Results and Problems in Cell Differentiation

Series Editor:
W. Hennig

38

Springer

Berlin
Heidelberg
New York
Barcelona
Hong Kong
London
Milan
Paris
Tokyo

Beate Brand-Saberi (Ed.)

Vertebrate Myogenesis

With 31 Figures

Springer

Professor Dr. Beate Brand-Saberi
Institute of Anatomy II
Albert Ludwigs University Freiburg
Albertstraße 17
79104 Freiburg
Germany

ISSN 0080-1844
ISBN 3-540-43178-0 Springer-Verlag Berlin Heidelberg New York

Library of Congress Cataloging-in-Publication Data

Vertebrate myogenesis / Beate Brand-Saberi (ed.). - Berlin ; Heidelberg :
New York : Barcelona ; Hong Kong : London ; Milan ; Paris : Tokyo :
Springer, 2002
(Results and problems in cell differentiation ; 38)
ISBN 3-540-43178-0

Springer-Verlag Berlin Heidelberg New York
a member of BertelsmannSpringer Science+Business Media GmbH
http://www.springer.de

Cover concept: Meta Design, Berlin, Germany
Cover production: Design & Production, Heidelberg, Germany
Typesetting: SNP Best-set Typesetter Ltd., Hong Kong
SPIN 10772625 39/3130 - 5 4 3 2 1 0 - Printed on acid-free paper

Preface

Muscle development of vertebrates has been a paradigm of cell differentiation for many years. Three types of muscle are found in the vertebrate body: skeletal, heart and smooth muscle, and there has been a gradient of concern about these different muscle types in the sequence they are mentioned here. Skeletal muscle has received much attention because it can be induced to differentiate in vitro and because of the clinical relevance of myopathies. The discovery of the muscle-specific members of the bHLH and MADS families of transcription factors must be regarded as a breakthrough not only in muscle research and have opened new insights into the genetic control of differentiation. Consequently, the effects of gene-targeting of the MyoD-related (myfs) and MEF transcription factors soon became objects of investigation. Along with the genetic control of skeletal and heart muscle development, the temporal-spatial appearance of cells fated to become myocytes has been of foremost interest.

The source of all skeletal muscle of the trunk is the paraxial mesoderm, which gives rise to metameric entities, the somites. The somite can be regarded as a turntable of mesodermal cell fates, chondrocytes, fibroblasts, angioblasts and skeletal muscle precursors. The coordinated development of these derivatives is tightly controlled by local tissue interactions between embryonic structures such as the neural tube, the notochord, the lateral plate and the ectoderm. In this connection, two populations of muscle precursor cells must be distinguished: those for the early differentiating epaxial muscle and those for the subsequently developing hypaxial muscle including limb muscle precursors. In the case of the latter, questions of cell migration, inhibition of differentiation and local patterning have been addressed, and molecules participating in each of these steps were characterized.

The histological grouping into skeletal, heart and smooth muscle is paralleled by an underlying difference in the expression of transcription factors. While, the *myfs* are restricted to the precursors of skeletal muscle, the *MEF*s are mutually expressed in the two striated groups, skeletal and heart muscle. Recent findings have indicated both common and distinct pathways of the two. Among the latter, Nkx 2.5 and GATA4 occupy prominent positions. Mutations in the evolutionary highly conserved *Nkx 2.5* gene result in heart malformations, congenital atrioventricular conduction defects and ventricular dysfunction in humans. Cardiomyocyte differentiation is intricately linked with heart morphogenesis and further analysis of heart development and malformation will be of high clinical relevance.

Likewise, clinical implications will be one driving force for enhancing our insights into the development of the third muscle type, smooth muscle. This type is found in the digestive, respiratory and urogenital organs and in the wall of blood vessels. However, its significance in pathological processes is balanced by the difficulties that this cell type presents to the investigator: First, the borderline between the undifferentiated and differentiated state in smooth muscle cells is blurred. Secondly, specific marker molecules are rare. Finally, smooth muscle has the highest plasticity as far as its phenotype is concerned. Therefore, I am particularly pleased that it has been possible to include into this volume recent findings concerning the control of smooth muscle development. We have reason to be confident that smooth muscle development is about to emerge from its gloomy existence as a stepchild of muscle research. Only when regarding the three, skeletal muscle, heart muscle and smooth muscle, will we begin to realize the process of muscle differentiation in all its facets, including the embryological and the evolutionary ones.

Furthermore, I wish to stress that the collection and discussion of novel functional data come to life especially when morphological documentation is given room. The combination of these aspects makes this volume intellectually and scientifically attractive. This is why I wish to express my thanks to all contributors for their efforts and enthusiasm, which is witnessed by their excellent articles.

I am grateful to Bodo Christ for participation in reviewing the contributions, and to Ernst-Martin Füchtbauer for helpful comments on the assembly of the volume. My thanks are due to Ulrike Uhl for assistance in the correspondence with the authors.

Beate Brand-Saberi
Freiburg, February 2002

Contents

The Genetics of Murine Skeletal Muscle Biogenesis
S. Tajbakhsh

Somite Patterning: a Few More Pieces of the Puzzle
C. Marcelle, C. Lesbros, and C. Linker

Transcription Factors in Skeletal Myogenesis of Vertebrates
P. Neuhaus and T. Braun

Hypaxial Muscle Development
G. Parkyn, R.C. Mootoosamy, L. Cheng, C. Thorpe, and S. Dietrich

Inhibition of Skeletal Muscle Development: Less Differentiation Gives More Muscle
E.-M. Füchtbauer

Control of Muscle Size During Embryonic, Fetal, and Adult Life
K. Patel, B. Christ, and F.E. Stockdale

Cadherins in Skeletal Muscle Development
Z. Waibler and A. Starzinski-Powitz

Development of Visceral Smooth Muscle

Giorgio Gabella[1]

The wide distribution of smooth muscle in the body is matched by patterns of development that differ in different organs. This review article deals with the origin, differentiation and growth of the smooth musculature of viscera. Only passing references will be made to the development of vascular musculature and to myoepithelial cells and myofibroblasts. The development of smooth muscle should be seen in the context of the special properties of this tissue. Smooth musculature is abundant (Table 1) and is found in all parts of the body; it performs with its contractions and its tone disparate functions, it grows while it is mechanically active, it is under the influence of local and systemic chemical factors and of mechanical factors, it produces the extracellular stroma (or matrix) that has the function of an intramuscular tendon and it adapts its growth and trophic condition to the functional demand imposed. The large assemblies of smooth muscle cells in the wall of viscera – as opposed to small groups or scattered muscle cells elsewhere in the body – undergo processes of development that consist not only in the cellular differentiation of a mesenchymal cell into a specialized contractile cell (cellular myogenesis), but also in the self-assembly and organization of the tissue, with the production of stroma and other features of supracellular organization (muscle differentiation).

1 Early Appearance of Smooth Muscles

The earliest morphological signs of the formation of a visceral smooth muscle are the elongation of the precursor cells and their grouping into a relatively dense cellular layer. The precursor cells arrange themselves parallel to each other, in contrast to the seemingly random orientation of the cells in the surrounding mesenchyme and of all the cells at an earlier stage. At the same time, the intercellular spaces of the primordial smooth muscle become narrower, a process related (it is not clear whether as a cause or as an effect) with the formation of more extensive intercellular contacts. In the chick, the process of elongation of mesenchymal cells, condensation of the cells and arrangement in a parallel array takes place at the end of the first week in ovo in the gizzard

[1] Department of Anatomy, University College of London, London WC1E 6BT, UK

Results and Problems in Cell Differentiation, Vol. 38
B. Brand-Saberi (Ed.): Vertebrate Myogenesis

Table 1. Total amounts of smooth muscle in the human body. These are rough estimates of the amount (in weight) of smooth musculature in the human body. These guesses give a total of 1,000–1,600 g in total or 1.5–2.2% of body weight

Organ	Weight (g)
Bladder	30–60
Ureter, vesicles, vas	30–60
Uterus	200–400
Gut	700–1,000
Airways	50–100
Vessels	150–300
Skin	10–30
Diffuse musculature	50–100

and intestine (Bennett and Cobb 1969; Gabella 1989) and results in the formation of a layer of musculature, which, in the case of the intestine, is the circular layer. In the rat embryo, similar events start in the gut wall around day 15 or soon afterwards.

Development of a smooth muscle is here taken as synonymous with development of the component smooth muscle cells. What constitutes a smooth muscle cell, in early development, is partly a matter of definition: here the term is intended morphologically and is taken to refer to an elongated cell containing a substantial amount of bundled myofilaments. At earlier stages in their development these cells are here referred to as smooth muscle cell precursors, these cells being committed to develop into muscle cells and beginning to express biochemical markers of smooth muscle cells. In histochemical studies, markers such as smooth muscle actin and smooth muscle myosin heavy chain are taken as identifiers of a smooth muscle cell from its earliest stage of development. Thus, smooth muscle differentiation is defined as the onset of smooth muscle myosin heavy chain mRNA expression (Miano et al. 1994), or the expression of alpha-smooth muscle actin, but not yet of gamma smooth muscle actin (McHugh 1995) (see section 5). The term anlage or anlagen refers to the assembly or assemblies that these cells form. The timing and the mechanisms of the commitment of primordial cells to turn into muscle cells are unknown; it is unclear to what extent the commitment is reversible, or the extent to which precursor cells retain some pluripotency (see Sect. 9).

The appearance of precursor smooth muscle cells within an organ and their differentiation into muscle cells is a local process: it does not appear to involve a distant migration of cells or their redeployment, and its timing is organ specific. Also, it seems that it is not a focal process: for example, in the primitive small intestine, the precursor cells are distributed around the entire circumference of the wall and along its entire length (although it is possible that there is a gradient along the length of the gut, or across the thickness of the wall, see p. 9).

The path of differentiation of smooth, skeletal and cardiac muscles from mesenchymal precursors is determined long before any morphological signs appear. The apposition between cardiac and smooth musculature (in the vena cava) and that between striated and smooth musculature (in the oesophagus and urethra) are already observed in development: in every case, the two tissues are morphologically distinct, even when, as is usually the case, there is a considerable admixture of two cell types (which can be properly resolved only be electron microscopy). However, there are several histochemical and ultrastructural studies describing a process of transdifferentation by which smooth muscles (or, rather, newly differentiated smooth muscle cells) differentiate into striated muscle fibres, in the chick embryo iris, in the rat urethra and in the mouse oesophagus (see p. 26).

The transition from precursor to smooth muscle cell involves marked changes in cell shape. The spatial orientation of these developing cells, within bundles, layers or the whole wall is well defined from the beginning and corresponds closely to that of the mature muscle cells in that particular organ. This early process of morphogenesis takes place at a time when there is very little extracellular materials and no collagen or other fibrils.

2 Timing of Smooth Muscle Development

The early development of smooth muscle varies in different organs. Only few accurate data are available, and the problem is often compounded by difficulty in the timing or staging of embryos and in the definition of early stages of development. For example, the 7-day-old embryos of Ko et al. (1996) were at the same Hamburger-and-Hamilton stage (26–28) as the 5-day-old embryos of Hirai and Hirabayashi (1983) and Paul et al. (1994).

In general, visceral smooth muscles are well differentiated soon after birth and are already similar to those of the adult; in contrast, vascular muscles (which had been the first to appear during embryonic life) undergo marked structural changes postnatally, a characteristic probably related to the occurrence of haemodynamic changes and to the late formation of conspicuous extracellular materials in large vessels. For example, incorporation of valin into elastic fibres of the media of the mouse aorta, as studied by autoradiography, continues up to 4 weeks after birth (Davis 1993) and, in the rat, the volume of the extracellular material in the media of the aorta increases tenfold from birth to day 11 (Olivetti et al. 1980). As to visceral muscles, there are clear differences in the time at which they appear in different viscera and probably also in their rate of growth. The intestinal musculature is more differentiated at birth than that of the vas deferens; the muscularis mucosae appears last in the wall of the intestine. Generally, it seems that the circular musculature develops earlier than the longitudinal musculature. Even in blood vessels there are marked difference in the timing of development; for example, the musculature of the coronary artery of the chick matures in advance of that of the aorta (Hood and

Rosenquist 1992). A special case is that of the chick amnion, in which a one-cell-thick layer of smooth musculature develops outside the embryo and reaches morphological maturity around the tenth day of incubation (Evans and Evans 1964), which is well in advance of any muscle of the embryo. The musculature of the amnion is contractile, producing some movement of the amniotic fluid at a time when the embryo has no spontaneous motility. (The early regression of the amnion musculature from the end of the second week is still unexplored.)

In chick embryos, smooth muscle appears in the gizzard (muscular stomach) at about day 7 (Bennett and Cobb 1969). Soon after that time a circular muscle layer becomes recognizable in the small intestine. A special layer of muscle cells at the inner border of the circular layer becomes recognizable at around day 11. A longitudinal muscle layer beneath the serosa is seen from around day 13. An inner longitudinal muscle layer (similar to a muscularis mucosae) beneath the mucosa becomes recognizable at around day 15 (Gabella 1992) and, at this stage, all the layers of the mature intestinal wall are recognizable and in place. The musculature of the rectum develops with a similar pattern (although there is no special layer of circular musculature), but some 2 days later than in the small intestine.

In the guinea-pig ileum the first muscle cells are seen in the presumptive circular layer from 25 days of gestation (and not at 20 days), and in the longitudinal muscle by day 32; between these two stages the myenteric plexus is formed, lying by the outer aspect of the circular muscle (Gershon et al. 1981). In another study of the guinea-pig ileum, a primordial circular musculature was found along the entire length of the small intestine in embryos of 8 mm CR length (Fig. 1A–C); at this stage, there was no musculature in the wall of the large intestine (Fig. 1D). Later in gestation, a longitudinal muscle appears (Fig. 2A,B) and, closer to term, a special layer of circular musculature and a muscularis mucosae becomes apparent (Fig. 2C,D). In the colon of 4-week-old dog embryos (mid-gestation) a circular muscle layer (~30 μm thick) and a minute longitudinal layer (~9 μm thick) were already visible (Ward and Torihashi 1995).

In the uterus of the swine at birth, a condensation of roundish cells into a subserosal layer is the primordial myometrium (Bal and Getty 1970). Muscle cells, elongated in a circular direction, are seen only from the age of 2 weeks and a longitudinal musculature from the age of 1 month. The process is more advanced in the body of the uterus than in the horns (Bal and Getty 1970). In the human uterus, primitive muscle cells (identified by the presence of myofilaments) are seen at 18 weeks of gestation (CR length 145 mm; and not at week 16 or earlier) in a cell layer close to the serosa; close to the mucosa the mesenchymal cells begin to acquire the appearance of fibroblasts. The distinction between the two regions of the wall is clear, and one gives rise to the myometrium the other to the endometrial stroma (Konishi et al. 1984). Differentiation of the myometrial muscle cells is more advanced at 31 weeks, but it is markedly less that in the bladder detrusor examined in the same study

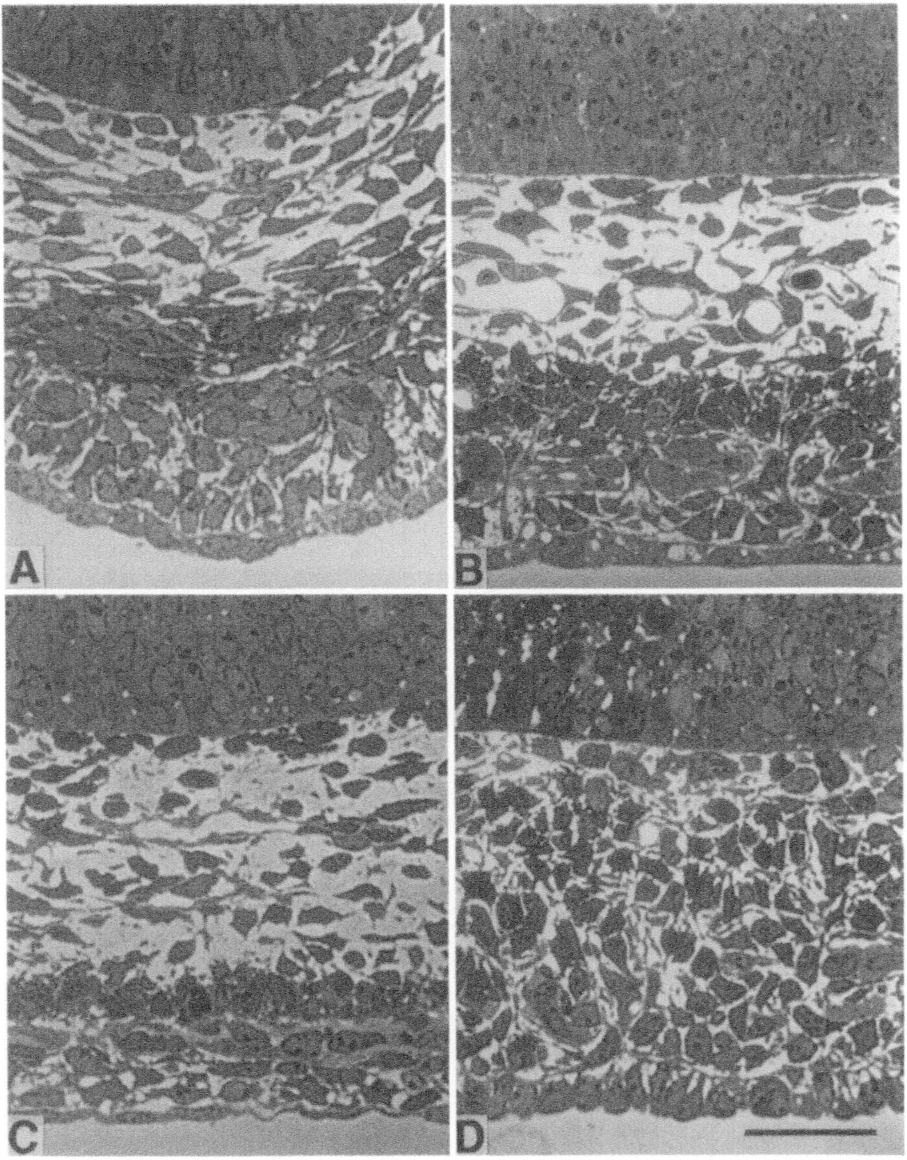

Fig. 1. Developing intestine from guinea-pigs in early gestation examined by light microscopy. **A** 8-mm CR embryo, transverse section of ileum. **B** 8-mm CR embryo, longitudinal section of ileum. **C** 8-mm CR embryo, longitudinal section of duodenum. **D** 8-mm CR embryo, longitudinal section of rectum. *Calibration bar* 30 µm

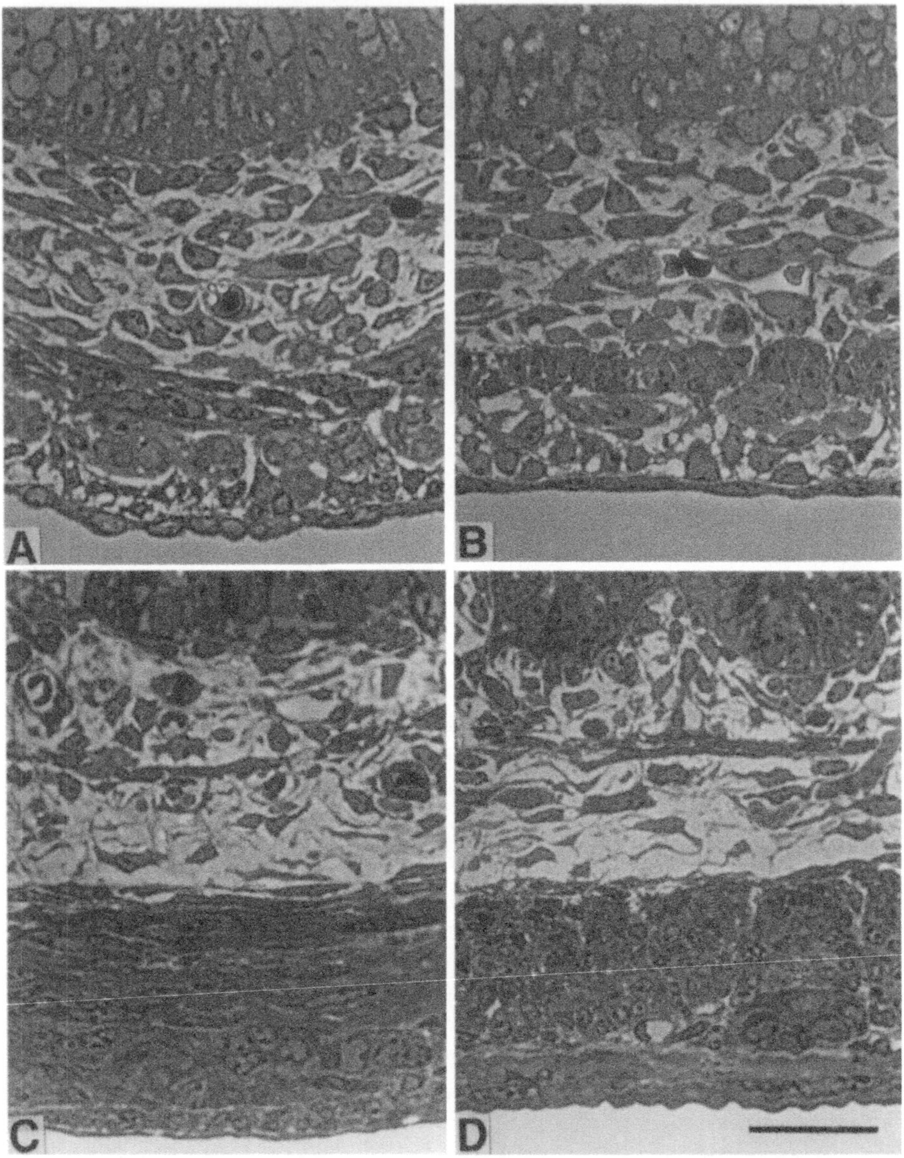

Fig. 2. Developing intestine from guinea-pigs in late gestation, examined by light microscopy. **A** 20-mm CR embryo, transverse section of ileum. **B** 20-mm CR embryo, longitudinal section of ileum. **C** 85-mm CR embryo, transverse section of ileum. **D** 85-mm CR embryo, transverse section of ileum. All sections display ganglia of the myenteric plexus. *Calibration bar* 30 μm

(Konishi et al. 1984). At the inner aspect of the myometrium there is, by week 31, a layer of cells with the features of immature smooth muscle cells (Konishi et al. 1984). The authors suggest that mesenchymal cells continue to generate muscle cells as well as endometrial stromal cells in this region. Generation of new smooth muscle cells from mesenchymal precursors is also observed in the human uterus in the adult (Fujii et al. 1989): by the inner surface of the myometrium, during the follicular phase of the oestrus cycle, there are cells described as resembling myofibroblasts (or immature muscle cells), which turn into differentiated muscle cells during the luteal phase or in early pregnancy (Fujii et al. 1989).

Oestrogen and progesterone receptors are found in the guinea-pig uterus starting from 34 and 50 days of gestation, respectively (Pasqualini et al. 1983). The sensitivity of myometrial muscle cells and of uterine mesenchymal cells to steroid hormones is well established (Ross and Klebanoff 1967; Bo et al. 1968), and the organ-specific nuclear receptors for oestrogen and progesterone are already present in the immature muscle cells of the rabbit myometrium (Hegele-Hartung et al. 1992). Treatment of newborn mice with oestrogens (diethylsilbestrol) has profound effects on the morphogenesis of the myometrium, including an increase in the middle layer of the muscle (the prospective circular musculature) and the early appearance and persistence of intense vimentin-immunoreactivity in muscle cells (Brody and Cunha 1989b). In the rat, the effects are similar but less severe (Brody and Cunha 1989b). Three layers - without distinct boundaries - are recognized in the uterus (uterine horn) of the newborn mouse (Brody and Cuhna 1989a). In the middle layer (the "prospective circular myometrial layer") the cells run circumferentially, are aggregated, and by day 5 after birth they show features of smooth muscle cells; at this stage, cells of the outer layer begin to elongate and to show a longitudinal orientation and are grouped into bundles and clearly recognized as longitudinal muscle cells by day 10. Similarly, in the rat uterus, the circular myometrium is clearly recognized by day 5 after birth, and the outer longitudinal myometrium by day 15 (Brody and Cuhna 1989a).

In the human bladder, studied in the electron microscope, small smooth muscle cells are first seen in embryos of 12 weeks (Newman and Antonakopoulos 1989) or 14 weeks (Konishi et al. 1984). The early musculature is subserosal and is initially confined to the caudal part of the bladder, but within 1 week it is found in the whole organ (Newman and Antonakopoulos 1989). Matsuno et al. (1984) gives much earlier ages for the appearance of smooth muscle, detected by light microscopy, in the intestine and the bladder at the sixth and seventh week of gestation, respectively (24 and 27 mm CR length). In the mouse bladder, smooth muscle cells are detected by alpha-actin histochemistry from gestation day 16, situated immediately beneath the serosa (Baskin et al. 1996; Wu et al. 1999). The case of the bladder is complicated by the fact that several authors describe the occurrence of myofibroblasts, a cell type whose presence is well characterized histochemically in other organs and in certain pathologic conditions. In the bladder of embryonic mice, myofibroblast, derived from mesenchymal cells, are morphologically distinct from muscle cells, but eventually they differentiate into muscle cells (Wu et al. 1999). In adult rabbits, subserosal fibroblasts and myofibroblasts differentiate into muscle cells during outlet obstruction hypertrophy of the bladder (Buoro et al. 1993). Myofibroblasts are not observed in the rat bladder (Gabella and Uvelius 1990).

In the rat, ureter cells identified as smooth muscle myoblasts are seen by electron microscopy at embryonic days 19 and 20. One day after birth there is a distinct layer of muscle cells with a predominant circular orientation; ultrastructurally these cells were well differentiated (Leeson and Leeson 1965).

In the mouse vas deferens, a circular musculature is just detectable on day 1 after birth, whereas longitudinal muscle is not seen until day 5 (Yamauchi and Burnstock 1969).

Developing muscle cells are seen by electron microscopy in the bronchi of rat embryos from day 16 of gestation (Collett and Des Beins 1974). The musculature of the airways is well differentiated in guinea-pig embryos at term (Gabella 1991); only cytological details distinguish it from that of the adult.

In the iris of the chick (which – in addition to a massive sphincter of striated musculature – has a ring of smooth muscle near the pupillary edge) cells with features of smooth muscle cells can be seen in 10–11-day embryos, but not at day 9. The muscle is well differentiated by day 17 (Gabella and Clarke 1983).

Muscle cells (myoepithelial cells) of the rat mammary gland can be detected only after birth (Deugnier et al. 1995).

3 Morphology of Developing Smooth Muscles

In the ileum (post-umbilical small intestine) of the chick embryo, smooth muscle cells – referred to by the authors as smooth muscle myoblasts and identified by alpha-actin immunofluorescence – are seen at day 8.5: these cells are apparently roundish in shape and constitute the presumptive circular muscle layer, whereas all the other cells in the wall are negative (Masumoto et al. 2000). At day 10.5, the immunopositive cells are elongated, parallel and circumferentially arranged; the outermost of them "began to stretch radially from the outermost layer of the developing circular musculature towards the presumptive longitudinal muscle layer". The radially arranged muscle cells form bundles, bend into an L-shape, lie close to the myenteric ganglia and acquire a longitudinal direction, giving origin to the longitudinal muscle layer at day 13.5 (Masumoto et al. 2000). Given the limited resolution of immunofluorescence on frozen sections, the origin of the longitudinal musculature is not entirely clear; however, in the view of these authors (Masumoto et al. 2000) the layer derives from the elongation – and presumably the migration – of muscle cells of the circular layer. The authors describe the muscularis mucosae as forming by a similar mechanism, protrusion, elongation, turning at a right angle and formation of a longitudinal layer at around day 15.5. At day 18.5, alpha-smooth muscle actin positive cells appear within the core of the mucosal villi and in the wall of large intramural blood vessels.

The early stages of development can be followed in more detail on thin sections for light and electron microscopy. This approach provides a better cellular resolution than histochemistry and gives a comprehensive view of the organ. However, embryonic tissues are particularly sensitive to the preparative procedures; they are also prone to muscle shrinkage and other artifacts that affect morphometry. Very few attempts have been made to correlate ultrastructural and immunohistochemical observations. In the intestine of chick and guinea-pig embryos, the aggregation and the elongation of mesenchymal cells into the muscle anlage occur approximately at the same time and they give rise to a distinct cellular layer (two to three cells thick at the start). At approximately the same time, the first evidence of a contractile apparatus is seen in the form of thin bundles of actin filaments with a small dense body halfway along the length of the bundle. Ultrastructural studies have provided no evidence as to whether the differentiation originates multi-focally before involving all the cells of the anlage; evidence of muscle cell migration is also absent (but any occurrence of cell migration would be difficult to document with ultrastructural data). Around the circumference of an intestinal loop there are no signs of maturational gradients, for example between the mesenteric and the antimesenteric regions. Along the length of the small intestine, the differentiation appears to occur synchronously, in contrast to the strong claim of the occurrence of cranial to caudal gradients based on alpha-smooth muscle actin histochemistry (McHugh 1995). At present, there is no ultrastructural evidence of such a gradient. It is possible, nevertheless, that there are differences in degree of maturation of just a few hours between different intestinal loops, which would not have been detected with the time resolution employed. However, ultrastructural differences in the early muscle differentiation between stomach, small intestine and large intestine are documented, and within the same embryo there is a stage when a musculature is present in the small intestine but not in the large intestine (Fig. 1). In contrast, within a given muscle, even a large one such as the gizzard muscle of the chick embryo, it has been claimed that the muscle cells develop synchronously; there are no undifferentiated cells remaining in the tissue among the developing muscle cells, blood vessels, interstitial cells and fibroblasts (Fig. 3; Gabella 1989).

An important aspect of the morphology of developing visceral muscles is that there is an early acquisition of those patterns of muscle cell grouping and stroma arrangement that are characteristic of each muscle. The layered structure of the intestinal musculature, with a uniform thickness and distinct arrangement of intramuscular septa, is acquired as soon as the early muscle cells are formed, and so is the distribution of intramuscular septa, which is characteristic of the rectum (Fig. 2), and the characteristic pattern of muscle bundles in the mammalian bladder.

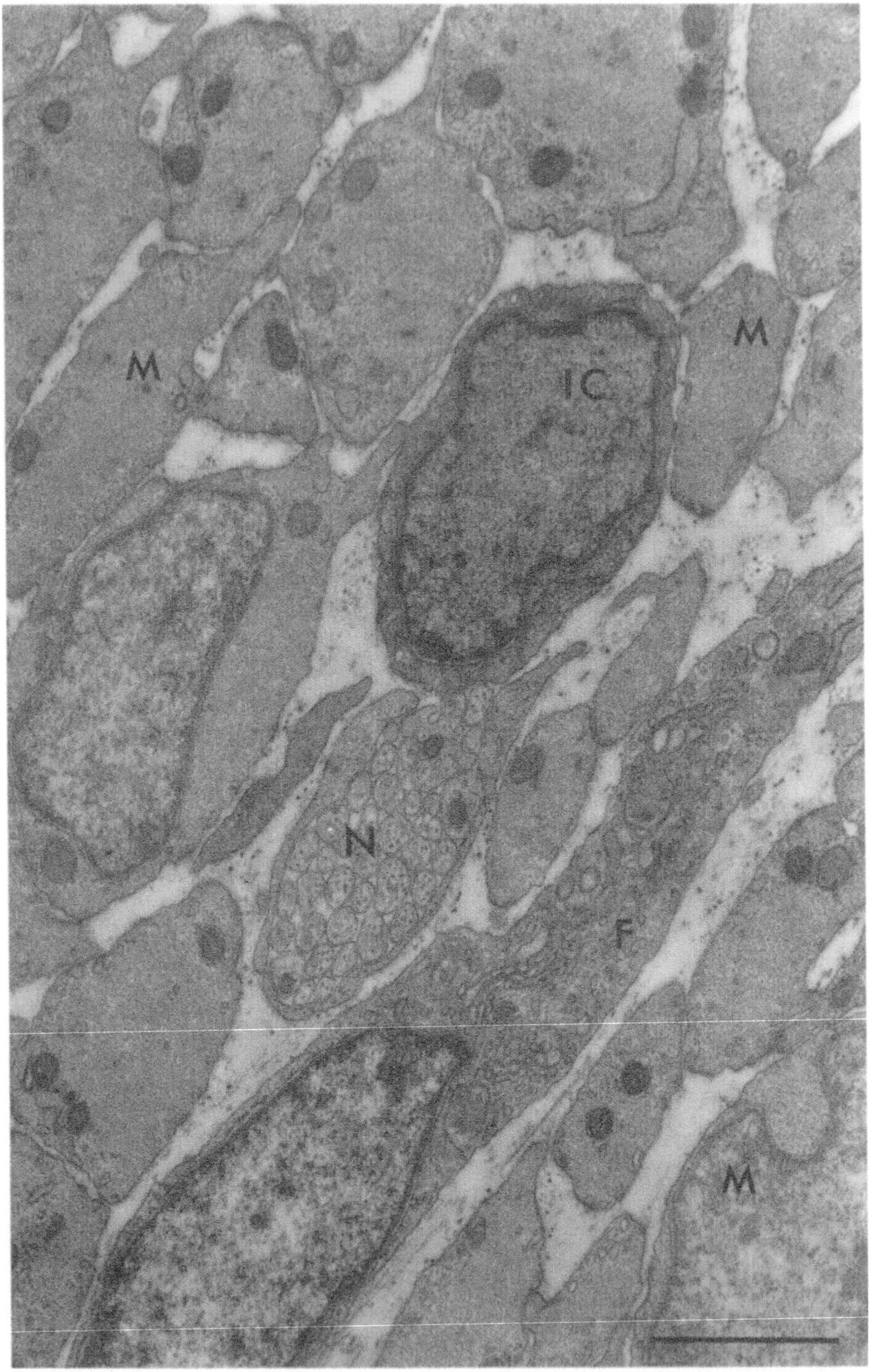

Fig. 3. Musculature of the gizzard of a 20-day-old chick embryo. In this electron micrograph a group of muscle cells are sectioned transversely (*M*). On the basis of their appearance these cells are regarded as being at the same degree of maturation. *N* Intramuscular nerve bundle; *IC* interstitial cell of Cajal; *F* fibroblast. *Calibration bar* 1 μm. (Modified from Gabella 1992)

4 Cytological Differentiation

Cell Shape. Elongation of the cell into a spindle takes place at the very beginning of differentiation, and it affects all the cells of the muscle anlage, which thus come to lie parallel to each other. At this stage, the muscle cells are smaller than the mesenchymal cells of the surrounding tissue. The shape of the cells and their orientation are initially more irregular than later in development, and the very orderly arrangement of the intestinal muscles and many other visceral muscles is acquired only progressively, and is not achieved until a substantial amount of extracellular fibrillar material is laid down.

The transverse outlines of the immature muscle cells are very irregular, showing the exuberance of cell membrane and cell surface over cytoplasm and cell volume (Fig. 4A). The slimmest of these expansions do not contain myofilaments or organelles, and they may be signs of exploratory movements of the cell, not yet encumbered by intercellular junctions or attachment to the stroma.

Cell Volume. The volume of the muscle cells is initially much smaller than at maturity (Fig. 4). In the canine colon at birth, muscle cells measure on average 4.0 μm in width and 183 μm in length; at 1 month of age the width is 5.2 μm and at 1 year of age the length is 372 μm (Ward and Torihashi 1995). In the chicken gizzard, the muscle cell volume is estimated at 600 μm^3 at day 14 in ovo and at 3,000 μm^3 in the adult, a fivefold growth in muscle cell size (Gabella 1992). Measurements have been carried out on the muscle cells of the guinea-pig taenia coli, but only from late embryonic life onwards. Shortly before birth, muscle cell volume is about 900 μm^3 and then grows to 2,000–3,000 μm^3 in the adult. In rats and guinea-pigs (and presumably in many other mammals) visceral muscle cells continue to grow through life, albeit at a diminishing rate; and in the taenia coli of ageing guinea-pigs (3 years) muscle cell volume is over 4,000 μm^3.

The nucleo-cytoplasmic ratio is a good indicator of muscle cell differentiation. During the postnatal development of the smooth muscle of the mouse vas deferens there is a substantial decrease of the nucleo-cytoplasmic ratio: from 0.79 at birth, to 0.41 at day 5, to 0.08 at day 25 and 0.03 in the adult (Yamauchi and Burnstock 1969).

Contractile Apparatus. The earliest sign of the development of the contractile apparatus is the appearance of isolated bundles of thin (actin) filaments, usually attached to a small dense body. In other studies it has been observed that dense bodies appear only subsequently to the development of actin filament bundles (e.g. Yamauchi and Burnstock 1969). However, in the case of the chick embryo gizzard and ileum, the earliest bundles of actin filaments already display a small dense band halfway along their length (Gabella 1992). These bundles are short (~1 μm) and minute, they are unattached to the cell membrane and they tend to be roughly aligned with the long axis of the cell.

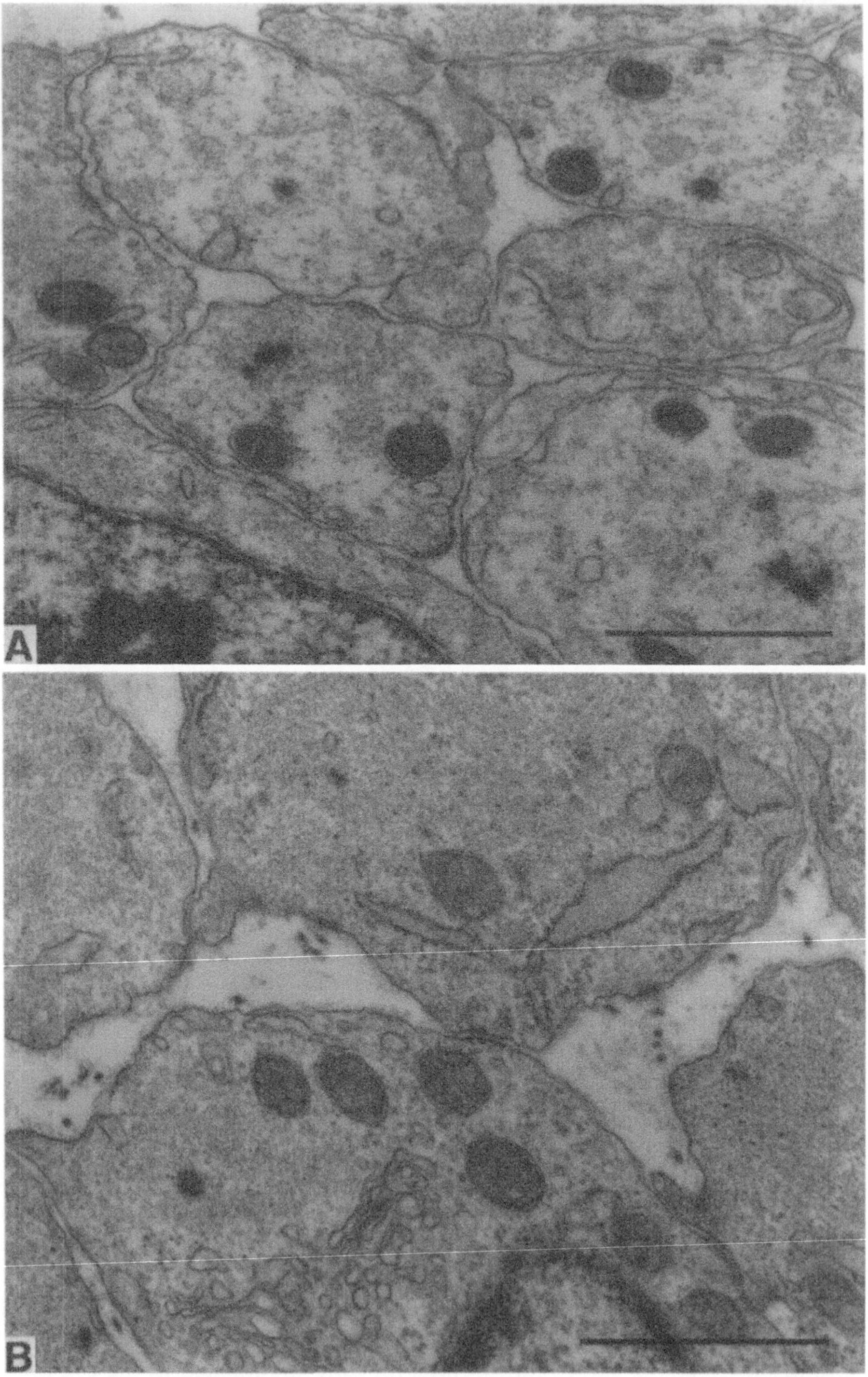
A
B

Thick (myosin) filaments are detected some days after the appearance of actin filaments (Fig. 4). In the absence of a good correlative study with histochemical data, it remains unclear to what extent the late appearance of myosin filaments is a technical problem of preservation (to which this type of filament is particularly sensitive). In their study of the mouse vas deferens, Yamauchi and Burnstock (1969) observed that the thick myosin ("coarse") filaments grow in width from 6 nm at day 10 after birth up to 14 nm in the adult.

Membrane-bound dense bands bearing insertion of bundles of actin filaments that are characteristic of mature smooth muscle cells are seen in the gizzard of 10-day chick embryos (Bennett and Cobb 1969). However, in the gizzard and ileum of chick embryos and in the intestinal musculature of rat and guinea-pig embryos, most of the dense bands appear relatively late in development (Gabella 1989, 1992). Initially, the actin bundles are mostly inserted into cytoplasmic dense bodies, and only a few are attached to the plasma membrane (Fig. 4). An extensive insertion of the contractile apparatus to the plasma membrane takes place rather rapidly, in the chick, during the first few days of hatching (Gabella 1989). A cytochemical study of vinculin and talin, two proteins associated with the dense bands of mature muscle cells (Geiger et al. 1981), shows that in the developing chicken gizzard these proteins remain primarily cytoplasmic up to embryonic days 16–18 (talin) or up to 1–3 days after hatching (vinculin; Volberg et al. 1986). Only at these times does their localization on the plasma membrane become apparent.

Endoplasmic Reticulum. The endoplasmic (or sarcoplasmic) reticulum of developing muscle cells is clearly in evidence and better developed, in spatial density terms, than in mature muscle cells. Single ribosomes, free rosettes of ribosomes and cisternae of rough reticulum are very common, and this correlates with the intense synthesis of cell proteins and of proteins for export. There are often large cisternae with an expanded lumen and an electron dense

Fig. 4. Muscle cells from chick embryo gizzard examined in the electron microscope. **A** Fourteen-day-old embryo. The muscle cell profiles have an undulated surface. There are no caveolae and there are no collagen fibrils in the intercellular spaces. Mitochondria are abundant and there are many thin (actin) filaments, but no obvious thick (myosin) filaments. Prominent dense bodies, surrounded by a few intermediate filaments, are visible in several cell profiles, while the attachment of filaments to the plasma membrane by means of dense bands is very limited. There are long cisternae of endoplasmic reticulum with few ribosomes attached. Part of a nucleus appears at the *bottom left*. *Calibration bar* 1 µm. **B** Nineteen-day-old embryo. The muscle cell profiles are larger than in the micrograph above. The cells surface is convoluted and shows only a few caveolae. The contractile apparatus (thin filaments, thick filaments and conspicuous dense bodies) is well recognizable, especially in the cell top of centre. There are few dense bands on the plasma membrane. Large cisternae of endoplasmic reticulum with electron-dense content and ribosomes attached. The nucleated muscle cell profile at bottom has a prominent Golgi apparatus. Some collagen fibrils can be seen in the spaces between the cells. *Calibration bar* 1 µm content (Fig. 4B). These cisternae reside not only in the cytoplasmic region near the poles of the nucleus, but also among myofilaments, often abutting with an edge on the plasma membrane. Their presence suggests that the cell is involved in synthesis of proteins destined for the extracellular space, but the extent to which the material in the cisternae are temporarily stored as opposed to being processed for secretion cannot be determined.

Golgi Apparatus. A large Golgi apparatus is found in developing muscle cells, especially in the early stages of differentiation.

Mitochondria. These organelles are abundant in developing smooth muscle cells. Their spatial density undergoes some reduction during development. For example, in the chicken gizzard muscle cells, mitochondrial spatial density is about 8% in 14-day-old embryos and it decreases progressively to about 5% in the adult (Gabella 1989). One would expect biochemical changes to take place in the mitochondria during development within one muscle, and also differences between different muscles, but these have not yet been investigated (these data would have a bearing on the question of the existence of subpopulations of muscle cells within a muscle, for which, in the case of visceral muscles, there is at the moment no firm evidence).

Microtubules. They are abundant in developing muscle cells, usually found beneath the plasma membrane and running parallel to the cell length.

Gap Junctions. Gap junctions are initially absent in developing smooth muscles. In the chicken gizzard, the first gap junctions are recognized in freeze-fracture preparations at day 16 in ovo (La Mantia and Shafiq 1982; Gabella 1989). In the circular muscle of the guinea-pig ileum, gap junctions are observed from around the time of birth. These early gap junctions of gut musculature are very small, formed by 5–10 intramembrane particles (connexons), and they grow progressively in size, presumably by confluence and by the addition of particles. The connexons – as seen in freeze-fracture preparations – do not change in appearance through development, although there is as yet no evidence on the types of connexins of the early junctions. The outline of the junctions is very sharp and in freeze-fracture preparations one does not observe particles in a transitional state between the membrane at large and the clustered connexons. In the tracheal muscle of the guinea-pig, a smooth muscle that is remarkably well differentiated at birth, gap junctions between muscle cells are not observed in foetuses or in newborn animals, but they are present at day 5 (Gabella 1991).

Adherens Junctions. As with dense bands of the plasma membrane, these junctions are less common in embryonic muscles than in fully differentiated muscles. There are several morphological types of adherens junctions in smooth muscles, and they are not well characterized. The main difference between developing and adult muscles is in the frequency of these junctions rather than in their ultrastructural appearance.

Caveolae. These organelles are rare in muscle cells during the early stages of development (Yamauchi and Burnstock 1969; La Mantia and Shafiq 1982; Gabella 1992). In the gizzard and ileum of the chick embryo occasional caveolae are observed as early as day 11, they become common by day 15; they increase substantially in number in the first few days after hatching and, on day 5, there are 14 caveolae per square micron (compared with 37 in the adult; La Mantia and Shafiq 1982). The size of caveolae, as judged in freeze-fracture preparations, is similar at all ages. If confirmed, this observation would indicate that caveolae are plasmalemmal derivatives of relatively fixed shape and size, probably having an invariant molecular structure.

Basal Lamina. A basal lamina becomes visible by electron microscopy only relatively late in development. Patchy lumps of faintly electron-dense material, adhering to the plasma membrane of muscle cells, precede the formation of a proper basal lamina.

5 Chemical Differentiation

Smooth muscle development is studied to a great advantage with histochemical and biochemical methods that detect components of contractile apparatus, cytoskeleton and extracellular matrix. These methods can be very specific, although limited by possible discrepancies between expression (mRNA levels, detected by riboprobes) and synthesis (protein levels, detected by immunohistochemistry or immunoassay; McHugh 1995; Takahashi et al. 1996), by the long half-life of proteins of the cytoskeleton and contractile apparatus, and by the limited spatial resolution and the distortion of thick frozen sections. However, Saborío et al. (1979) comment that in the case of actin isoforms in the chick embryo gizzard, expression and synthesis proceed in parallel, and Eddinger and Meer (1997) showed that this is true for myosin isoforms in single, isolated vascular muscle cells.

Histochemical methods are strong tools for the analysis of the chemical development of a muscle and the expression of different genes (Saborío et al. 1979; Kuroda 1985; Eddinger and Murphy 1991; McHugh et al. 1991). They can also be useful in localizing substances as markers or identifiers of a smooth muscle, in order to detect the earliest appearance or the extent of a muscle. However, few markers, if any, are entirely specific, because transient expression of non-typic genes is common in development and, conversely, a developing striated muscle briefly expressing a smooth muscle "marker" should not be considered a transformed smooth muscle. The best value of histochemical markers is given when they are used in conjunction with other methods that provide strong morphological evidence.

Some isoforms of actin (Skalli et al. 1986), smooth muscle myosin heavy chain (Miano et al. 1994), fibronectin variants (Glukhova et al. 1990), SM22 (Duband et al. 1993), calponin (Miano and Olson 1996), a smooth muscle alpha-actinin-like molecule (Hungerford et al. 1997) intermediate filament proteins, are some of the markers used. Like alpha-smooth muscle actin, the specific marker of mature smooth muscle SM22, which is a calponin-related protein of unknown function (Lees-Miller et al. 1987; Duband et al. 1993), is expressed very early in development (Duband et al. 1993; Li et al. 1996); however, it is also transiently expressed in the early development of cardiac and skeletal muscles (Li et al. 1996). In contrast, smooth muscle myosin heavy chain mRNA is expressed exclusively in smooth muscle cells, in both adult and developing muscles and the promoter of smooth muscle myosin heavy chain is the most specific among the smooth muscle markers (Miano et al. 1994). Myosin heavy chain has isoforms specific for vascular and for visceral smooth muscles (Kelley et al. 1993). Studied by in-situ hybridization, expression of mRNA for smooth muscle myosin heavy chain is detected at day 10.5 (but not at day 9.5 or earlier) in the mouse embryo; at day 12.5 it appears in the developing gut and bronchi, and by day 17.5 in all major viscera, including oesophagus, bladder and ureter (Miano et al. 1994), while there is no labelling of any other tissue.

Alpha-smooth muscle actin is an early marker of developing smooth muscle; the molecule is not completely specific because it is present in embryonic skeletal and cardiac muscles (Ruzicka and Schwartz 1988; Woodcock-Mitchell et al. 1988; Sawtell and Lessard 1989) and in myofibroblasts. A day or two after the first expression of alpha-smooth muscle actin, smooth muscles begin to express gamma-smooth muscle actin; this actin isoform is not expressed at all in striated muscles (Sawtell and Lessard 1989) and, therefore, is a more specific marker of smooth muscle. McHugh (1995) proposes the name of smooth muscle myoblasts for the cells of the smooth muscle lineage during the period, which lasts from a few hours to 48 h, in which they are alpha-smooth muscle actin positive and gamma-smooth muscle actin negative. Myoblasts of this type are also identified in cell cultures (Brittingham and McHugh 1992), where they are also known as synthetic smooth muscle cells (Chamley-Campbell et al. 1979). In the amphibian embryo, alpha-smooth muscle actin is expressed at the time of the induction of mesoderm and is a specific marker of the ventral mesoderm, which is destined to give origin to vascular and visceral smooth muscle (Saint-Jeannet et al. 1992; see Sect. 9).

In the small intestine of the chick embryo, expression of alpha-smooth muscle actin is observed at day 8.5 (the earliest stage studied) in the circular muscle anlage, at day 11.5 in the longitudinal muscle; at day 16.5 in the muscularis mucosae and at day 18.5 in the core of the villi (Masumoto et al. 2000). These times correspond approximately with the times of appearance of actin filaments in the various muscle layers (Gabella 1992). In the adult, the musculature of the gizzard, examined by electrophoresis, contains predominantly gamma-enteric smooth muscle actin; there are also small amounts of alpha-vascular smooth muscle actin (Rubenstein 1981), but these are not detected by immunohistochemistry (Skalli et al. 1986). The proportion of alpha- and gamma-actins varies in different parts of the chicken gut (Yamamoto et al. 1996), and in the main circular muscle of the small intestine where alpha-smooth muscle actin is first expressed – and later on becomes a very strong marker – this actin isoform is no longer detectable in the adult (Masumoto et al. 2000). It is common in most visceral smooth muscles in adults of all species to contain, predominantly, the gamma isoform of actin, whereas alpha-smooth muscle actin predominated in vascular muscles (Fatigati and Murphy 1984).

In the anlage of the musculature of the chicken gizzard, very faint immunoreactivity for myosin heavy chain, desmin and beta-tropomyosin appears during the fifth day of incubation and is distinct at day 7 (Hirai and Hirabayashi 1983). This is confirmed by Stuewer and Gröschel-Stewart (1985) and Zimmermann et al. (1995) who add the important observations that expression of smooth muscle myosin in the gizzard anlage proceeds, wave-like, from a region close to the serosa towards the lumen, at the same time as nerve fibres from the Auerbach's plexus grow into the muscle. This chemical differentiation proceeds from small localized foci (already detected at day 5), which expand and coalesce; immunoreactivity spreads in all directions, becomes more intense, and by day 13 is detected in the entire muscle. The gizzard musculature in birds is characterized by a high content of the respiratory pigment myoglobin (Gröschel-Stewart et al. 1971; Blessing and Müller 1973): immunoreactivity for haemoglobin does not appear in chicken gizzard muscle cells until embryonic day 18 (Stuewer and Gröschel-Stewart 1985).

In the mouse, smooth muscle cells that are alpha-actin positive are detected at embryonic day 11 in the oesophagus and lung buds, and they spread along a distinct cranio-caudal gradient (McHugh 1996), and by day 13 they are found in the entire alimentary tract. This chemical differentiation of the circular muscle, detected with in-situ riboprobes for the isoforms alpha – and beta-smooth muscle actin, is followed 48–72 h later by that of the longitudinal muscle and then by the muscularis mucosae (McHugh 1995). In the bladder, actin-positive cells appear at embryonic day 11, and at day 13 in the ureter, and in the vagina at day 15 (McHugh 1995). In the bladder, the early muscle cells are located within the mesenchyme of the outer half of the wall, whereas in the ureter they appear in the mesenchyme immediately underlying the epithelium in the same manner as in developing blood vessels (McHugh 1995). Alpha-smooth muscle actin is detected in the rat urethral sphincter from day 14 of gestation onwards, but it is probable that it is localized in developing striated musculature (Borirakchanyavat et al. 1997).

Anti-actin immunoreactive cells in the rat and mouse uterus are first observed at day 3 after birth, and at day 5 the immunoreactivity outlines the presumptive circular muscle layer of the myometrium, in addition to the blood vessels. The developing outer longitudinal muscle stains from day 10 (Brody and Cunha 1989a). During this period of life, mesenchymal cells (unlike the muscle cells of the circular layer) are vimentin-positive, and so are the precursors of the muscle cells of the longitudinal layer. Vimentin-immunoreactivity, therefore, discriminated between mesenchymal and connective tissue cells (positive) and muscle cells (negative; Brody and Cunha 1989a).

An important observation reported by McHugh – albeit only briefly discussed there – is that "immature smooth muscle myocytes" (here referred to as smooth muscle cell precursors) are also observed in embryos near term and in new-born mice. On this basis, these studies, in their authors' words, "have demonstrated that smooth muscle tissues are composed of a heterogeneous population of phenotypically distinct smooth muscle myocytes" (1995, p. 280). Heterogeneous expression of actin genes in the adult intestine of the rat supports this view (Liddell et al. 1993), if one assumes that co-expression of two isoforms by a tissue indicates existence of two subpopulations of cells.

McHugh (1996) stresses that gastrointestinal smooth muscle develops along a cranio-caudal gradient (see a discussion of this point in the light of structural evidence in Sect. 3). In addition to the early localization in the foregut, McHugh (1995) observes expression of alpha-smooth muscle actin at the caudal extremity of the hindgut (and in the allantois and the placenta) and on this basis concludes that the development of the primitive gut tube initiates at two distinct sites, which include a cranial and a caudal centre of organization.

In the canine colon, in embryos about midway through the gestation period (approximately at 30 days after fertilization), both the longitudinal and the circular developing muscle layers are vimentin-immunoreactive; the circular layer, but not the longitudinal layer, is also positive for desmin at this stage (Ward and Torihashi 1995). By 1 week prior to birth, intermediate filament immunoreactivity changes and becomes similar to that found in the adult: both muscle layers express desmin-immunofluorescence, and vimentin-immunoreactivity has disappeared in all muscle cells, but it is found in cells regarded as precursors of interstitial cells of Cajal.

Fibronectin variants, laminins and laminin-binding integrin subunits, are expressed according to specific patterns, which have been investigated mainly in vascular muscles (Glukhova et al. 1990, 1993). Similar patterns are found in some visceral muscles, such as that of the human colon, and the myoepithelial cells of the mammary gland (Deugnier et al. 1995). Developing muscle cells in the bladder, intestine and aorta express neural cell adhesion molecule (NCAM) polypeptides in the rat at around embryonic day 17: this NCAM immunoreactivity becomes undetectable soon after birth (Akeson at al. 1988).

6 Growth of Visceral Smooth Muscles

A vast increase in volume occurs in developing smooth muscles from the stage at which they are identified to adulthood. The gizzard (some 90% of which is constituted by smooth muscle) increases in weight from 60 mg in 10-day-old chick embryos to about 60 g in 6-month-old chickens (Fig. 5). The 1,000-fold

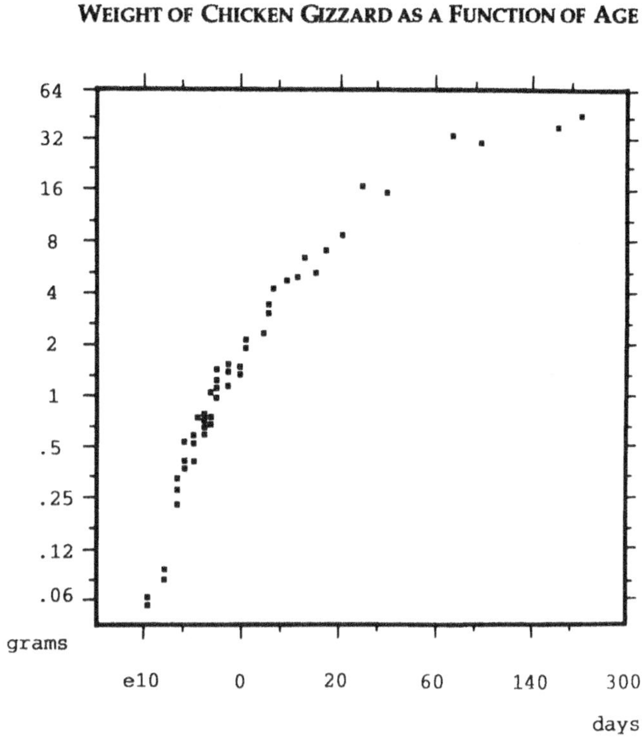

Fig. 5. Weight of chicken gizzard as a function of age. Relation of the weight of the gizzard (in grams) and the age of the animal, from the 10th day of the embryonic life (*e10*) through hatching (*0*) to 6 months of age, plotted on a base 2 log-log scale (each division represents a doubling either of the age or of the weight). Each *dot* is one measurement. (Gabella 1989)

increase is accounted for by an increase in muscle cell size and by an increase in muscle cell number, the latter factor being the more conspicuous. A perinatal increase in the size of the muscle is also well documented by Ward and Takahashi (1995) in the canine colon: from mid-gestation to 30 days of age, the thickness of the muscle coat increases from 42 to 317 μm.

As discussed in section 4, the muscle cell volume in viscera grows by a factor of three to five from the embryo to the adult. Therefore, a large part of the increase in the volume of the whole muscle must be accounted for by an increase in the number of muscle cells, apart from an increase in the amount of extracellular space, which, in the case of visceral smooth muscles, is very limited. There are four possible ways for an increase in the muscle cell number: (1) there may be differentiation of intramuscular non-muscle cells into muscle cells; (2) there may be migration of cells into the muscle, and differentiation; (3) there may be addition of muscle cells at the surface of a muscle by differentiation of surrounding cells; and (4) there may be mitosis of existing muscle cells. In the opinion of this writer the last of the four factors is by far the most significant (see Sect. 6).

As to the increase by late differentiation of intramuscular non-muscle cells (point 1 above), studies by Northern blot analysis have identified immature smooth muscle myoblasts in late embryogenesis/early postnatal development, amidst relatively mature muscle cells (McHugh et al. 1991; Liddell et al. 1993) in the gut of rodents; in the opinion of these authors a smooth muscle, therefore, is composed of a heterogeneous muscle cell population, showing a "wide range of cellular phenotypes". The migration of cells into and out of a developing smooth muscle (point 2 above) is vaguely implicit in some accounts, but it has not been clearly observed in any muscle. The addition of muscle cells at the surface or at one of the surfaces of a muscle is an interesting possibility. This form of growth, by accretion or deposition (as opposed to by intussusception, see p. 23), would produce arrays of muscle cells of different age, which are possibly phenotypically distinct. Data from the myometrium and bladder detrusor (Konishi et al. 1984; Fujii et al. 1989; Wu et al. 1999) may fit with this mode of growth (as discussed on p. 19).

7 Cell Division and Increase in Cell Number

The bulk of primordial smooth muscle is produced by mitotic division of precursor cells. In the well-studied chick embryo gizzard (and similarly in other visceral muscles) there is intense mitotic activity in the mesenchymal cells at day 6 of incubation, which is during the period preceding the start of smooth muscle differentiation and synthesis of smooth muscle proteins (Bennett and Cobb 1969).

At later stages of development, the muscle cells of many viscera (intestine, airways, bladder) undergo mitosis on a significant scale (Fig. 6A–D). Characteristically, the dividing cells are not undifferentiated muscle cells or muscle

Fig. 6. Dividing muscle cells in the gizzard of chick embryos examined in the electron microscope. **A** Transverse section of the musculature of 20-day-old embryo. The large cell profile in the centre is a muscle cell undergoing mitotic division and showing chromosomes (*g*), paired cisternae of endoplasmic reticulum (*p*) and a centriole (*o*); at the periphery of the profile there are bundles of myofilaments and other organelles. To the *bottom left* is a fibroblast; the other cell profiles are muscle cells. *Calibration bar* 2 μm. **B** Longitudinal section of a muscle cell in mitosis from a 19-day-old embryo. The ovoid part of the cell (*left*), where the division occurs, shows chromosomes (*g*), round mitochondria and long cisternae of endoplasmic reticulum (*r*). The remaining part of the cell shown to the *right* is packed with myofilaments, elongated mitochondria and other organelles, and resembles in structure the adjacent muscle cells. *Calibration bar* 2.5 μm. **C** Two muscle cells linked end-to-end by a characteristic cytoplasmic bridge, at the site where the constriction ring formed between the daughter cells at the end of mitosis. From the ring microtubules are seen to project into both cells. Lipid vacuoles (*v*) are visible in the cell at bottom. *Calibration bar* 1 μm. **D** A post-mitotic intercellular bridge in transverse section, showing the regular array of microtubule and the incrustation of electron-dense material on the inner side of the plasma membrane; there are also a few tubules of endoplasmic reticulum. *Calibration bar* 0.5 μm. (Modified from Gabella 1989)

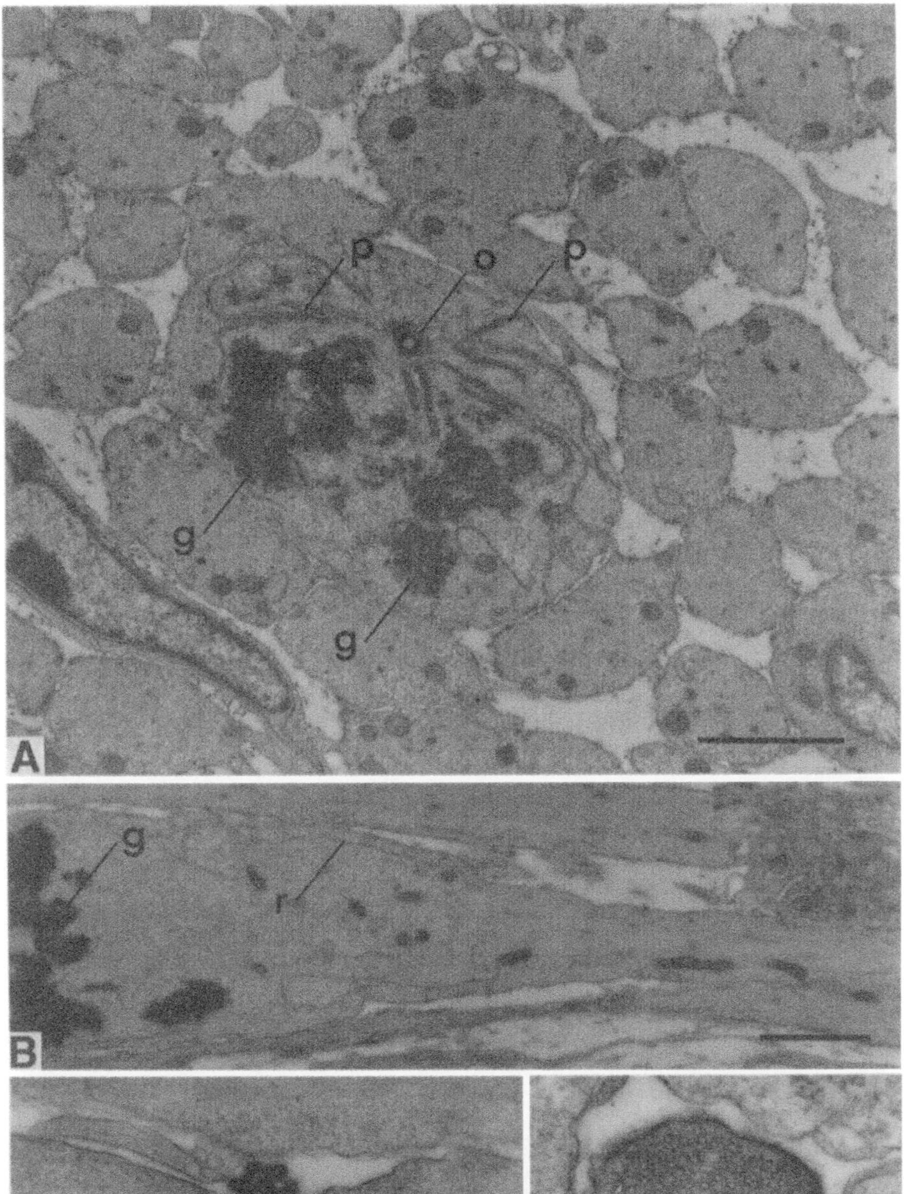
p
o
p
g
g
A
g
r
B
v
C
D

cells that first de-differentiate; they are well-developed muscle cells (Bennett and Cobb 1969; Cobb and Bennett 1970; Gabella 1989). De-differentiating muscle cells or mesenchymal cells undergoing late differentiation is not observed. However, in other developing smooth muscles, such as the ureter and vas deferens, muscle cells in mitosis have not been observed (Leeson and Leeson 1965; Yamauchi and Burnstock 1970), and in cultures of gizzard muscle cells in vitro mitoses occur only among the muscle cells that have de-differentiated (Campbell et al. 1974). If mitoses are absent in the developing muscle cells of the ureter and vas deferens (as yet there are no thymidine studies of these muscles), and if the number of muscle cells increases (which seems likely, but not yet documented), then migration and differentiation of mesenchymal cells, within or at the edges of an already existing muscle, as discussed in section 6, is a possible alternative mechanism that should be investigated.

Mitosis of smooth muscle cells without de-differentiation is seen on a large scale in the chicken gizzard towards the end of the embryonic life (Gabella 1989; Fig. 6). A similar process is observed in the small and large intestine of the guinea-pig, where the mitotic peak for muscle cells occurs around the time of birth or soon afterwards. Studies with tritiated thymidine have shown intense uptake (during a 30-min pulse) in muscle cells of both the longitudinal and the circular layer (Gabella 1992; Fig. 7). The label is entirely localized in the nuclei. The silver granules outline the elongated muscle cell nuclei and this confirms that the dividing cells are the differentiated muscle cells of the muscle

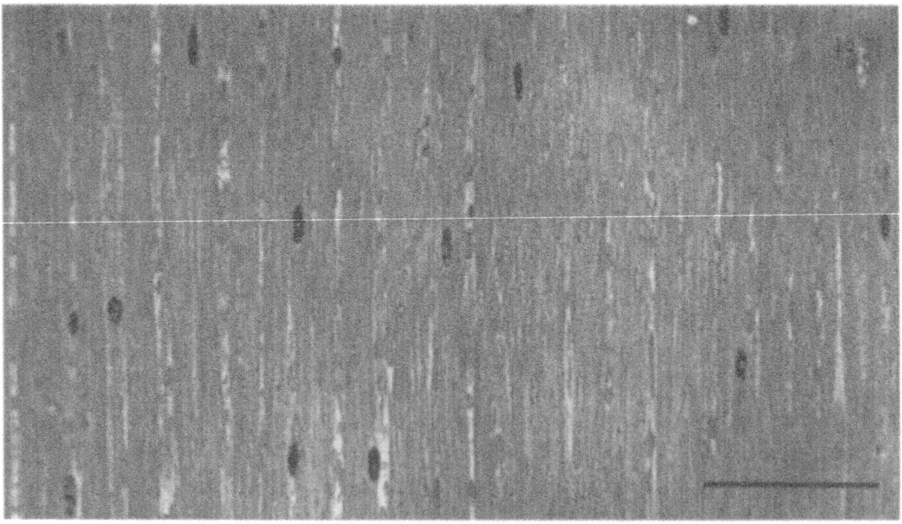

Fig. 7. Circular musculature of the proximal colon of a 1-day-old guinea-pig, sectioned on a plane parallel to the serosal surface. The animal was injected with tritiated thymidine; 30 min later the tissue was fixed and processed for autoradiography. The *dark elongated patches* are clusters of silver grain on some muscle cell nuclei and they indicate uptake of thymidine by cells preparing to divide. *Calibration bar* 100 µm. (Gabella 1992)

bundles. Muscle cells actually undergoing division are not labelled with tritiated thymidine, because the DNA replication had been completed before the injection of the tracer. Tritiated thymidine extensively labels fibroblasts and/or interstitial cells (they cannot be distinguished from one another in these preparations) and endothelial cells. Some Schwann cells in the intramuscular nerves and glial cells in the myenteric plexus are labelled, but not any enteric neurons.

In both avian and mammalian intestinal muscles, the dividing muscle cells are located well within a muscle bundle (Figs. 6A and 7); during the process of mitosis the cells maintain their elongated shape and remain connected to their neighbours by cell junctions, including gap junctions (Cobb and Bennett 1970; Gabella 1989; Fig. 6B). Except in the middle region, where the actual division takes place, the dividing muscle cells are indistinguishable from their neighbours, as regards their organelles, filament complement and membrane features. The middle region of the cell, where the chromosomes appear, becomes wider than the rest of the cell. Paired cisternae of reticulum, probably originating from the break-up of the nuclear envelope (Taura 1978) appear near the chromosomes. The equatorial plate of chromosomes is invariably oriented across the cell length, and the cell divides transversely. For some time afterwards, the daughter cells remain connected lengthwise by a cytoplasmic bridge occupied by microtubules (Fig. 6C,D).

There is no apparent pattern in the distribution of dividing muscle cells. Muscle cells in mitosis and muscle cells labelled with tritiated thymidine are observed in any part of a muscle layer. There is no preferential localization of dividing cells near either surface of a muscle layer or, in the case of the gizzard, near the insertion on the tendons (Gabella 1989, 1992). Cobb and Bennett (1970) observed that, in the developing chicken gizzard, smooth muscle cells in mitosis occurred in small groups, surrounded by cells in interphase, rather than isolated dividing cells, but there is no mention that these groups might be localized only in some regions of the very bulky musculature of the gizzard.

This type of growth by division of muscle cells in place is termed intussusceptive as it involves the addition of new cells within the texture of the muscle rather than at its surface or its ends. In some way, the muscle thus increases in thickness, length and width by growth from within rather than by addition to its borders. Intussusceptive growth implies that the muscle has enough plasticity in its cell-to-cell and cell-to-stroma junctions to accommodate new cells generated within a tight array of differentiated and working cells.

The occurrence of mitosis in differentiated muscle cells shows that, in visceral muscles, even after the unfolding of differentiation and the expression of major markers of a muscle cell, including the synthesis of myosin and the formation of cell-to-cell junctions, the muscle cell remains capable of dividing and do so. The same is true of developing cardiac muscle cells (Zak 1973) whereas, in skeletal muscles, myoblasts stop dividing at the onset of myosin expression (Moss and Leblond 1970).

If one takes the data on chicken gizzard as representative of visceral muscle development (because similar results are obtained from chicken and rat ileum and from guinea-pig taenia coli), then the 1,000-fold increase in muscle volume is accounted for unevenly by muscle cell size increase and increase in muscle cell number. Given that cell volume increases by a factor of three-to five, there must be an increase in cell number by a factor of 200–330. If the increase in muscle cell number is accounted for by mitoses and if the mitotic load is born uniformly by all muscle cells, then each of the muscle cells at embryonic day 10 must divide at least eight times to produce the final population.

Mitoses among muscle cells are found throughout the entire period of development and, on a small scale, they occur even in muscle cells of the adult. Smooth muscle cells in mitosis are common in hypertrophying intestinal musculature in adult animals (Gabella 1990).

8 Extracellular Materials and Vascularization

The stroma (or matrix) is the material filling the space between the muscle cells; it includes fibrils, of collagen, elastin and fibronectin, and amorphous materials. Smooth muscle cells are capable of, and are involved in, the synthesis of these materials, which usually assemble extracellularly in vast molecular complexes. The stroma has specific features in many different smooth muscles, and they have a major significance in the muscle's mechanical properties. For example, changes in the amount of extracellular materials (in addition to changes in muscle cell tone) account for the increase in compliance of the bovine bladder during embryonic life (Coplan et al. 1994). As to its origin, fibroblasts, located in the interstices between muscle bundles, contribute to the production of the stroma, but the muscle cells themselves are probably the main contributors. In the media of large blood vessels, muscle cells are probably the only source of the stroma, and this process is well characterized, for example in the rat aorta (Davis 1993).

The production of stroma goes hand in hand with the synthesis of the contractile apparatus by the same cells, and the two processes must be co-ordinated in the sense that the mechanical properties of these two components of the muscle must be finely matched. The process involves co-ordinated activations of two sets of genes: genes for the contractile apparatus proteins and genes for the stromal proteins, linked ultimately through the mechanical effects of their products. Commonly the early collagen fibrils that appear in the muscle lie close to a point of the plasma membrane bearing a dense band and are associated with it by some dense material (Fig. 4B). This observation suggests that there is a mechanical link between the two structures, which allows for the force generated by the contractile apparatus to be discharged onto the stroma and along the muscle. In addition, it suggests the possibility that the cell's dense bands and the stromal fibrils somehow appear and develop in parallel.

The importance of the stroma rests also on the formation of receptor–ligand links between intrinsic membrane proteins (e.g. integrins) and specific/

selective components of the stroma; the establishment of these links generate transmembrane signals that activate pathways to genes, surely a key mechanism in the unfolding of the muscle structural development.

Most visceral muscles are associated with blood vessels (capillaries) from early development. As soon as the muscle reaches an adequate thickness, blood vessels are found in it, running close and parallel to muscle cells. In the gizzard of 6-day-old chick embryos, mesenchymal cells at the mucosal side of the muscle are identified histochemically as putative angiogenetic cells; at day 8 these cells begin to cluster, and then they merge and migrate to form capillaries that invade the muscle in a mucosal to serosal direction (Paul et al. 1994).

9 Origin of Smooth Muscle Precursors

The precursors of most smooth muscles are cells of the lateral plate mesoderm. Both the somatic or somatopleural sheet of the lateral plate and the splanchnic or splanchnopleural sheet contribute the precursors of vascular and visceral muscles (Bellairs et al. 1992; Christ and Ordahl 1995). The muscle cells of the gut and of its blood vessels originate from the splanchnic mesoderm. Vascular smooth muscles originate also from the segmented para-axial mesoderm (Christ and Ordahl 1995).

An additional major source of smooth muscle cell precursors (mainly for vascular muscles) is the neural crest. The neuroectodermal structure gives rise to mesoectodermal derivatives, including the muscles of the large arteries derived from the aortic arches and the smooth musculature in the skin and face and ventral part of neck (Le Lièvre and Le Douarin 1975; Le Douarin and Kalcheim 1999). The nerves for smooth muscles are all of neural crest origin. It is possible that some of the myofibroblasts in the seminiferous tubules are of neural crest origin (Holstein et al. 1996).

The iris, with its dilator and sphincter pupillae muscles, contains myoepithelial cells, smooth muscle cells and striated muscle fibres in various proportions in different species. In the rat iris, where all the musculature is of the smooth type, muscle cells derive from epithelial cells of the optic cup (Lai 1972), and the small component of smooth muscle of the iris of amphibians and birds also derives from the epithelium (Tonosaki and Kelly 1971). In the chick embryo, cells can be clearly seen emerging from the anterior epithelium and differentiating into smooth muscle cells near the pupillary margin in 10-day-old embryos (Gabella and Clarke 1983); the earliest signs of this process are seen on embryonic day 9 in the chick (hatching at 20 days), on day 8 in the quail (hatching at 17 days) and on day 12 in the duck (hatching at 27 days; Yamashita and Sohal 1986). The striated muscle fibres of the iris begin to appear 2 days later and they originate from mesenchymal cells within the entire stroma of the iris (Gabella and Clarke 1983; Yamashita and Sohal 1986). Earlier studies, however, have concluded that the iridial striated musculature originates by transformation of smooth muscle cells or myoepithelial cells

(Reger 1966; Lucchi et al. 1974; Mussini et al. 1976; Pilar et al. 1981) or directly from the iridial epithelium (Ferrari and Koch 1984).

Transdifferentiation. The question of a possible transformation of smooth muscle into striated muscle has attracted much attention. This issue remains controversial, in spite of firm claims that transdifferentiation does occur. The most abundant observations on smooth-to-striated differentiation comes from the chick embryo iris where modern authors (Pilar et al. 1987; Scapolo 1988; Volpe et al. 1993; Link and Nishi 1998) describe smooth muscle cells (which are of epithelial origin) transforming into striated fibres. These observations are in direct contrast with the suggestion that the striated musculature of the iris differentiate directly from mesenchymal cells (Gabella and Clarke 1983), and that the two types of muscle intermix but have separate origin (Fig. 8).

Borirakchanyavat and colleagues (1997) studied the parallel development of smooth and striated muscle in the urethral sphincters of the rat. In this light microscope study the authors observed transient co-expression of alpha-smooth muscle actin and alpha-sarcomeric actin in the same tissue, but they point out that alpha-smooth muscle actin, and other smooth muscle isoactins, are expressed in many non-muscle tissues during development. Transdifferentation is mentioned as one possible hypothesis, but the authors are very careful not to suggest that transdifferentiation does occur, a view, however, that has been attributed to them in the subsequent literature.

The strongest observations on transdifferentiation were made on the mouse oesophagus (Patopoutian et al. 1995; Stratton et al. 2000), where ultrastructural and histochemical evidence is provided to support the hypothesis of a direct transformation of smooth muscle cells into striated muscle fibres.

In spite of the considerable weight of evidence and the amount of experimental work, scepticism on this issue is not entirely unreasonable and the present reviewer regards the case for a transdifferentiation of smooth muscle cells into striated fibres (visceral striated fibres, rather than skeletal, that is fibres which are not inserted via a tendon to either bone or cartilage) as still unproven.

10 Influence of Endothelium, Epithelium, Connective Tissue and Nerves on Smooth Muscle Development

10.1 Role of Endothelium and Epithelium

The early vascular muscle cells derive from mesenchymal cells that have come to lie very close to the endothelium. A role for the endothelium in vascular muscle differentiation is suggested not only by its spatial relationship but also by the ability of endothelial cells to secrete a variety of factors that influence proliferation and growth of muscle as reviewed by Owen (1995). As regards visceral muscles, a role of the mucosal epithelium or the serosal mesothelium

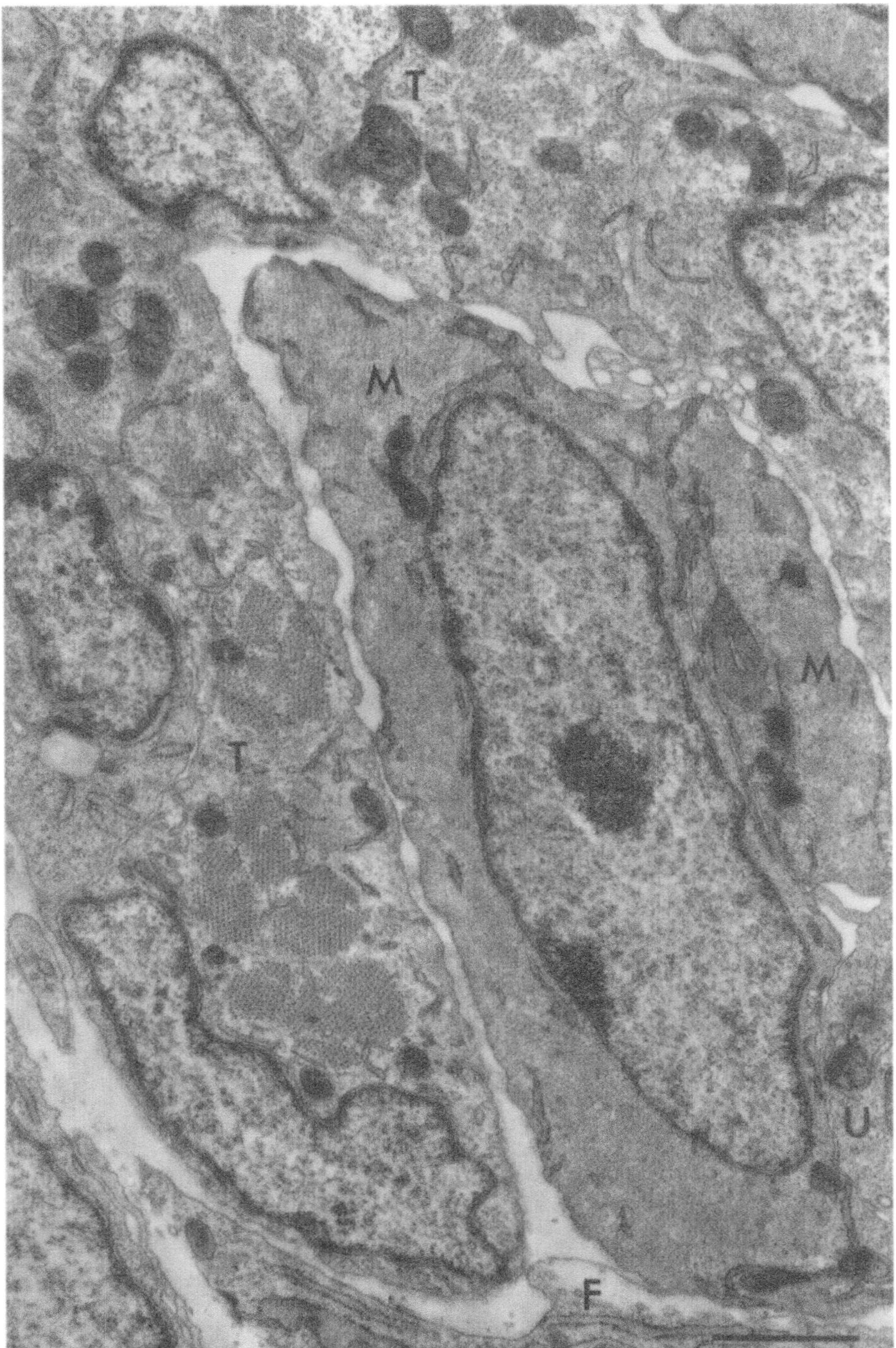

Fig. 8. Sphinter pupillae of a 14-day-old chick embryo, sectioned transversely (radial section of the iris). The electron micrograph shows smooth *M* muscle cells; *T* striated muscle fibres or myotubes at various stages of maturation; *F* fibroblasts; *U* undifferentiated cells. *Calibration bar* 1 µm (18.4 mm.) (Modified from Gabella and Clarke 1983)

is difficult to ascertain. In the wall of the intestine of the mouse embryo there is cell signalling, via members of the Hedgehog family of signals, between the epithelium, endodermally derived, and mesodermal derivatives such as the musculature (Ramalho-Santos et al. 2000). Both *Sonic Hedgehog* (*Shh*) and *Indian Hedgehog* (*Ihh*) are expressed in the gut epithelium of controls, and both *Shh* and *Ihh* mutant mice have large malformations of the intestinal wall and a reduction of the amount of smooth muscle (Ramalho-Santos et al. 2000); *Ihh* mutant mice lack smooth muscle in their lungs (Pepicelli et al. 1998).

The uterine epithelium may play an important role in the differentiation of the myometrium. Grafts of mesenchyme from the uterus of neonate mice differentiate into smooth muscle when grafted under the renal capsule of syngenic female hosts – but only if the graft contains some epithelium (Cunha et al. 1989). The epithelium remains effective in its promotional role on the differentiation of smooth muscle from mesenchyme of the neonate mouse, even when it derives from an adult mouse. However, in the uterus, the mesenchymal cells in the region adjacent to the epithelium differentiate into fibroblasts that are part of the endometrial stroma. The earliest muscle cells differentiate from mesenchymal cells lying close to the serosa of the organ (Brody and Cunha 1989a). In the intestine and the bladder too, the early layer of musculature lies closer to the serosa than to the mucosa. Only in the bronchi does the musculature develop close to the epithelium, a position that it maintains through life.

10.2 Role of Mesenchymal Cells and the Extracellular Matrix

A role for fibroblasts in smooth muscle development is proposed by Ko et al. (1996). In explants of fragments of gizzard of 7-day-old chick embryos there is migration of fibroblast-like cells, followed by migration of round cells that spread over the fibroblasts (identified by fibronectin-immunoreactivity) and differentiate into muscle cells (identified by a ribbon shape and smooth-muscle-myosin immunoreactivity).

Developmental interaction between muscle cells and fibroblasts is more in evidence in vascular muscles (reviewed by Carey 1991; Owen 1995).

10.3 Role of Nerves

Nerves have a strong influence on many adult smooth muscles. Denervation of the bladder detrusor muscle or the intrinsic ocular muscles leads to paralysis. In other smooth muscles the effect of nerves, and the consequences of denervation, are more subtle. Myogenic activity, extensive ionic and mechanical coupling between the muscle cells, sensitivity to physical factors such as temperature and stretch, presence of receptors for hormones and other active substances, all these factors make smooth muscles less acutely dependent on the activity of nerve than for example skeletal muscles.

In development, smooth muscle differentiation starts before the arrival of nerves. Colonization of the gut by neural crest neurons takes place when some muscle has already differentiated. There is some evidence that the same is true not only for the gut, but also for the airways, uterus and bladder. However, in the sphincter pupillae of the chick, axons running by the smooth muscle cells are observed from the earliest stage of development (Gabella and Clarke 1983).

In chick embryos, when segments of the gut are excised before nerve colonization (aneural hindgut of embryonic day 4) and are transplanted onto chorio-allantoid membrane and followed for 10 days, they display an apparently normal development (Lecoin et al. 1996; Fig. 9). In this case, the differentiation and growth of the intestinal musculature clearly continues and progresses in a similar manner in the aneural intestine as in the control, richly-innervated gut. These experiments show that in the avian intestine, suppression of the arrival of nerves into the muscle does not stop muscle growth (although the exact extent of growth was not determined) and differentiation, nor does it affect the organogenesis of the intestinal wall, its layered structure and admixing of several tissues (Lecoin et al. 1996). In contrast, in the lethal spotted mouse, which has a congenital aganglionosis of part of the large intestine, there is malformation of both circular and longitudinal muscle layers (Tennyson et al. 1986), an observation that may suggest that normal innervation plays a role in the development of that musculature. In chicken gizzard, smooth muscle myosin immunoreactive muscle cells are first found close to neuroblasts of the Auerbach's plexus; with further development, cells expressing myosin are found progressively deeper into the wall in a wave that appears to follow the growth of nerve fibres emerging from the Auerbach's plexus (Zimmerman et al. 1995).

11 Development of Mechanical Activity

Smooth muscles contract and produce mechanical work from early in their development. Donahue and Bowen (1972) observed spontaneous contractions in the gizzard of 8-day-old chick embryos, a time at which the differentiation of the contractile apparatus has barely begun, there is no stroma around the muscle cells and the muscle is not innervated. Freshly isolated muscle cells of the chicken gizzard were contractile when obtained from embryos of 7 days or older, but not when obtained from 5-day-old embryos (Hirai and Hirabayashi 1986). There is peristaltic activity in the intestine of mammalian embryos (Neu 1989).

It appears that in all smooth muscles mechanical activity (probably including tone as well as phasic contractions) is present very early in development. Therefore, growth and differentiation take place while the muscle is already

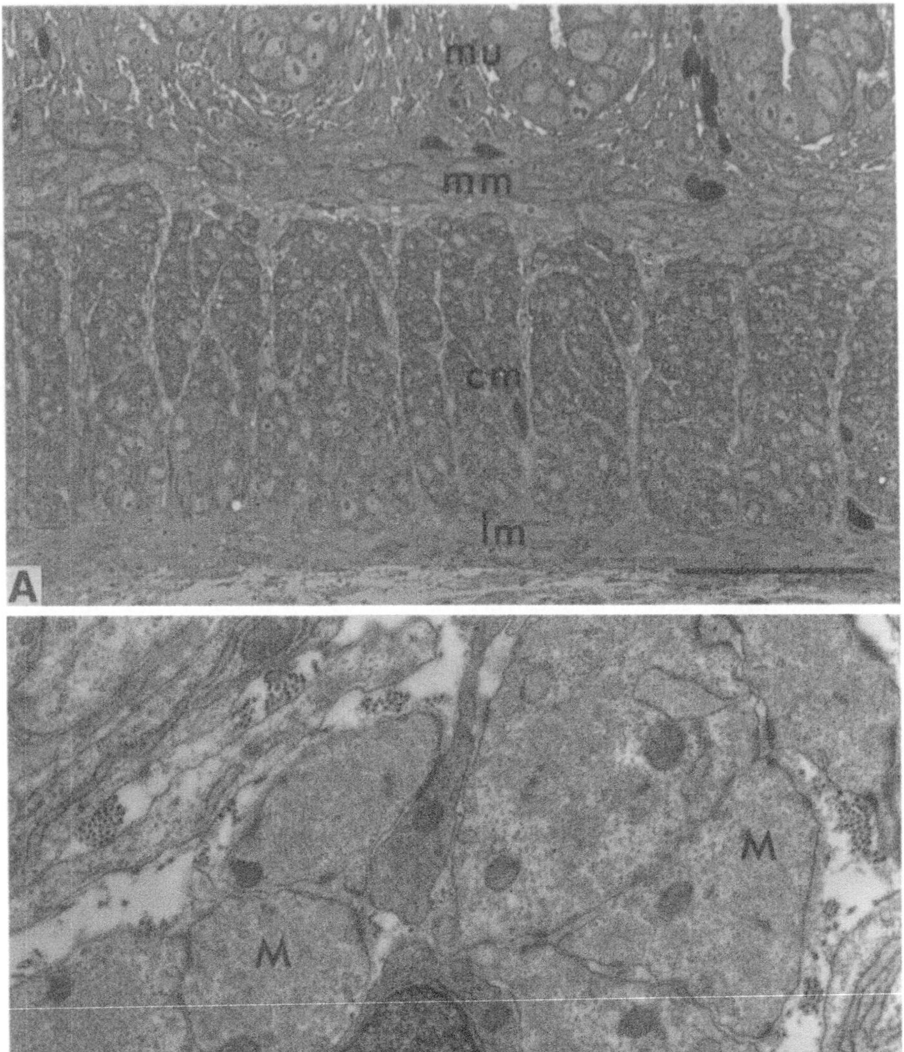
mu
mm
cm
lm
A
M
M
IC
F
B

doing mechanical work. On the one hand, the intrinsic growth of a contractile machine that is already performing work poses complex structural demands: addition of new cells, secretion of extracellular materials, growth in cell size and formation of attachment to the stroma. On the other hand, it is probably the mechanical activity (which is also a way of sensing the physical environment of the muscle) that feeds back onto the muscle to trigger and regulate the cell's synthetic processes in order to match growth with functional demands.

12 Related Processes of Development and Growth

The process of smooth muscle development, which occurs early in life, has features in common with smooth muscle tissue repair in the adult, with hypertrophic growth (as is found in the intestine, bladder, uterus), and with pathological growth (myoepithelioma, myoma). Similarities are also found with muscle cells that differentiate and grow in vitro (Campbell et al. 1974) and with in vivo transplants of smooth muscles after excision and mincing (Burnstock 1981).

A clear distinction between differentiation and simple growth is impossible. Furthermore, biological growth is usually allometric, and this involves qualitative changes reminiscent of differentiation. Also, a clear distinction between a life period of development and the stage of maturity is difficult. In rodents, body growth continues throughout life and is mirrored in individual tissues. Smooth muscles such as the taenia coli of the guinea-pig maintain a slow increase in cell number by mitosis and a continuous increase in cell size (volume) through ageing.

In addition to standard growth, smooth muscles can undergo – in adults and even in ageing subjects – hypertrophic growth (reviewed in Gabella 1991). Hypertrophic growth is induced by increased functional demands on the muscle. For example, partial obstruction of a segment of intestine is followed by hypertrophy of the ileal loops on the oral side; with partial outlet obstruction of the bladder, the musculature hypertrophies. The massive enlargement of the muscle (a tenfold increase in muscle volume) is accounted for by an

Fig. 9. Chick intestine developing in the absence of nerves. A segment of hindgut explanted from a 4-day-old chick embryo and cultivated without nerve tissue on the chorio-allantoic membrane of a chick host. **A** Longitudinal section of the wall examined by light microscopy. Longitudinal (*lm*) and circular (*cm*) muscle layers have an appearance similar to controls. A muscularis mucosae (*mm*) is present and the mucosa (*mu*) shows the bottom of villi. Note the absence of nerve ganglia. *Calibration bar* 25 μm. **B** In the electron microscope a transverse section of the circular muscle layer shows muscle cells (*M*) with the standard appearance for the age. An interstitial cell of Cajal (*IC*) and a fibroblast (*F*) are visible. A few collagen fibrils are present in the intercellular spaces. *Calibration bar*: 1 μm. (Modified from Lecoin et al. 1996, with permission of the publishers) increase in cell size and by mitotic division of the existing muscle cells (Gabella 1991).

In the bladder with outlet partial obstruction, connective tissue cells of the serosa turn into myofibroblasts and subsequently into smooth muscle cells (all cell types identified with histochemical markers; Buoro et al. 1993). The development in the rabbit bladder of new muscle cells from multi-potent mesenchymal cells was suggested also by Johnson et al. (1962) and Brent and Stephens (1975).

13 Synopsis

The development of the smooth musculature of viscera has attracted the interest of only relatively few investigators, and thus the field appears somewhat underexplored. The major emphasis on histochemical evidence - at the expense of ultrastructural and functional studies - may have limited the progress in this area.

Mature tissue is formed through the differentiation of precursors into muscle cells and through the organization of these cells into a complex tissue where distribution and orientation of muscle cells, deployment of abundant extracellular materials and addition of other cellular elements (interstitial cells, fibroblasts, nerves, blood vessels) are characteristic and specific features. The precursor cells are found at sites where a muscle develops, and they derive predominantly from the mesoderm, but also from the neuroectoderm and from the endoderm. The process starts at different times in different organs.

The earliest stages of differentiation are characterized by the precursor cells aggregating and becoming elongated; their longitudinal axis lies in a position similar to the one they will have in the mature muscle. Both the cytological and the histochemical differentiation follow distinct patterns in various muscles, with characteristic temporal sequences in the appearance of key features. This process must impart distinct functional properties to a muscle cell at each stage of its development. However, the chronological correspondence between ultrastructural and histochemical development is poorly understood. Histochemical studies have detected gradients of maturation of the muscle cells, for example, across the thickness of the gizzard musculature and along the length of the small intestine; ultrastructural studies have not yet confirmed the existence of these gradients.

Muscle growth is accounted for by muscle cell enlargement (without nucleus duplication) and an increase in muscle cell number by mitosis of pre-existing differentiated muscle cells. De-differentiation and division of muscle cells, migration of muscle cells and late development of muscle cell precursors have all also been considered as possible mechanisms for muscle growth. Several authors have described the presence of precursor cells within developing smooth muscles, and they have described late differentiation of some muscle cells or waves of differentiation that would give rise to phenotypic heterogeneity of the mature muscle cell population. In contrast, other studies, mainly by electron microscopy, have suggested that, within large visceral muscles, the muscle cells differentiate synchronously.

There are interesting data on the influence of adjacent tissues on the development of a smooth muscle, but the interplay of these and other factors has not been fully investigated. Smooth muscles contract from early in their development, hence mechanical factors are likely to influence development: on the one hand, passive stresses imposed on the muscle by other tissues, such as adjacent muscles or the contents of the viscera and, on the other hand, active forces generated by the muscle itself. The very attraction of visceral smooth muscles in the study of cellular morphogenesis - an attraction that has not yet been highlighted or exploited in scientific studies, either descriptively or experimentally - is that, onto a single type of cell, a large range of factors interact, such as the genetic expression, chemical influences (from other muscles, endocrine glands, nerves, other intramuscular cells) and mechanical factors.

References

Akeson RA, Wujek JR, Roe S, Warren SL, Small SJ (1988) Smooth muscle cells transiently express NCAM. Mol Brain Res 4:107–120

Bal HS, Getty R (1970) Postnatal growth of the swine uterus from birth to six months. Growth 34:15–30

Baskin LS, Hayward SW, Young PF, Cunha GR (1996) Ontogeny of the rat bladder: smooth muscle and epithelial differentiation. Acta Anat 155:163–171

Bellairs R, Sanders EF, Lash JW (eds) (1992) Formation and differentiation of early embryonic mesoderm. Plenum, New York

Bennett T, Cobb JLS (1969) Studies of the avian gizzard: the development of the gizzard and its innervation. Z Zellforsch Mikrosk Anat 98:599–621

Blessing MH, Müller G (1973) Myoglobin concentration in the chicken, especially in the gizzard. Comp Biochem Physiol [A] 47:534–540

Bo WJ, Odor DL, Rothrock ML (1968) Ultrastructure of uterine smooth muscle following progesterone or progesterone-estrogen treatment. Anat Rec 163:121–132

Borirakchanyavat S, Baskin LS, Kogan BA, Cunha GR (1997) Smooth and striated muscle development in the intrinsic urethral sphincter. J Urol 158:1119–1122

Brent L, Stephens FD (1975) The response of smooth muscle cells in the rabbit urinary bladder to outflow obstruction. Invest Urol 12:494–502

Brittingham J, McHugh K (1992) Characterization of isoactin gene expression in primary cultures of rat gastrointestinal smooth muscle cells. Mol Biol Cell 5:158a

Brody JR, Cunha GR (1989a) Histologic, morphometric, and immunocytochemical analysis of myometrial development in rats and mice. I. Normal development. Am J Anat 186:1–20

Brody JR, Cunha GR (1989b) Histologic, morphometric, and immunocytochemical analysis of myometrial development in rats and mice. II. Effect of DES on development. Am J Anat 186: 21–42

Buoro S, Ferrarese P, Chiavegato A, Roelofs M, Scatena M, Pauletto P, Passerini-Glazel G, Pagano F, Sartore S (1993) Myofibroblast-derived smooth muscle cells during remodelling of rabbit urinary bladder wall induced by partial outflow obstruction. Lab Invest 69:589–602

Burnstock G (1981) Development of smooth muscle and its innervation. In: Bülbring E, Brading AF, Jone AW, Tomita T (eds) Smooth muscle: an assessment of current knowledge. Arnold, London, pp 431–457

Campbell GR, Chamley JH, Burnstock G (1974) Development of smooth muscle cells in tissue culture. J Anat (Lond) 117:295–312
Carey DJ (1991) Control of growth and differentiation of vascular cells by extracellular matrix proteins. Annu Rev Physiol 53:61–177
Chamley-Campbell J, Campbell G, Ross R (1979) The smooth muscle cell in culture. Physiol Rev 59:1–61
Christ B, Ordahl CP (1995) Early stages of chick somite development. Anat Embryol 191:381–396
Cobb JLS, Bennett T (1970) An ultrastructural study of mitotic division in differentiated gastric smooth muscle cells. Z Zellforsch Mikrosk Anat 108:177–189
Collett AJ, Des Beins GD (1974) Fine structure of myogenesis and elastogenesis in the developing rat lung. Anat Rec 179:343–360
Coplan DE, Macarak EJ, Levin RM (1994) Developmental changes in normal fetal bovine whole bladder physiology. J Urol 151:1391–1395
Cunha GR, Young P, Brody JR (1989) Role of uterine epithelium in the development of myometrial smooth muscle cells. Biol Reprod 40:861–871
Davis EC (1993) Stability of elastin in the developing mouse aorta: a quantitative radioautographic study. Histochemistry 100:17–26
Deugnier M-A, Moiseyeva EP, Thiery JP, Glukhova M (1995) Myoepithelial cell differentiation in the developing mammary gland: progressive acquisition of smooth muscle phenotype. Dev Dyn 204:107–117
Donahoe JR, Bowen JM (1972) Analysis of the spontaneous motility of the avian embryonic gizzard. Am J Vet Res 33:1835–1848
Duband J-L, Gimona M, Scatena M, Sartore S, Small JV (1993) Calponin and SM 22 as differentiation markers of smooth muscle: spatiotemporal distribution during avian embryonic development. Differentiation 55:1–11
Eddinger TJ, Meer DP (1997) Myosin isoform heterogeneity in single smooth muscle cells. Comp Biochem Physiol 117B:29–38
Eddinger TJ, Murphy RA (1991) Developmental changes in actin and myosin heavy chain isoform expression in smooth muscle. Arch Biochem Biophys 284:232–237
Evans DH, Evans EM (1964) The membrane relationships of smooth muscles: an electron microscope study. J Anat (Lond) 98:37–46
Fatigati V, Murphy RA (1984) Actin and tropomyosin variants in smooth muscles. J Biol Chem 259:14384–14388
Ferrari PA, Koch WE (1984) Development of the iris in the chicken embryo. I. A study of growth and histodifferentiation utilizing immunocytochemistry for muscle differentiation. J Embryol Exp Morphol 81:153–167
Fujii S, Konishi I, Mori T (1989) Smooth muscle differentiation at endometrio-myometrial junction. Virchows Archiv A Pathol Anat 414:105–112
Gabella G (1989) Development of smooth muscle: ultrastructural study of the chick embryo gizzard. Anat Embryol 180:213–226
Gabella G (1990) Hypertrophy of visceral smooth muscle. Anat Embryol 182:409–424
Gabella G (1991) Ultrastructure of tracheal muscle in developing, adult and ageing guinea-pigs. Anat Embryol 183:71–79
Gabella G (1992) Intestinal smooth muscle development. In: Holle GE (ed) Advances in the innervation of the gastrointestinal tract. Elsevier, Amsterdam, pp 35–47
Gabella G, Clarke E (1983) Embryonic development of the smooth and striated musculatures of the chicken iris. Cell Tissue Res 229:37–59
Gabella G, Uvelius B (1990) Urinary bladder fine structure: normal and hypertrophic musculature. Cell Tissue Res 262:67–79
Geiger B, Dutton AH, Tokuyasu KT, Singer SJ (1981) Immunoelectron microscope studies of membrane–microfilament interaction. The distribution of α-actinin, tropomyosin and vinculin in intestinal epithelial brush border and chicken gizzard smooth muscle cells. J Cell Biol 91:614–628
Gershon MD, Sherman P, Gintzler AR (1981) An ultrastructural analysis of the developing enteric nervous system of the guinea-pig small intestine. J Neurocytol 10:271–296

Glukhova MA, Frid MG, Shekhonin BV, Balabanov YV, Koteliansky VE (1990) Expression of fibronectin variants in vascular and visceral smooth muscle cells in development. Dev Biol 141:193–202

Glukhova M, Koteliansky V, Fondacci C, Marotte F, Rappaport L (1993) Laminin variants and integrin laminin receptors in developing and adult human smooth muscle. Dev Biol 157:437–447

Gröschel-Stewart U, Jaroschik U, Schwalm H (1971) Chicken gizzard, a myosin-containing smooth muscle. Experientia 27:512

Hegele-Hartung C, Chwalisz K, Beier HM (1992) Distribution of estrogen and progesterone receptors in the uterus: an immunohistochemical study in the immature and adult pseudopregnant rabbit. Histochemistry 97:97–50

Hirai S, Hirabayashi T (1983) Developmental changes of protein constituents in chicken gizzards. Dev Biol 97:483–493

Hirai S, Hirabayashi T (1986) Development of myofibrils in the gizzard of chicken embryos. Intracellular distribution of structural proteins and development of contractility. Cell Tissue Res 243:487–493

Holstein AF, Maekawa M, Nagano T, Davidoff MS (1996) Myofibroblasts in the lamina propria of human seminiferous tubules are dynamic structures of heterogeneous phenotype. Arch Histol Cytol 59:109–125

Hood LS, Rosenquist TH (1992) Coronary artery development in the chick: origin and deployment of smooth muscle cells, and the effects of neural crest ablation. Anat Rec 234:291–300

Hungerford JE, Hoeffler JP, Bowers CW, Dahm LM, Falchetto R, Shabanowitz J, Hunt DF, Little CD (1997) Identification of a novel marker for primordial smooth muscle and its differential expression pattern in contractile vs noncontractile cells. J Cell Biol 137:925–937

Johnson AJ, Kinsey DL, Rehm RA (1962) Observations on bladder regeneration. J Urol 88:494–499

Kelley CA, Takahashi M, Yu JH, Adelstein RS (1993) An insert of seven amino acids confers functional differences between smooth muscle myosins from the intestine and vasculature. J Biol Chem 268:12848–12854

Ko J-A, Murahashi S, Arata T, Inoue A (1996) Differentiation of smooth muscle cells from undifferentiated cells of chicken gizzard occurs on the layer of fibroblast-like cells. Cell Tissue Res 285:395–401

Konishi I, Fujii S, Okamura H, Mori T (1984) Development of smooth muscle in the human fetal uterus: an ultrastructural study. J Anat 139:239–252

Kuroda M (1985) Change of actin isomers during differentiation of smooth muscle. Biochem Biophys Acta 843:208–213

La Mantia J, Shafiq SA (1982) Developmental changes in the plasma membrane of gizzard smooth muscle of the chicken. A freeze-fracture study. J Anat 134:243–253

Lai Y-L (1972) The development of the sphincter muscle in the iris of the albino rat. Exp Eye Res 14:196–202

Le Douarin N, Kalcheim C (1999) The neural crest. Cambridge University Press, Cambridge

Le Lièvre CS, Le Douarin NM (1975) Mesenchymal derivatives of the neural crest: analysis of chimaeric quail and chick embryos. J Embryol Exp Morphol 34:125–154

Lecoin L, Gabella G, Le Douarin N (1996) Origin of the *c*-kit-positive interstitial cells in the avian bowel. Development 122:725–733

Lees-Miller JP, Heeley DH, Smillie LB (1987) An abundant and novel protein of 22 kDa (SM22) is widely distributed in smooth muscles: purification from bovine aorta. Biochem J 244:705–709

Leeson TS, Leeson CR (1965) The rat ureter. Fine structural changes during its development. Acta Anat 62:60–70

Li L, Miano JM, Cserjesi P, Olson EN (1996) SM22α, a marker of adult smooth muscle, is expressed in multiple myogenic lineages during embryogenesis. Circ Res 78:188–195

Liddell RA, Syms M, McHugh KM (1993) Heterogeneous isoactin gene expression in the adult rat gastrointestinal tract. Gastroenterology 105:347–356

Link BA, Nishi R (1998) Development of the avian iris and ciliary body: mechanisms of cellular differentiation during the smooth-to-striated muscle transition. Dev Biol 203:163–176

Liu H-C (1962) The comparative structure of the ureter. Am J Anat 111:1–15

Lucchi ML, Bortolami R, Callegari E (1974) Fine structure of intrinsic eye muscles of birds: developmental and postnatal changes. J Submicrosc Cytol 6:205–218
Masumoto K, Nada O, Suita S, Taguchi T, Guo R (2000) The formation of the chick ileal muscle layers as revealed by α-smooth muscle actin immunohistochemistry. Anat Embryol 201: 121–129
Matsuno T, Tokunaka S, Koyanagi T (1984) Muscular development in the urinary tract. J Urol 132:148–152
McHugh KM (1995) Molecular analysis of smooth muscle development in the mouse. Dev Dyn 204:278–290
McHugh KM (1996) Molecular analysis of gastrointestinal muscle development. J Pediatr Gastroenterol Nutr 23:379–394
McHugh KM, Crawford K, Lessard JL (1991) A comprehensive analysis of the developmental and tissue-specific expression of the isoactin multigene family in the rat. Dev Biol 148:442–458
Miano JM, Olson EN (1996) Expression of the smooth muscle cell calponin gene marks the early cardiac and smooth muscle cell lineages during mouse embryogenesis. J Biol Chem 271: 7095–7103
Miano JM, Cserjesi P, Ligon KL, Periasamy M, Olson EN (1994) Smooth muscle myosin heavy chain exclusively marks the smooth muscle lineage during muscle embryogenesis. Circ Res 75:803–812
Moss F, Leblond C (1970) Nature of dividing nuclei in skeletal muscle of growing rats. J Cell Biol 44:459–462
Mussini I, Aloisi M, Lucke RS (1976) Transient smooth muscle features in developing chick striated muscle. J Submicrosc Cytol 8:256
Neu J (1989) Functional development of the fetal gastrointestinal tract. Semin Perinatol 13: 224–235
Newman J, Antonakopoulos GN (1989) The fine structure of the human fetal urinary bladder. Development and maturation. A light, transmission and scanning electron microscope study. J Anat 166:135–150
Olivetti G, Anversa P, Melissari M, Loud AV (1980) Morphometric study of early postnatal development of the thoracic aorta in the rat. Circ Res 47:417–424
Owen GK (1995) Regulation of differentiation of vascular smooth muscle cells. Physiol Rev 75: 487–517
Pasqualini JR, Sumida C, Gulino A, Tardy J, Nguyen BL, Gelly C, Cosquer-Clavreul C (1983) Progesterone receptors during fetal development. In: Bardin CW, Milgröm E, Mauvais-Jarkis P (eds) Progesterone and progestrins. Raven Press, New York
Patapoutian A, Wold BJ, Wagner RA (1995) Evidence for developmentally programmed transdifferentiation in mouse esophageal muscle. Science 270:1818–1821
Paul ER, Christian A-L, Franke R, Gröschel-Stewart U (1994) Embryonic chicken gizzard: smooth muscle and non-muscle myosin isoforms. Cell Tissue Res 276:381–386
Pepicelli CV, Lewis PM, McMahon AP (1998) Sonic hedgehog regulates branching morphogenesis in the mammalian lung. Curr Biol 8:1083–1086
Pilar G, Tuttle J, Vaca K (1981) Functional maturation of motor nerve terminals in the avian iris: ultrastructure, transmitter metabolism and synaptic reliability. J Physiol (Lond) 321: 175–193
Pilar G, Nunez R, McLennan ISS, Meriney SD (1987) Muscarinic and nicotinic synaptic activation of the developing chicken iris. J Neurosci 7:3813–3826
Ramalho-Santos M, Melton DA, Macmahon AP (2000) Hedgehog signals regulate multiple aspects of gastrointestinal development. Development 127:2763–2772
Reger JF (1966) The fine structure of iridial constrictor pupillae muscle of *Alligator mississipiensis.* Anat Rec 155:197–216
Ross R, Klebanoff SJ (1967) The structural changes in uterine smooth muscle and fibroblasts in response to estrogens. J Cell Biol 32:155–169
Rubenstein PA (1981) Differential behavior of gizzard isoactins. Arch Biochem Biophys 43: 151–161

Ruzicka D, Schwartz R (1988) Sequential activation of α-actin genes during avian cardiogenesis: vascular smooth muscle α-actin gene transcripts mark the onset of cardiomyocyte differentiation. J Cell Biol 107:2575–2586
Saborío JL, Segura M, Flores M, García R, Palmer E (1979) Differential expression of gizzard actin genes during chick embryogenesis. J Biol Chem 254:1119–1125
Saint-Jeannet J-P, Levi G, Girault J-M, Koteliansky V, Thiery J-P (1992) Ventrolateral regionalization of *Xenopus laevis* mesoderm is characterized by the expression of α-smooth muscle actin. Development 115:1165–1173
Sawtell N, Lessard J (1989) Cellular distribution of smooth muscle actins during mammalian embryogenesis: expression of the α-vascular but not the γ-enteric isoform in differentiating striated myocytes. J Cell Biol 109:2929–2937
Scapolo PA, Peirone SM, Filogamo G, Veggetti A (1988) Histochemical, immunohistochemical, and ultrastructural observation on the iris muscles of *Gallus gallus*. Anat Rec 221:687–699
Skalli O, Ropez P, Trzeciak A, Benzonana G, Gillessen D, Gabbiani G (1986) A monoclonal antibody against α-smooth muscle actin: a new probe for smooth muscle differentiation. J Cell Biol 103:2787–2796
Stratton CJ, Bayguinov Y, Sanders KM (2000) Ultrastructural analysis of the transdifferentation of smooth muscle to skeletal muscle in the murine esophagus. Cell Tissue Res 301:283–298
Stuewer D, Gröschel-Stewart U (1985) Expression of immunoreactive myosin and myoglobin in the developing chicken gizzard. Roux's Arch Dev Biol 194:417–424
Takahashi Y, Imanaka T, Takano T (1996) Spatial and temporal pattern of smooth muscle cell differentiation during development of the vascular system in the mouse embryo. Anat Embryol 194:515–526
Taura M (1978) Origin and fate of paired cisternae in mitotic aortic cells of swine. J Electron Micr 27:283–291
Tennyson V, Duc Pham T, Rothman T, Gershon M (1986) Abnormalities of smooth muscle, basal laminae, and nerves in the aganglionic segments of the bowel of lethal spotted mutant mice. Anat Rec 215:267–281
Tonosaki A, Kelly DE (1971) Fine structural study on the origin and development of the sphincter pupillae muscle in the West Coast newt (*Taricha torosa*). Anat Rec 170:57–74
Volberg T, Sabanay H, Geiger B (1986) Spatial and temporal relationships between vinculin and talin in developing chicken gizzard muscle. Differentiation 32:34–43
Volpe P, Biral D, Pizzo P, Salviati G, Margreth A (1993) Ontogenesis of chick iris intrinsic muscle: evidence for a smooth-to-striated transition. Dev Biol 159:441–449
Ward SM, Torihashi S (1995) Morphological changes during ontogeny of the canine proximal colon. Cell Tissue Res 282:93–108
Woodcock-Mitchell J, Mitchell J, Low R, Kieny M, Sengel P, Rubbia L, Skalli O, Jackson B, Gabbiani G (1988) α-Smooth muscle actin is transiently expressed in embryonic rat cardiac and skeletal muscles. Differentiation 39:161–166
Wu H-S, Baskin LS, Blakey C, Goodman J, Cunha GR (1999) Ultrastructural smooth muscle ontogeny of the rat bladder. In: Baskin LS, Cunha GR (eds) Advances in bladder research. Kluwer and Plenum, New York, pp 93–102
Yamamoto Y, Kubota T, Atoji Y, Suzuki Y (1996) Distribution of α-vascular smooth muscle actin in the smooth muscle cells of the gastrointestinal tract of the chicken. J Anat (Lond) 189: 623–630
Yamashita T, Sohal GS (1986) Development of smooth and skeletal muscle cells in the iris of the domestic duck, chick and quail. Cell Tissue Res 244:121–131
Yamauchi A, Burnstock G (1969) Post-natal development of smooth muscle cells in the mouse vas deferens. J Anat (Lond) 104:1–15
Zak R (1973) Cell proliferation during cardiac growth. Am J Cardiol 3:211–219
Zimmermann A, Haine A, Gröschel-Stewart U (1995) Neural and smooth muscle development in the chicken gizzard. Roux's Arch Dev Biol 204:271–274

Mammalian Smooth Muscle Differentiation: Origins, Markers and Transcriptional Control

Joseph M. Miano[1]

1 Introduction

Three principal muscle types have evolved in essentially all vertebrate species to carry out functions related to cellular contraction. Traditionally, the three muscle types - cardiac, skeletal and smooth - have been distinguished by their unique structural and functional attributes. In recent years, great strides have been made with respect to the molecular characteristics of each muscle type and the regulatory pathways and factors governing muscle cell lineage determination and differentiation. This is especially true with skeletal and cardiac muscle where several transcription factors have been assigned critical roles in orchestrating developmental programmes unique to these two sarcomeric muscle types. The myogenic regulatory factors in skeletal muscle (myf5, MyoD, myogenin, and MRF4) have, indeed, formed the basis of a paradigm of cellular differentiation (Olson 1990). This paradigm has subsequently been extended to cardiac muscle where related transcription factors (e.g., dHAND) have been cloned and shown, through genetic means, to be essential for normal cardiogenesis (Olson and Srivastava 1996). The relative ease in which progress has been made with the two sarcomeric muscle types probably reflects their well-circumscribed points of origin and their terminal differentiation.

In marked contrast, our understanding of the molecular decision-making processes involved with smooth muscle differentiation is only just beginning. The lag in progress with respect to elucidating the molecular underpinnings governing smooth muscle cell (SMC) differentiation is because of several unique aspects of this muscle type's biology. First, essentially no convenient in vitro model systems exist to manipulate SMC differentiation in a manner analogous to several skeletal muscle models (e.g., the C2C12 cell line). This is, in part, because of the inability of most SMC to undergo terminal differentiation, which has often complicated the very definition of SMC, particularly in vitro. Second, we now appreciate the fact that SMC have diverse origins of ontogeny and display remarkable heterogeneity even within a particular tissue (e.g., vessel wall). Related to this point are the many cell types throughout the mammalian body plan that have phenotypic features resembling SMC, including

[1] Center for Cardiovascular Research, Box 679, University of Rochester Medical Center, 601 Elmwood Avenue, Rochester, New York 14642, USA

Results and Problems in Cell Differentiation, Vol. 38
B. Brand-Saberi (Ed.): Vertebrate Myogenesis

capillary pericytes, Ito cells of the liver, mesangial cells of the renal glomerulus, interstitial cells of Cajal in the intestine, and myoepithelial cells in glandular tissues. Third, because most SMC do not terminally differentiate, they have the capacity to modulate their phenotype and can, in some instances, transdifferentiate to completely different cell types. Nowhere is the phenotypic plasticity of SMC more important from a clinical standpoint than in the many diseases whose pathogenesis appears to be intricately linked to the differentiated status of SMC. For example, many vascular disorders including atherosclerosis and hypertension exhibit defects in the differentiated status of SMC (Owens 1995). The same can be said for certain visceral SMC (e.g. respiratory SMC in asthma). Somewhat surprisingly however, is the rarity in which SMC sarcomas occur, particularly those associated with the vasculature (Giangola et al. 1995).

In this chapter, recent progress made with respect to vascular smooth muscle cell ontogeny and differentiation will be summarized. The evolution of SMC with respect to the other muscle types will be considered first. The diverse origins of SMC and the model systems available to further characterize SMC development will then be outlined. An up-to-date appraisal of the markers used to define SMC lineages in vitro and in vivo will then be summarized. Many of these markers have been used as model genes to begin characterizing the transcriptional circuitry involved in SMC differentiation. The latter will be discussed with emphasis placed on transgenic mouse model systems. Finally, some perspectives for future analysis of this mysterious, if not challenging, muscle cell type will be presented.

2 Smooth Muscle Cell Ontogeny

2.1 Evolutionary Concepts

We begin with a rather difficult, yet fascinating question to consider: which muscle type evolved first? If we consider ordered structure as a criterion for evolution, then one might conclude that the highly organized sarcomere in striated muscle evolved earlier than the less structurally organized SMC. Moreover, proteins with homology to MyoD have been discovered in several invertebrates including *Caenorhabditis elegans* and *Drosophila melanogaster* (Olson 1990). Studies with a smooth muscle alpha actin antibody showed that the invertebrate, planaria, contain SMC (Pascolini et al. 1992). However, as will be discussed below, this particular protein is not a stringent marker for SMC lineages. In a recent study using a panel of antibodies to several muscle-restricted proteins including heavy caldesmon and smooth muscle calponin, the invertebrate, *Helix* (snail), was shown to possess SMC in its body plan (Royuela et al. 2000). Thus, the available data would suggest that certain invertebrates probably evolved a muscle type resembling vertebrate SMC. A broader analysis of chordates using additional markers for SMC should assist in determining how ancient SMC lineages really are. A related question is whether smooth muscle evolved from striated muscle or vice versa. Evidence for the latter is supported by the fact that several SMC marker genes are expressed in early skeletal and cardiac muscle lineages whereas the reverse is not true. In a rigorous phylogenetic analysis of the smooth muscle myosin heavy chain gene, Goodson and Spudich provided evidence that SMC evolved independently of striated muscle via convergent evolution (Goodson and Spudich 1993; Goodson 1994). Clearly, the emergence of complex multi-organ systems in birds and mammals necessitated the evolution of a variety of smooth muscle cell types, many of which have distinct physiological features that are critical to the function of such organs as the intestinal tract, uterus and bladder. The widespread distribution of smooth muscle cell types throughout mammalian body plans highlights its heterogeneity and complexity. This is particularly noteworthy with respect to embryological origins of SMC, which is considered next.

2.2 Embryological Origins of SMC

Unlike cardiac and skeletal muscle, which arise from well-defined regions of the developing embryo, SMC have multiple sites of origin. Most SMC lineages arise from splanchnic mesoderm. Chick-quail chimera studies, however, have revealed a neural crest origin for SMC in the great vessels and cranial vasculature (Le Lievre and Le Douarin 1975; Bergwerff et al. 1998). Thus, aortic SMC, which are inherently heterogeneous (Lemire et al. 1994), have populations of SMC arising from two distinct germ layers: mesoderm (splanchnic mesoderm) and ectoderm (neural crest). Topouzis and Majesky characterized the molecular markers and growth properties of these two aortic SMC lineages and found that although both lineages express essentially the same panoply of SMC markers, ectomesenchymal SMC grew autonomously and exhibited a much more robust TGF-β1 response (Topouzis and Majesky 1996). Interestingly, ablation of the neural crest has profound effects on great vessel and heart development with little effects on normal coronary vessel development (Kirby et al. 1983; Waldo et al. 1994). We now know that coronary SMC arise from a specialized tissue from the dorsal aspect of the developing heart, the proepicardial organ (PEO; Landerholm et al. 1999; Vrancken Peeters et al. 1999). The epithelial–mesenchymal transformation that occurs when cells of the PEO migrate outwardly is accompanied by the sequential expression of a battery of SMC markers and the loss of epithelial cell-restricted markers (e.g. cytokeratin) (Landerholm et al. 1999). Importantly, transfecting PEO cells with a dominant negative serum response factor (SRF) cDNA abolishes key SMC markers such as SM-Calponin (Landerholm et al. 1999). As will be discussed below, SRF is a central mediator of SMC marker gene transcription both in vitro and in vivo. Because coronary arteries are commonly associated with atherosclerosis and SMC differentiation is compromised in such vessels (Aikawa et al. 1993, 1997; Chen et al. 1997), there is intense interest in understanding the developmental cues and transcriptional circuitry underlying coronary SMC differentiation. Recently, the hemangioblast has been demonstrated to give rise to SMC based on in vitro and embryological studies (Yamashita et al. 2000). SMC may also arise from endothelial cells (Arciniegas et al. 1992, 2000; DeRuiter et al. 1997) and adventitial fibroblasts during vascular remodeling (Shi et al. 1996; Miano et al. 1998). From this discussion, we can see that vascular SMC have multiple points of origin during development including cells of completely different germ layers. The variable points of origin for vascular SMC highlights their presence throughout the entire body plan and implies that there exists heterogeneity with respect to the functional properties of vascular SMC, even within a seemingly homogeneous vessel segment (Bevan 1979). Only when we unravel the signaling pathways and transcriptional mechanisms that specify SMC lineage determination and differentiation, will we begin understanding the mechanisms underlying their de-differentiation in disease states.

2.3 Models for Studying SMC Differentiation

As mentioned in the Introduction, a major problem in SMC biology is the notable lack of easily manipulative model systems for studying the differentiation process. This is largely because the phenotypic characteristics of differentiated SMC are quickly lost once they are placed into culture, a phenomenon that has been referred to as phenotypic modulation (Chamley-Campbell et al. 1979). Recently, however, several cell lines have emerged that should help investigators in the study of certain aspects of SMC differentiation (e.g., transcriptional regulation of a SMC differentiation marker). For example, Rothman and colleagues (1992) first described a clonal rat pulmonary artery SMC line (called PAC1) that does not exhibit the heterogeneity of primary derived vascular SMC (Lemire et al. 1994). PAC1 SMC initially appeared to have a "fixed" phenotype as evidenced by the stable expression of several SMC markers over many passages (Rothman et al. 1992; Firulli et al. 1998). Recently, however, the PAC1 SMC line has been shown to transdifferentiate into skeletal muscle upon culturing at confluency for prolonged periods (Graves and Yablonka-Reuveni 2000). Another cell line that has properties of both skeletal and smooth muscle is the BC_3H1 muscle line (Schubert et al. 1974). These cells exhibit molecular characteristics of SMC when actively growing, but then modulate to a sarcomeric muscle phenotype upon withdrawal from the cell cycle (Taubman et al. 1989; Miano et al. 2001). This process is fully reversible, making the BC_3H1 muscle cell line a potentially useful model system to define the conditions and factors that specify a more SMC-like phenotype. A number of other reports have shown that various pluripotent cell lines (e.g., P-19 cells) can adopt a SMC fate under certain culture conditions (e.g., retinoic acid treatment; Rudnicki et al. 1990; Blank et al. 1995; Suzuki et al. 1996; Drab et al. 1997). Finally, a neural crest cell line (Monc-1) was shown to differentiate into a SMC lineage upon stimulation with either TGF-β1 or bone morphogenic protein 2 (Shah et al. 1996). This cell line has been used to show induction of genes uniquely expressed in vascular SMC (Layne et al. 1998; Chin et al. 2000). Although these cell lines have some utility in characterizing a differentiated SMC phenotype, the final analysis will require an appraisal of SMC promoters and signalling pathways in transgenic mice (see below).

3 Molecular Definitions of Smooth Muscle Cell Lineages

3.1 SMC-Restricted Markers

Vascular SMC express a gene set whose encoded proteins facilitate the unique aspects of contraction in this muscle type (Stull et al. 1991). In situ hybridization, immunohistochemistry and RNase protection assays have been performed in developing and postnatal animals to allow for a hierarchical classification of SMC-restricted markers (Table 1). Several SMC markers represent specific splice variants of a more widely expressed gene. These include α-tropomyosin, heavy caldesmon and meta-vinculin. Because the specificity of these markers is not conferred by a transcriptional event, we will not consider them here. Several of the genes listed in Table 1, however, are expressed specifically in SMC at the transcriptional level. Here, a brief description of the functional and expression characteristics of some of the more widely studied SMC markers will be highlighted.

One of the first SMC markers to be analyzed was the smooth muscle isoform of alpha actin (SM α-actin), which is considered to be the most abundant SMC-restricted protein in adult SMC (Owens 1995). SM α-actin first emerges in the developing heart and somites of chickens and rats (Ruzika and Schwartz 1988; Woodcock-Mitchell et al. 1988; McHugh 1995). The protein product of SM α-actin was subsequently shown in these muscle types as well (Sugi and Lough 1992). SM α-actin expression in sarcomeric muscle is transient with levels virtually extinguished in heart and skeletal muscle by late gestation. Thus, SM α-actin is considered a fetal marker in cardiac tissue and, as with other fetal cardiac genes (e.g., atrial naturetic peptide), SM α-actin mRNA and protein are reactivated in the adult heart subjected to stress (e.g., pressure overload; Black et al. 1991; Leslie et al. 1991). As will be discussed below, several other SMC markers are expressed in the developing heart raising interesting questions as to the physiological and evolutionary significance of such cardiac gene expression. Although SM α-actin has been historically used as a "gold standard" marker for SMC, it is expressed in a variety of non-SMC lineages including endothelial cells (Borrione et al. 1990), astrocytes (Lecain et al. 1991) and fibroblasts (Desmoulière et al. 1992). Thus, when characterizing cells that are considered to be of SMC origin, it is best to couple SM α-actin with at least one additional marker of higher specificity (Owens 1995). The SM α-actin gene

Table 1. Salient characteristics of a SMC differentiation gene set

SMC gene in order of specificity	Function	Expression in vascular disease	CArG/MEF2 dependent?	Promoter specificity in vitro?	Promoter specificity in vivo?
SM-MHC (Miano et al. 1994)	Contraction	Down (Aikawa et al. 1999)	Y (Madsen et al. 1997; Katoh et al. 1998)	Y (Katoh et al. 1994)	Y (Madsen et al. 1998)
SM γ-Actin (McHugh and Lessard 1988)	Contraction	?	Y (Browning et al. 1998)	Y (Kovacs and Zimmer 1998)	Y (Qian et al. 1996)
Telokin (Herring and Smith 2001)	Contraction	?	Y (Herring and Smith 1997)	Y (Herring and Smith 1996)	Y (Herring and Smith 1996)
SM-CALP (Miano and Olson 1996; Samaha et al. 1996)	Contraction	Down (Shanahan et al. 1994)	Y (Miano et al. 2000)	N (Miano and Olson 1996)	Y (Miano et al. 2002)
Smoothelin (van der Loop et al. 1996)	Cytoskeleton	Up (van der Loop et al. 1997)	?	?	?
SM22 (Li et al. 1996a)	Unknown	Down (Yamamura et al. 1997)	Y (Kim et al. 1997; Li et al. 1997)	Y (Solway et al. 1995; Li et al. 1997)	Y (Moessler et al. 1996; Li et al. 1997)[a]
SM α-actin (Li et al. 1996a)	Contraction	Down Gabbiani et al. 1984)	Y (Shimizu et al. 1995)	Y (Shimizu et al. 1995)	Y (Wang et al. 1997; Mack and Owens 1999)
SM-Lim/CRP-1 (Jain et al. 1996; Henderson et al. 1999)	Cytoskeleton	Down (Jain et al. 1996)	?	N (Yet et al. 1998)	Y (Yet et al. 1998)
Sox-13 (Roose et al. 1998)	Transcription	?	?	?	?
MaxiK β-subunit (Jiang et al. 1999)	Ion flux	?	?	?	?
Nov (Ellis et al. 2000)	Adhesion	Down (Ellis et al. 2000)	N	?	?

APEG (Hsieh et al. 1996)	Unknown	Down (Hsieh et al. 1996)	?	?	?
Gax (Gorski et al. 1993)	Transcription	Down (White et al. 1993)	Y (Andrés et al. 1995)	N (Andrés et al. 1995)	?
α_8-Integrin (Schnapp et al. 1995; Chen et al. 2001)	Adhesion	?	?	?	?
α_1-Integrin (Obata et al. 1997)	Adhesion	?	Y (Obata et al. 1997)	Y (Obata et al. 1997)	?
Notch3 (Joutel et al. 2000)	Signaling	?	?	?	?
h-Caldesmon (Ueki et al. 1987)	Contraction	Down (Glukhova et al. 1988)	Y (Momiyama et al. 1998)	Y (Yano et al. 1995)	?
Desmin (van Groningen et al. 1994)	Cytoskeleton	Down (Osborn et al. 1987)	Y (Kuisk et al. 1996)	Y (Li and Paulin 1991)	Y (Kuisk et al. 1996; Raguz et al. 1998)
PGM-RP (Moiseeva et al. 1996)	Cytoskeleton	?	?	Y (Moiseeva and Critchley 1997)	?
Tropoelastin (Fazio et al. 1988)	Structural	Up (Majesky et al. 1992; Belknap et al. 1996)	?	N (Kahari et al. 1990)	Y (Hsu-Wong et al. 1994)
β-Tropomyosin (Nakamura et al. 2001)	Contraction	?	Y (Nakamura et al. 2001)	?	?

[a] The promoter was active in only a subset of cell types that normally express the endogenous gene (see text for details)

has been knocked out and its phenotype is associated with an attenuation in gastrointestinal motility and an apparent upregulation of the SM-γ actin gene (Schildmeyer et al. 2000).

SM22 was first isolated from chicken gizzard (Lees-Miller et al. 1987a) and like SM α-actin it is an abundant protein found in various SMC lineages (Lees-Miller et al. 1987b). The SM22 gene was independently cloned and named transgelin (Prinjha et al. 1994) and p27 (Almendral et al. 1989). The latter studies point to SM22 having a role in actin cross-linking. During development, SM22 first emerges in the heart and then the somites with virtually the same temporal pattern of expression as SM α-actin (Duband et al. 1993; Li et al. 1996a). As with SM α-actin and other SMC markers (Table 1), SM22 becomes highly restricted to SMC lineages in adults (Li et al. 1996a). Interestingly, genetic inactivation of the SM22 has not revealed a remarkable phenotype (Zhang et al. 2001). The promoters of both SM22 and SM α-actin are very similar suggesting similar modes of gene regulation, although transgenic studies reveal salient differences (see Sect. 3.2).

There are no troponins expressed in SMC lineages. Takahashi and colleagues, however, purified a protein of 34,000 molecular weight from chicken gizzard that bound *cal*cium and had homology to tro*ponin* T, hence its designation as calponin (Takahashi et al. 1986, 1988). The calponin gene family is comprised of three distinct loci on different chromosomes. Neutral or h2 calponin (Strasser et al. 1993) and acidic calponin (Applegate et al. 1994; Ferhat et al. 1996) are widely expressed (Cal and Levine 1995; Masuda et al. 1996; Fukui et al. 1997), but their cellular function is unknown. In contrast, the smooth muscle calponin (SM-Calp) gene is more tightly restricted in its expression and has been studied extensively. Indeed, SM-Calp appears to be a multifunctional protein involved in contraction (Winder and Walsh 1990), cytoskeletal remodelling (North et al. 1994; Parker et al. 1998) and cell growth inhibition (Jiang et al. 1997; Horiuchi et al. 1999). SM-Calp was thought initially to be restricted to SMC lineages during development in chickens and in adult mammalian tissues (Gimona et al. 1990; Duband et al. 1993). In situ hybridization studies, however, revealed that this gene is also expressed in the developing mouse heart (Miano and Olson 1996; Samaha et al. 1996). An independent report verified SM-Calp protein expression in developing rat hearts (Ya et al. 1997). Moreover, there are reports of SM-Calp being expressed in developing bone (Yoshikawa et al. 1998) and a variety of sarcomas (Sugimoto et al. 1999). Unlike SM α-actin and SM22, which are expressed in both sarcomeric muscle types during development (Li et al. 1996a), SM-Calp is not expressed in skeletal muscle (Miano and Olson 1996; Samaha et al. 1996). The other non-SMC lineage exhibiting SM-Calp expression during development is normal bone (Yoshikawa et al. 1998); some osteosarcomas also appear to express SM-Calp (Yamamura et al. 1998). Interestingly, the SM-Calp knockout mouse shows a bone defect that is associated with accelerated healing of a fracture (Yoshikawa et al. 1998). Thus, SM-Calp appears to be a negative regulator of osteogenesis during development. For simplicity, we wish to refer to the more restricted calponin as SM-Calp, and acidic and neutral calponin as non-muscle calponin A and B, respectively, following the convention for the non-muscle myosin heavy chains.

Smoothelin was first cloned from a human colon SMC cDNA library using an antibody raised from mice immunized with chicken gizzard extract (van der Loop et al. 1996). Smoothelin is highly restricted to differentiated SMC with levels precipitously dropping once SMC are placed in culture (van der Loop et al. 1996). As with SM α-actin and SM22, smoothelin transcripts are present in developing cardiac and skeletal muscle (G. van Eys, pers. comm.). In adults, smoothelin appears to be totally restricted to SMC lineages. Interestingly, there appears to be a vascular-specific isoform (117 kDa) and a visceral-specific isoform (59 kDa) that putatively arise from alternative promoters (Krämer et al. 2001). The function of smoothelin is not clear, but its co-localization with the cytoskeleton suggests a role in signal transduction or intracellular mechanics. Telokin is a 17-kDa protein encoded by a transcription unit embedded within the myosin light chain kinase locus (Gallagher and Herring 1991). Telokin shows a high degree of specificity for SMC lineages during development and in postnatal tissues with no expression in sarcomeric muscle (Herring et al. 2001). It is transcribed from an internal promoter that shows SMC-restricted activity in vivo (see next section). The function of telokin is not entirely clear, but it has been shown to bind unphosphorylated myosin filaments and stimulates myosin assembly in vitro suggesting an important role in the contractile activity of SMC (Herring and Smith 1996).

Another highly restricted SMC gene is the smooth muscle isoform of myosin heavy chain (SMMHC). This gene generates four splice variants, which show differential expression between visceral and vascular SMC (Babij 1993; Eddinger and Wolf 1993; Kelley et al. 1993; White et al. 1993). Such differential expression of SMMHC splice variants probably confers the distinct contractile properties of visceral versus vascular SMC. For example, a 7-amino acid insert at the N-terminus of SMMHC (visceral SMC-restricted) correlates with a higher velocity of movement of actin filaments in an in vitro assay as well as a higher actin-activated Mg^{2+}-ATPase activity in visceral versus vascular SMC (Kelley et al. 1993). It is tempting to speculate that this visceral-restricted SMMHC isoform interacts with other visceral SMC-restricted proteins such as telokin, smoothelin (59 kDa isoform) and SM γ-actin to establish the unique contractile properties of visceral SMC. As with telokin and SM γ-actin (Sawtell and Lessard 1989; McHugh 1995), the SMMHC gene is not expressed in developing heart or skeletal muscle (Miano et al. 1994). In fact, with the exception of myofibroblasts (Lazard et al. 1993), which arguably are SMC, SMMHC is only found in SMC. Genetic inactivation of SMMHC results in perinatal death by starvation (from severely reduced intestinal motility), cardiomyopathy (from delayed closure of the ductus arteriosus), and possible renal failure (from an over-distended bladder and failed micturition; Morano et al. 2000).

Two additional SMC markers are worthy of comment. Using differential display, a gene was isolated that, upon Northern blotting, showed highly enriched expression in the aorta. The gene was aptly named aortic preferentially expressed gene (APEG-1; Hsieh et al. 1996). The function of APEG-1 is unknown, but it appears to localize in the nucleus and, like so many other SMC markers, is rapidly down-regulated upon cell culture and after balloon catheter injury of the vessel wall (Hsieh et al. 1996). Interestingly, another aortic SMC restricted gene was identified in a screen for retinoid-inducible genes (Chen et al. 2001). The gene was identified as α_8-integrin, which encodes for a protein that heterodimerizes exclusively with the β_1-integrin subunit (Schnapp et al. 1995). The fact that these genes do not show conspicuous expression in visceral SMC suggests for a unique role in signalling within vascular (and particularly aortic) SMC. It is also possible that the transcriptional regulation of these two markers, whose promoters have yet to be functionally characterized, is distinct from other SMC markers. Several additional genes that show preferential accumulation in SMC have been defined and are listed in Table 1.

In summary, several SMC markers show embryonic expression in both developing heart and skeletal muscle. It is interesting to speculate that this early sarcomeric expression of SMC markers is a vestige of evolution. Indeed, there is some evidence, albeit weak, that some SMC markers (including SM α-actin and SM-Calp) are expressed in invertebrates (Royuela et al. 1997). It will be important to determine whether cardiac expression of SMC markers has physiological relevance and, if so, whether this is an adaptive response in the failing heart where the contractile function is compromised and may very well approximate the peristaltic-like contraction of the early embryonic heart. Finally, there appears to be hierarchical specificity of expression for SMC between the genes listed in Table 1. Investigators wishing to define their cells as SMC should exercise care when choosing the marker for study. Clearly, genes such as telokin, SMMHC and smoothelin are more stringent markers for SMC lineages than SM α-actin, desmin or Gax. In the final analysis, it is prudent to consider using more than one marker to define a cell type (e.g. SM α-actin and SMMHC).

3.2 SMC-Restricted Promoter Activity

Virtually all of the promoters to the above-described SMC markers have been cloned and characterized in vitro and, in many cases, in transgenic mice as well. Here, data with respect to SMC-restricted promoter activity is briefly summarized. A common thread among virtually all SMC genes is the presence of one or more CArG boxes (Minty and Kedes 1986) either in the 5′ flanking promoter or within introns of the transcription unit itself. The CArG box binds serum response factor (SRF), which was first cloned from serum stimulated fibroblasts (Norman et al. 1988). SRF is a target of several signaling pathways that are operative in developing SMC. The SM α-actin and SM22 genes both have two proximal promoter CArG boxes in virtually the same spatial arrangement relative to the pre-initiation complex. Interestingly, however, although both promoters can exhibit SMC-restricted activity in vitro, the situation in transgenic mice is more complex. SM α-actin shows appropriate promoter activity during development with widespread SMC activity in postnatal tissues; however, inclusion of the first intron (which contains a consensus CArG box element) specifies vascular SMC-specific activity (Mack and Owens 1999). In sharp contrast, the SM22 promoter only partially recapitulates the developmental profile of SM22 expression during development with activity in somites, arterial SMC and the bulbous cordis (future right ventricle). No SM22 promoter activity was noted in visceral, venular or microvascular SMC (Li et al. 1996b; Moessler et al. 1996; Kim et al. 1997), although we have recently observed SM22 promoter activity in the arterial microvascular bed using a more discriminating nuclear-targeted lacZ reporter gene (unpubl. observ.). In adults, SM22 promoter activity appears to be restricted only to arterial SMC. Thus, there are other more distal regulatory cassettes in or around the SM22 locus that are necessary for the complete expression of SM22 during development and in adult SMC tissues. One approach to defining such distal elements is through the use of large genomic constructs. Such an approach has been successfully used for the analysis of the SM-Calp promoter.

The SM-Calp gene has the unique attribute of being expressed in the heart, but not the somites of developing mouse embryos (Miano and Olson 1996; Samaha et al. 1996; Ya et al. 1997). Cloning and analysis of the SM-Calp promoter revealed non-specific activity in various cell lines (Miano and Olson 1996; Kitami et al. 1999). Surprisingly, similar promoter constructs used in transgenic mice failed to show any activity (unpubl. observ.). This led to the idea of there being distal control elements that were not contained in the original promoter constructs. To facilitate the capture of distal regulatory elements that might confer SM-Calp-specific expression, our laboratory turned to a bacterial artificial chromosome (BAC) that harbors the human SM-Calp gene (Miano et al. 1997). The success of such an approach has recently been shown with respect to the desmin promoter, where a 250-kb BAC containing the human desmin locus completely recapitulated the endogenous desmin expression profile (Raguz et al. 1998). We have recently found that a 103-kb BAC containing the human SM-Calp locus faithfully expressed the human SM-Calp transgene in a SMC-restricted manner both in vitro and in vivo (Miano et al. 2002). An important question is where, within this 103-kb of DNA sequence, the SM-Calp regulatory elements lie. Using phylogenetic fingerprinting, we discovered several evolutionarily conserved CArG boxes in the first intron of SM-Calp that enhance SM-Calp promoter activity in a SMC-specific manner (Miano et al. 2000). Whether the intronic CArG boxes are necessary for SMC-specific expression of SM-Calp in transgenic mice awaits further study.

Several other SMC promoters have been characterized, including SMMHC, telokin, SM γ-actin and α_1-integrin (see Table 1). Remarkably, one or more proximal CArG box controls each of these promoters in vitro and in vivo. A major question that has baffled investigators is how a transcription factor such as SRF, which also controls the expression of sarcomeric muscle genes as well as growth-promoting genes (Johansen and Prywes 1995), confers cell-specific expression of certain genes. What prohibits the expression of SMMHC in the heart and conversely the expression of cardiac α-actin in vascular SMC? A number of theories have been advanced including the recruitment of cell-specific co-activators or the interaction of SRF with neighboring transacting factors such as MyoD (Sartorelli et al. 1990). In SMC, there have been reports showing an enhanced binding/activity of SRF with such transcription factors as Mhox (Hautmann et al. 1997) and the mouse bagpipe homologue, mNkx3.1 (Carson et al. 2000). The latter study revealed a direct interaction between SRF and mNkx3.1. More recently, Olson and colleagues cloned (through in silico means) a cardiac and smooth muscle-restricted SRF coactivator called myocardin (Wang et al. 2001). Myocardin forms a ternary complex with SRF on a SM22 CArG template and activates the SM22 promoter more than a 1,000-fold, making it perhaps the most potent transactivator of SMC-restricted genes described to date. It is likely that myocardin activates other SMC-restricted genes that depend upon a functional CArG element. Moreover, it will be important to determine whether myocardin expression or activity is compromised in vascular disease states where expression of many CArG-dependent SMC markers is reduced. From this discussion, it is clear there exists critical protein–protein interactions that involve SRF and other transacting factors that function to define a SMC-restricted transcriptome. Alternatively, cell-specific SRF splice variants (Belaguli et al. 1999) or nuclear trafficking (Camoretti-Mercado et al. 2000) may co-ordinate cell-specific SRF-dependent gene transcription in SMC. An intriguing and recently advanced hypothesis is the conformation of chromatin around cell-specific SRF-dependent gene loci (Miano et al. 2000). In this model, a SMC-restricted gene (e.g. SMMHC) would have an open chromatin template in SMC, but not in non-SMC. Recently, elegant studies from the Owens' lab have demonstrated precisely this scenario (Manabe and Owens 2001). Using the chromatin immunoprecipitation (ChIP) assay, they showed that CArG boxes from the SMMHC locus were in an "open" state (accessible to SRF binding) in SMC but "closed" (inaccessible to SRF binding) in non-SMC (Manabe and Owens 2001). It will be important to define the mechanisms that underlie chromatin remodelling in CArG boxes as well as other regulatory elements that are of importance in SMC (e.g., MEF2 binding sites).

4 Future Perspectives

In this chapter, I have briefly highlighted some of the more timely molecular aspects of SMC differentiation that are helping us define their embryological origin as well as transcriptional control mechanisms. The importance of understanding the molecular underpinnings of SMC determination and differentiation are underscored by the profound changes in phenotype that occur in this cell type in various disease settings. The next task will be to begin exploring how some of the signaling and transcriptional regulatory pathways governing SMC differentiation are perturbed in disease states such as atherosclerosis and in-stent restenosis. Early insight into this issue has recently been obtained in transgenic mice carrying various SMC-restricted promoters. Upon injury to the vessel wall, the activity of the SMMHC, SM22 and SM α-actin promoters was compromised (Regan et al. 2000). Mutation of a putative GC-repressor element in the SM22 promoter prevented injury-induced down-regulation of this promoter's activity, which suggests that this element binds a factor(s) that inhibits transcription of the SM22 gene following injury (Regan et al. 2000). As more and more promoter/enhancer elements are defined in SMC differentiation markers, similar types of assays may be performed to piece together a map of the critical elements involved in the disease-associated phenotypic modulation of SMC. Another critical (if not Herculean) task will be to test the importance of critical regulatory elements in their native genomic context through Cre-lox-mediated knock-out technology (Oosthuyse et al. 2001). Finally, various genome projects (both completed and contemplated) will bring a wealth of data to the wet-lab bench for analyzing evolutionarily conserved regulatory elements. For example, interspecies DNA sequence comparisons (i.e. phylogenetic fingerprinting) between orthologous genes may reveal sequence elements beyond the CArG box important for SMC-restricted gene expression.

Acknowledgements. The author is funded by the National Institutes of Health (HL-62572) and wishes to thank N.J.C. for her support and encouragement.

References

Aikawa M, Sivam PN, Kuro-o M, Kimura K, Nakahara K, Takewaki S, Ueda M, Yamaguchi H, Yazaki Y, Periasamy M, Nagai R (1993) Human smooth muscle myosin heavy chain gene isoforms as molecular markers for vascular development and atherosclerosis. Circ Res 73:1000–1012

Aikawa M, Sakomura Y, Ueda M, Kimura K, Manabe I, Ishiwata S, Komiyama N, Yamaguchi H, Yazaki Y, Nagai R (1997) Redifferentiation of smooth muscle cells after coronary angioplasty determined via myosin heavy chain expression. Circulation 96:82–90

Almendral JM, Santaren JF, Perera J, Zerial M, Bravo R (1989) Expression, cloning and cDNA sequence of a fibroblast serum-regulated gene encoding a putative actin-associated protein (p27). Exp Cell Res 181:518–530

Andrés V, Fisher S, Wearsch P, Walsh K (1995) Regulation of *Gax* homeobox gene transcription by a combination of positive factors including myocyte-specific enhancer factor 2. Mol Cell Biol 15:4272–4281

Applegate D, Feng W, Green RS, Taubman MB (1994) Cloning and expression of a novel acidic calponin isoform from rat aortic vascular smooth muscle. J Biol Chem 269:10683–10690

Arciniegas E, Sutton AB, Allen TD, Schor AM (1992) Transforming growth factor beta 1 promotes the differentiation of endothelial cells into smooth muscle cells in vitro. J Cell Sci 103: 521–529

Arciniegas E, Ponce L, Hartt Y, Graterol A, Carlini RG (2000) Intimal thickening involves transdifferentiation of embryonic endothelial cells. Anat Rec 258:47–57

Babij P (1993) Tissue-specific and developmentally regulated alternative splicing of a visceral isoform of smooth muscle myosin heavy chain. Nucleic Acids Res 21:1467–1471

Belaguli NS, Zhou W, Trinh T-HT, Majesky MW, Schwartz RJ (1999) Dominant negative murine serum response factor: alternative splicing within the activation domain inhibits transactivation of serum response factor binding targets. Mol Cell Biol 19:4582–4591

Belknap JK, Grieshaber NA, Schwartz PE, Orton EC, Reidy MA, Majack RA (1996) Tropoelastin gene expression in individual vascular smooth muscle cells: relationship to DNA synthesis during vascular development and after arterial injury. Circ Res 78:388–394

Bergwerff M, Verberne ME, DeRuiter MC, Poelmann RE, Gittenberger-de Groot AC (1998) Neural crest cell contribution to the developing circulatory system: implications for vascular morphology? Circ Res 82:221–231

Bevan JA (1979) Sites of transition between functional systemic and cerebral arteries of rabbits occur at embryological junctional sites. Science 204:635–637

Black FM, Packer SE, Parker TG, Michael LH, Roberts R, Schwartz RJ, Schneider MD (1991) The vascular smooth muscle α-actin gene is reactivated during cardiac hypertrophy provoked by load. J Clin Invest 88:1581–1588

Blank RS, Swartz EA, Thompson MM, Olson EN, Owens GK (1995) A retinoic acid-induced clonal cell line derived from multipotential P19 embryonal carcinoma cells expresses smooth muscle characteristics. Circ Res 76:742–749

Borrione AC, Zanellato AMC, Giuriato L, Scannapieco G, Pauletto P, Sartore S (1990) Nonmuscle and smooth muscle myosin isoforms in bovine endothelial cells. Exp Cell Res 190:1–10

Browning CL, Culberson DE, Aragon IV, Fillmore RA, Croissant JD, Schwartz RJ, Zimmer WE (1998) The developmentally regulated expression of serum response factor plays a key role in the control of smooth muscle-specific genes. Dev Biol 194:18–37

Cal H, Levine M (1995) Modulation of enhancer-promoter interactions by insulators in the *Drosophila* embryo. Nature 376:533–536

Camoretti-Mercado B, Liu H-W, Halayko AJ, Forsythe SM, Kyle JW, Li B, Fu Y, McConville J, Kogut P, Vieira JE, Patel NM, Hershenson MB, Fuchs E, Sinha S, Miano JM, Parmacek MS, Burkhardt JK, Solway J (2000) Physiological control of smooth muscle-specific gene expression through regulated nuclear translocation of serum response factor. J Biol Chem 275:30387–30393

Carson JA, Fillmore RA, Schwartz RJ, Zimmer WE (2000) The smooth muscle γ-actin gene promoter is a molecular target for the mouse *bagpipe* homologue, mNkx3-1, and serum response factor. J Biol Chem 275:39061–39072

Chamley-Campbell J, Campbell GR, Ross R (1979) The smooth muscle cell in culture. Physiol Rev 59:1–61

Chen J, Maltby KM, Miano JM (2001) A novel retinoid-response gene set in vascular smooth muscle cells. Biochem Biophys Res Comm 281:475–482

Chen YH, Chen YL, Lin SJ, Chou CY, Mar GY, Chang MS, Wang SP (1997) Electron microscopic studies of phenotypic modulation of smooth muscle cells in coronary arteries of patients with unstable angina pectoris and postangioplasty restenosis. Circulation 95:1169–1175

Chin MT, Maemura K, Fukumoto S, Jain MK, Layne MD, Watanabe M, Hsieh CM, Lee ME (2000) Cardiovascular basic helix loop helix factor 1, a novel transcriptional repressor expressed preferentially in the developing and adult cardiovascular system. J Biol Chem 275: 6381–6387

DeRuiter MC, Poelmann RE, Van Munsteren JC, Mironov V, Markwald RR, Gittenberger-de Groot AC (1997) Embryonic endothelial cells transdifferentiate into mesenchymal cells expressing smooth muscle actins in vivo and in vitro. Circ Res 80:444–451

Desmoulière A, Rubbia-Brandt L, Abdiu A, Walz T, Macieira-Coelho A, Gabbiani G (1992) α-smooth muscle actin is expressed in a subpopulation of cultured and cloned fibroblasts and is modulated by γ-interferon. Exp Cell Res 201:64–73

Drab M, Haller H, Bychkov R, Erdmann B, Lindschau C, Haase H, Morano I, Luft FC, Wobus AM (1997) From totipotent embryonic stem cells to spontaneously contracting smooth muscle cells: a retinoic acid and db-cAMP in vitro differentiation model. FASEB J 11:905–915

Duband JL, Gimona M, Scatena M, Sartore S, Small JV (1993) Calponin and SM 22 as differentiation markers of smooth muscle: spatiotemporal distribution during avian embryonic development. Differentiation 55:1–11

Eddinger TJ, Wolf JA (1993) Expression of four myosin heavy chain isoforms with development in mouse uterus. Cell Motil Cytoskeleton 25:358–368

Ellis PD, Chen Q, Barker PJ, Metcalfe JC, Kemp PR (2000) *Nov* gene encodes adhesion factor for vascular smooth muscle cells and is dynamically regulated in response to vascular injury. Arterioscler Thromb Vasc Biol 20:1912–1919

Fazio MJ, Olsen DR, Kuivaniemi H, Chu ML, Davidson JM, Rosenbloom J, Uitto J (1988) Isolation and characterization of human elastin cDNAs, and age-associated variation in elastin gene expression in cultured skin fibroblasts. Lab Invest 58:270–277

Ferhat L, Charton G, Represa A, Ben-Ari Y, der Terrossian E, Khrestchatisky M (1996) Acidic calponin cloned from neural cells is differentially expressed during rat brain development. Eur J Neurosci 8:1501–1509

Firulli AB, Han D, Kelly-Roloff L, Koteliansky VE, Schwartz SM, Olson EN, Miano JM (1998) A comparative molecular analysis of four rat smooth muscle cell lines. In Vitro Cell Dev Biol 34:217–226

Fukui Y, Masuda H, Takagi M, Takahashi K, Kiyokane K (1997) The presence of h2-calponin in human keratinocyte. J Dermatol Sci 14:29–36

Gabbiani G, Kocher O, Bloom WS, Vandekerckhove J, Weber K (1984) Actin expression in smooth muscle cells of rat aortic intimal thickening, human artheromatous plaque, and cultured rat aortic media. J Clin Invest 73:148–152

Gallagher PG, Herring BP (1991) The carboxyl terminus of the smooth muscle myosin light chain kinase is expressed as an independent protein, telokin. J Biol Chem 266:23945–23952

Giangola G, Migaly J, Crawford B, Moskowitz P, Sebenick M (1995) Leiomyosarcoma of the subclavian artery. J Vasc Surg 22:496–500

Gimona M, Herzog M, Vandekerchkhove J, Small JV (1990) Smooth muscle specific expression of calponin. FEBS Lett 274:159–162

Glukhova M, Kabakov AE, Frid MG, Ornatsky OI, Belkin AM, Mukhin DN, Orekhov AN, Koteliansky VE, Smirnov VN (1988) Modulation of human aorta smooth muscle cell phenotype: a study of muscle-specific variants of vinculin, caldesmon, and actin expression. Proc Natl Acad Sci USA 85:9542–9546

Goodson HV (1994) Molecular evolution of the myosin superfamily: application of phylogenetic techniques to cell biological questions. Society of General Physiologists Series 49. Rockefeller University Press, New York, pp 141–157

Goodson HV, Spudich JA (1993) Molecular evolution of the myosin family: relationships derived from comparisons of amino acid sequences. Proc Natl Acad Sci USA 90:659–663

Gorski DH, LePage DF, Patel CV, Copeland NG, Jenkins NA, Walsh K (1993) Molecular cloning of a diverged homeobox gene that is rapidly down-regulated during the G_0/G_1 transition in vascular smooth muscle cells. Mol Cell Biol 13:3722–3733

Graves DC, Yablonka-Reuveni Z (2000) Vascular smooth muscle cells spontaneously adopt a skeletal muscle phenotype: a unique $Myf5^-/MyoD^+$ myogenic program. J Histochem Cytochem 48:1173–1193

Hautmann MB, Thompson MM, Swartz EA, Olson EN, Owens GK (1997) Angiotensin II-induced stimulation of smooth muscle α -actin expression by serum response factor and the homeodomain transcription factor MHox. Circ Res 81:600–610

Henderson JR, Macalma T, Brown D, Richardson JA, Olson EN, Beckerle MC (1999) The LIM protein, CRP1, is a smooth muscle marker. Dev Dyn 214:229–238

Herring BP, Smith AF (1996) Telokin expression is mediated by a smooth muscle cell-specific promoter. Am J Physiol 270:C1656-C1665

Herring BP, Smith AF (1997) Telokin expression in A10 smooth muscle cells requires serum response factor. Am J Physiol 272:C1394-C1404

Herring BP, Lyons GE, Hoggatt AM, Gallagher PG (2001) Telokin expression is restricted to smooth muscle tissues during mouse development. Am J Physiol 280:C12–C21

Horiuchi A, Nikaido T, Taniguchi S, Fujii S (1999) Possible role of calponin h1 as a tumor suppressor in human uterine leiomyosarcoma. J Natl Cancer Inst 91:790–796

Hsieh CM, Yoshizumi M, Endege WO, Kho CJ, Jain MK, Kashiki S, Santos R, Lee WS, Perrella MA, Lee ME (1996) *APEG-1*, a novel gene preferentially expressed in aortic smooth muscle cells, is down-regulated by vascular injury. J Biol Chem 271:17354–17359

Hsu-Wong S, Katchman SD, Ledo I, Wu M, Khillan J, Bashir MM, Rosenbloom J, Uitto J (1994) Tissue-specific and developmentally regulated expression of human elastin promoter activity in transgenic mice. J Biol Chem 269:18072–18075

Jain MK, Fujita KP, Hsieh CM, Endege WO, Sibinga NES, Yet SF, Kashiki S, Lee WS, Perrella MA, Haber E, Lee ME (1996) Molecular cloning and characterization of SmLIM, a developmentally regulated LIM protein preferentially expressed in aortic smooth muscle cells. J Biol Chem 271: 10194–10199

Jiang Z, Grange RW, Walsh MP, Kamm KE (1997) Adenovirus-mediated transfer of the smooth muscle cell calponin gene inhibits proliferation of smooth muscle cells and fibroblasts. FEBS Lett 413:441–445

Jiang Z, Wallner M, Meera P, Toro L (1999) Human and rodent MaxiK channel β-subunit genes: cloning and characterization. Genomics 55:57–67

Johansen FE, Prywes R (1995) Serum response factor: transcriptional regulation of genes induced by growth factors and differentiation. Biochem Biophys Acta 1242:1–10

Joutel A, Andreux F, Gaulis S, Domenga V, Cecillion M, Battail N, Piga N, Chapon F, Godfrain C, Tournier-Lasserve E (2000) The ectodomain of the Notch3 receptor accumulates within the cerebrovasculature of CADASIL patients. J Clin Invest 105:597–605

Kahari VM, Fazio MJ, Chen YQ, Bashir MM, Rosenbloom J, Uitto J (1990) Deletion analysis of 5′-flanking region of the human elastin gene: delineation of functional promoter and regulatory *cis*-elements. J Biol Chem 265:9485–9490

Katoh Y, Loukianov E, Kopras E, Zilberman A, Periasamy M (1994) Identification of functional promoter elements in the rabbit smooth muscle myosin heavy chain gene. J Biol Chem 269: 30538–30545

Katoh Y, Molkentin JD, Dave V, Olson EN, Periasamy M (1998) MEF2B is a component of a smooth muscle-specific complex that binds an A/T-rich element important for smooth muscle myosin heavy chain gene expression. J Biol Chem 273:1511–1518

Kelley CA, Takahashi M, Yu JH, Adelstein RS (1993) An insert of seven amino acids confers functional differences between smooth muscle myosins from the intestines and vasculature. J Biol Chem 268:12848–12854

Kim S, Ip HS, Lu MM, Clendenin C, Parmacek MS (1997) A serum response factor-dependent transcriptional regulatory program identifies distinct smooth muscle cell sublineages. Mol Cell Biol 17:2266–2278

Kirby ML, Gale TF, Stewart DE:1983. Neural crest cells contribute to normal aorticopulmonary septation. Science 220:1059–1061

Kitami Y, Maguchi M, Nishida W, Okura T, Kohara K, Hiwada K (1999) The unique 5′-flanking region of the human basic calponin gene. Hypertens Res 22:187–193

Kovacs A, Zimmer WE (1998) Cell-specific transcription of the smooth muscle gamma-actin gene requires both positive- and negative-acting *cis* elements. Gene Exp 7:115–129

Krämer J, Quensel C, Meding J, Cardosa MC, Leonhardt H (2001) Identification and characterization of novel smoothelin isoforms in vascular smooth muscle. J Vasc Res 38:120–132

Kuisk IR, Li H, Tran D, Capetanaki Y (1996) A single MEF2 site governs desmin transcription in both heart and skeletal muscle during mouse embryogenesis. Dev Biol 174:1–13

Landerholm TE, Dong X-R, Lu J, Belaguli NS, Schwartz RJ, Majesky MW (1999) A role for serum response factor in coronary smooth muscle differentiation from proepicardial cells. Development 126:2053–2062

Layne MD, Endege WO, Jain MK, Yet SF, Hsieh CM, Chin MT, Perrella MA, Blanar MA, Haber E, Lee ME (1998) Aortic carboxypeptidase-like protein, a novel protein with discoidin and carboxypeptidase-like domains, is up-regulated during vascular smooth muscle cell differentiation. J Biol Chem 273:15654–15660

Lazard D, Sastre X, Frid MG, Glukhova MA, Thiery JP, Koteliansky VE (1993) Expression of smooth muscle-specific proteins in myoepithelium and stromal myofibroblasts of normal and malignant human breast tissue. Proc Natl Acad Sci USA 90:999–1003

Le Lievre CS, Le Douarin NM (1975) Mesenchymal derivatives of the neural crest: analysis of chimaeric quail and chick embryos. J Embryol Exp Morph 34:125–154

Lecain E, Alliot F, Laine MC, Calas B, Pessac B (1991) α isoform of smooth muscle actin is expressed in astrocytes in vitro and in vivo. J Neurosci Res 28:601–606

Lees-Miller JP, Heeley DH, Smillie LB, Kay CM (1987a). Isolation and characterization of an abundant and novel 22-kDa protein (SM22) from chicken gizzard smooth muscle. J Biol Chem 262:2988–2993

Lees-Miller JP, Heeley DH, Smillie LB (1987b) An abundant and novel protein of 22 kDa (SM22) is widely distributed in smooth muscles. Biochem J 244:705–709

Lemire JM, Covin CW, White S, Giachelli CM, Schwartz SM, (1994) Characterization of cloned aortic smooth muscle cells from young rats. Am J Pathol 144:1068–1081

Leslie KO, Taatjes DJ, Schwarz J, von Turkovich M, Low RB (1991) Cardiac myofibroblasts express alpha smooth muscle actin during right ventricular pressure overload in the rabbit. Am J Pathol 139:207–216

Li L, Miano JM, Cserjesi P, Olson EN (1996a) SM22α, a marker of adult smooth muscle, is expressed in multiple myogenic lineages during embryogenesis. Circ Res 78: 188–195

Li L, Miano JM, Mercer B, Olson EN (1996b) Expression of the *SM22α* promoter in transgenic mice provides evidence for distinct transcriptional regulatory programs in vascular and visceral smooth muscle cells. J Cell Biol 132:849–859

Li L, Liu ZC, Mercer B, Overbeek P, Olson EN (1997) Evidence for serum response factor-mediated regulatory networks governing SM22α transcription in smooth, skeletal, and cardiac muscle cells. Dev Biol 187:311–321

Li Z, Paulin D (1991) High level desmin expression depends on a muscle-specific enhancer. J Biol Chem 266:6562–6570

Mack CP, Owens GK (1999) Regulation of smooth muscle α-actin expression in vivo is dependent on CArG elements within the 5′ and first intron promoter regions. Circ Res 84:852–861

Madsen CS, Hershey JC, Hautmann MB, White SL, Owens GK (1997) Expression of the smooth muscle myosin heavy chain gene is regulated by a negative-acting GC-rich element located between two positive-acting serum response factor-binding elements. J Biol Chem 272: 6332–6340

Madsen CS, Regan CP, Hungerford JE, White SL, Manabe I, Owens GK (1998) Smooth muscle-specific expression of the smooth muscle myosin heavy chain gene in transgenic mice requires 5′-flanking and first intronic DNA sequence. Circ Res 82:908–917

Majesky MW, Giachelli CM, Reidy MA, Schwartz SM (1992) Rat carotid neointimal smooth muscle cells reexpress a developmentally regulated mRNA phenotype during repair of arterial injury. Circ Res 71:759–768

Manabe I, Owens GK (2001) CArG elements control smooth muscle subtype-specific expression of *smooth muscle myosin* in vivo. J Clin Invest 107:823–834

Masuda H, Tanaka K, Takagi M, Ohgami K, Sakamaki T, Shibata N, Takahashi K (1996) Molecular cloning and characterization of human non-smooth muscle calponin. J Biochem 120: 415–424

McHugh KM (1995) Molecular analysis of smooth muscle development in the mouse. Dev Dyn 204:278–290

McHugh KM, Lessard JL (1988) The developmental expression of the rat α-vascular and gamma-enteric smooth muscle isoactins; isolation and characterization of a rat gamma-enteric actin cDNA. J Biol Chem 8:5224–5231

Miano JM, Olson EN (1996) Expression of the smooth muscle cell calponin gene marks the early cardiac and smooth muscle cell lineages during mouse embryogenesis. J Biol Chem 271:7095–7103

Miano JM, Cserjesi P, Ligon KL, Periasamy M, Olson EN (1994) Smooth muscle myosin heavy chain exclusively marks the smooth muscle lineage during mouse embryogenesis. Circ Res 75:803–812

Miano JM, Krahe R, Garcia E, Elliott JM, Olson EN (1997) Expression, genomic structure and high resolution mapping to 19p13.2 of the human smooth muscle cell calponin gene. Gene 197:215–224

Miano JM, Kelly LA, Artacho CA, Nuckolls TA, Piantedosi R, Blaner WS (1998) all-*trans*-retinoic acid reduces neointimal formation and promotes favorable geometric remodeling of the rat carotid artery after balloon withdrawal injury. Circulation 98:1219–1227

Miano JM, Carlson MJ, Spencer JA, Misra RP (2000) Serum response factor-dependent regulation of the smooth muscle calponin gene. J Biol Chem 275:9814–9822

Miano JM, Thomas S, Disteche CM (2001) Expression and chromosomal mapping of the mouse smooth muscle calponin gene. Mamm Genome 12:187–191

Miano JM, Kitchen CM, Chen J, Maltby KM, Kelly LA, Weiler H, Krahe R, Ashworth LK, Garcia E (2002) Human smooth muscle calponin expression in transgenic mice revealed with a bacterial artificial chromosome. In Press, Am J Physiol Heart & Circulatory

Minty A, Kedes L (1986) Upstream regions of the human cardiac actin gene that modulate its transcription in muscle cells: presence of an evolutionarily conserved repeated motif. Mol Cell Biol 6:2125–2136

Moessler H, Mericskay M, Li Z, Nagl S, Paulin D, Small JV (1996) The SM 22 promoter directs tissue-specific expression in arterial but not in venous or visceral smooth muscle cells in transgenic mice. Development 122:2415–2425

Moiseeva EP, Critchley DR (1997) Characterisation of the promoter which regulates expression of a phosphoglucomutase-related protein, a component of the dystrophin/utrophin cytoskeleton predominantly expressed in smooth muscle. Eur J Biochem 248:634–643

Moiseeva EP, Belkin AM, Spurr NK, Koteliansky VE, Critchley DR (1996) A novel dystrophin/utrophin-associated protein is an enzymatically inactive member of the phosphoglucomutase superfamily. Eur J Biochem 235:103–113

Momiyama T, Hayashi K, Obata H, Chimori Y, Nishida T, Ito T, Kamiike W, Matsuda H, Sobue K (1998) Functional involvement of serum response factor in the transcriptional regulation of caldesmon gene. Biochem Biophys Res Comm 242:429–435

Morano I, Chai G-X, Baltas LG, Lamounier-Zepter V, Lutsch G, Kott M, Haase H, Bader M (2000) Smooth-muscle contraction without smooth-muscle myosin. Nat Cell Biol 2:371–375

Nakamura M, Nishida W, Mori S, Hiwada K, Hayashi K, Sobue K (2001) Transcriptional activation of β-tropomyosin mediated by serum response factor and a novel Barx homologue, Barx1b, in smooth muscle cells. J Biol Chem 276:18313–18320

Norman C, Runswick M, Pollock R, Treisman R (1988) Isolation and properties of cDNA clones encoding SRF, a transcription factor that binds to the c-fos serum response element. Cell 55:989–1003

North AJ, Gimona M, Cross RA, Small JV (1994) Calponin is localised in both the contractile apparatus and the cytoskeleton of smooth muscle cells. J Cell Sci 107:437–444

Obata H, Hayashi K, Nishida W, Momiyama T, Uchida A, Ochi T, Sobue K (1997) Smooth muscle cell phenotype-dependent transcriptional regulation of the α1 integrin gene. J Biol Chem 272: 26643–26651

Olson EN (1990) MyoD family: a paradigm for development. Genes Dev 4:1454–1461

Olson EN, Srivastava D (1996) Molecular pathways controlling heart development. Science 272:671–676

Oosthuyse B, Moons L, Storkebaum E, Beck H, Nuyens D, Brusselmans K, Van Dorpe J, Hellings P, Gorselink M, Heymans S, Theilmeier G, Dewerchin M, Laudenbach V, Vermylen P, Raat H, Acker T, Vleminckx V, Van Den Bosch L, Cashman N, Fujisawa H, Drost MR, Sciot R, Bruyninckx F, Hicklin DJ, Ince C, Gressens P, Lupu F, Plate KH, Robberecht W, Herbert J-M, Collen D, Carmeliet P (2001) Deletion of the hypoxia-response element in the vascular endothelial growth factor promoter causes motor neuron degeneration. Nat Genet 28:131–138

Osborn M, Caselitz J, Puschel K, Weber K (1987) Intermediate filament expression in human vascular smooth muscle and in arteriosclerotic plaques. Virchows Arch A Pathol Anat Histol 411:449–458

Owens GK (1995) Regulation of differentiation of vascular smooth muscle cells. Physiol Rev 75: 487–517

Parker CA, Takahashi K, Tang JX, Tao T, Morgan KG (1998) Cytoskeletal targeting of calponin in differentiated, contractile smooth muscle cells of the ferret. J Physiol 508:187–198

Pascolini R, Di Rosa I, Fagotti A, Panara F, Gabbiani G (1992) The mammalian anti-α-smooth muscle actin monoclonal antibody recognizes an α-actin-like protein in *planaria* (*Dugesia lugubris s.l.*). Differentiation 51:177–186

Prinjha RK, Shapland CE, Hsuan JJ, Totty NF, Mason IJ, Lawson D (1994) Cloning and sequencing of cDNAs encoding the actin cross-linking protein transgelin defines a new family of actin-associated proteins. Cell Motil Cytoskeleton 28:243–255

Qian J, Kumar A, Szucsik JC, Lessard JL (1996) Tissue and developmental specific expression of murine smooth muscle gamma-actin fusion genes in transgenic mice. Dev Dyn 207:135–144

Raguz S, Hobbs C, Yague E, Ioannou PA, Walsh FS, Antoniou M (1998) Muscle-specific locus control region activity associated with the human desmin gene. Dev Biol 201:26–42

Regan CP, Adam PJ, Madsen CS, Owens GK (2000) Molecular mechanisms of decreased smooth muscle differentiation marker expression after vascular injury. J Clin Invest 106:1139–1147

Roose J, Korver W, Oving E, Wilson A, Wagenaar G, Markman M, Lamers W, Clevers H (1998) High expression of the HMG box factor Sox-13 in arterial walls during embryonic development. Nucleic Acids Res 26:469–476

Rothman A, Kulik TJ, Taubman MB, Berk BC, Smith CWJ, Nadal-Ginard B (1992) Development and characterization of a cloned rat pulmonary arterial smooth muscle cell line that maintains differentiated properties through multiple subcultures. Circulation 86:1977–1986

Royuela M, Fraile B, Picazo L, Paniagua R (1997) Immunocytochemical electron microscopic study and Western blot analysis of caldesmon and calponin in striated muscle of the fruit fly *Drosophila melanogaster* and in several muscle cell types of the earthworm *Eisenia foetida*. Eur J Cell Biol 72:90–94

Royuela M, Fraile B, Arenas MI, Paniagua R (2000) Characterization of several invertebrate muscle cell types: a comparison with vertebrate muscles. Microsc Res Tech 48:107–115

Rudnicki MA, Sawtell NM, Reuhl KR, Berg R, Craig JC, Jardine K, Lessard JL, McBurney MW (1990) Smooth muscle actin expression during P19 embryonal carcinoma differentiation in cell culture. J Cell Physiol 142:89–98

Ruzika DL, Schwartz RJ (1988) Sequential activation of α-actin genes during avian cardiogenesis: vascular smooth muscle α-actin gene transcripts mark the onset of cardiomyocyte differentiation. J Cell Biol 107:2575–2586

Samaha FF, Ip HS, Morrisey EE, Seltzer J, Tang Z, Solway J, Parmacek MS (1996) Developmental pattern of expression and genomic organization of the calponin-h1 gene: a contractile smooth muscle cell marker. J Biol Chem 271:395–403

Sartorelli V, Webster KA, Kedes L (1990) Muscle-specific expression of the cardiac α-actin gene requires MyoD1, CArG-box binding factor, and Sp1. Genes Dev 4:1811–1822

Sawtell NM, Lessard JL (1989) Cellular distribution of smooth muscle actins during mammalian embryogenesis: expression of the α-vascular but not the γ-enteric isoform in differentiating striated myocytes. J Cell Biol 109:2929–2937

Schildmeyer LA, Braun R, Taffet G, Debiasi M, Burns AE, Bradley A, Schwartz RJ (2000) Impaired vascular contractility and blood pressure homeostasis in the smooth muscle α-actin null mouse. FASEB J 14:2213–2220

Schnapp LM, Breuss JM, Ramos DM, Sheppard D, Pytela R (1995) Sequence and tissue distribution of the human integrin α_8 subunit: a β_1-associated α subunit expressed in smooth muscle cells. J Cell Sci 108:537–544

Schubert D, Harris AJ, Devine CE, Heinemann S (1974) Characterization of a unique muscle cell line. J Cell Biol 61:398–413

Shah NM, Groves AK, Anderson DJ (1996) Alternative neural crest cell fates are instructively promoted by TGFβ superfamily members. Cell 85:331–343

Shanahan CM, Cary NRB, Metcalfe JC, Weissberg PL (1994) High expression of genes for calcification-regulating proteins in human atherosclerotic plaques. J Clin Invest 93:2393–2402

Shi Y, O'Brien JE, Fard A, Mannion JD, Wang D, Zalewski A (1996) Adventitial myofibroblasts contribute to neointimal formation in injured porcine coronary arteries. Circulation 94:1655–1664

Shimizu RT, Blank RS, Jervis R, Lawrenz-Smith SC, Owens GK (1995) The smooth muscle α-actin gene promoter is differentially regulated in smooth muscle versus non-smooth muscle cells. J Biol Chem 270:7631–7643

Solway J, Seltzer J, Samaha FF, Kim S, Alger LE, Niu Q, Morrisey EE, Ip HS, Parmacek MS (1995) Structure and expression of a smooth muscle cell-specific gene, SM22α. J Biol Chem 270:13460–13469

Strasser P, Gimona M, Moessler H, Herzog M, Small JV (1993) Mammalian calponin: identification and expression of genetic variants. FEBS Lett 330:13–18

Stull JT, Gallagher PG, Herring BP, Kamm KE (1991) Vascular smooth muscle contractile elements: cellular regulation. Hypertension 17:723–732

Sugi Y, Lough J (1992) Onset of expression and regional deposition of alpha-smooth and sarcomeric actin during avian heart development. Dev Dyn 193:116–124

Sugimoto T, Hosoi H, Horii Y, Ishida H, Mine H, Takahashi K, Abe T, Ohta S, Sawada T (1999) Malignant rhabdoid-tumor cell line showing neural and smooth-muscle-cell phenotypes. Int J Cancer 82:678–686

Suzuki T, Kim HS, Kurabayashi M, Hamada H, Fujii H, Aikawa M, Watanabe M, Watanabe N, Sakomura Y, Yazaki Y, Nagai R (1996) Preferential differentiation of P19 mouse embryonal carcinoma cells into smooth muscle cells: Use of retinoic acid and antisense against the central nervous system-specific POU transcription factor Brn-2. Circ Res 78:395–404

Takahashi K, Hiwada K, Kokubu T (1986) Isolation and characterization of a 34,000-dalton calmodulin- and F-actin-binding protein from chicken gizzard smooth muscle. Biochem Biophys Res Comm 141:20–26

Takahashi K, Hiwada K, Kokubu T (1988) Vascular smooth muscle calponin. A novel troponin T-like protein. Hypertension 11 (Pt 2):620–626

Taubman MB, Smith CWJ, Izumo S, Grant JW, Endo T, Andreadis A, Nadal-Ginard B (1989) The expression of sarcomeric muscle-specific contractile protein genes in BC_3H1 cells: BC_3H1 cells resemble skeletal myoblasts that are defective for commitment to terminal differentiation. J Cell Biol 108:1799–1806

Topouzis S, Majesky MW (1996) Smooth muscle lineage diversity in the chick embryo: two types of aortic smooth muscle cell differ in growth and receptor-mediated transcriptional responses to transforming growth factor-β. Dev Biol 178:430–445

Ueki N, Sobue K, Kanda K, Toshikazu H, Higashino K (1987) Expression of high and low molecular weight caldesmons during phenotypic modulation of smooth muscle cells. Proc Natl Acad Sci USA 84:9049–9053

van der Loop FTL, Schaart G, Timmer EDJ, Ramaekers FCS, van Eys GJJM (1996) Smoothelin, a novel cytoskeletal protein specific for smooth muscle cells. J Cell Biol 134:401–411

van der Loop FTL, Gabbiani G, Kohnen G, Ramaekers FCS, van Eys GJJM (1997) Differentiation of smooth muscle cells in human blood vessels as defined by smoothelin, a novel marker for the contractile phenotype. Arterioscler Thromb Vasc Biol 17:665–671

van Groningen JJM, Bloemers HPJ, Swart GWM (1994) Rat desmin gene structure and expression. Biochim Biophys Acta 1217:107–109

Vrancken Peeters M-PFM, Gittenberger-de Groot AC, Mentink MMT, Poelmann RE (1999) Smooth muscle cells and fibroblasts of the coronary arteries derive from epithelial-mesenchymal transformation of the epicardium. Anat Embryol 199:367–378

Waldo KL, Kumiski DH, Kirby ML (1994) Association of the cardiac neural crest with development of the coronary arteries in the chick embryo. Anat Rec 239:315–331

Wang D-Z, Chang PS, Wang Z, Sutherland L, Richardson JA, Small E, Krieg PA, Olson EN (2001) Activation of cardiac gene expression by myocardin, a transcriptional cofactor for serum response factor. Cell 105:851–862

Wang J, Niu W, Nikiforov Y, Naito S, Chernausek S, Witte D, LeRoith D, Strauch A, Fagin JA (1997) Targeted overexpression of IGF-I evokes distinct patterns of organ remodeling in smooth muscle cell tissue beds of transgenic mice. J Clin Invest 100:1425–1439

Weir L, Chen D, Pastore C, Isner JM, Walsh K (1995) Expression of *gax*, a growth arrest homeobox gene, is rapidly down-regulated in the rat carotid artery during the proliferative response to balloon injury. J Biol Chem 270:5457–5461

White S, Martin AF, Periasamy M (1993) Identification of a novel smooth muscle myosin heavy chain cDNA: isoform diversity in the S1 head region. Am J Physiol 264:C1252-C1258

Winder S, Walsh M (1990) Inhibition of the actomyosin MgATPase by chicken gizzard calponin. Prog Clin Biol Res 327:141–148

Woodcock-Mitchell JL, Mitchell JJ, Low RB, Kieny M, Sengel P, Rubbia L, Skalli O, Jackson B, Gabbiani G (1988) α-smooth muscle actin is transiently expressed in embryonic rat cardiac and skeletal muscles. Differentiation 39:161–166

Ya J, Markman MW, Wagenaar GT, Blommaart PJ, Moorman AF, Lamers WH (1997) Expression of the smooth-muscle proteins alpha-smooth-muscle actin and calponin, and of the intermediate filament protein desmin are parameters of cardiomyocyte maturation in the prenatal rat heart. Anat Rec 249:495–505

Yamamura H, Masuda H, Ikeda W, Tokuyama T, Takagi M, Shibata N, Tatsuta M, Takahashi K (1997) Structure and expression of the human SM22α gene, assignment of the gene to chromosome 11, and repression of the promoter activity by cytosine DNA methylation. J Biochem 122:157–167

Yamamura H, Yoshikawa H, Tatsuta M, Akedo H, Takahashi K (1998) Expression of the smooth muscle calponin gene in human osteosarcoma and its possible association with prognosis. Int J Cancer 79:245–250

Yamashita J, Itoh H, Hirashima M, Ogawa M, Nishikawa S, Yurugi T, Naito M, Nakao K, Nishikawa S-I (2000) Flk1-positive cells derived from embryonic stem cells serve as vascular progenitors. Nature 408:92–96

Yano H, Hayashi K, Momiyama T, Saga H, Haruna M, Sobue K (1995) Transcriptional regulation of the chicken caldesmon gene: activation of gizzard-type caldesmon promoter requires a CArG box-like motif. J Biol Chem 270:23661–23666

Yet SF, Folta SC, Jain MK, Hsieh CM, Maemura K, Layne MD, Zhang D, Marria PB, Yoshizumi M, Chin MT, Perrella MA, Lee ME (1998) Molecular cloning, characterization, and promoter analysis of the mouse *Crp2/SMLim* gene: preferential expression of its promoter in the vascular smooth muscle cells of transgenic mice. J Biol Chem 273:10530–10537

Yoshikawa H, Taniguchi S, Yamamura H, Mori S, Sugimoto M, Miyado K, Nakamura K, Nakao K, Katsuki M, Shibata N, Takahashi K (1998) Mice lacking smooth muscle calponin display increased bone formation that is associated with enhancement of bone morphogenetic protein responses. Genes Cells 3:685–695

Zhang JCL, Kim S, Helmke BP, Yu WW, Du KL, Lu MM, Strobeck M, Yu Q-C, Parmacek MS (2001) Analysis of SM22α-deficient mice reveals unanticipated insights into smooth muscle cell differentiation and function. Mol Cell Biol 21:1336–1344

The Genetics of Murine Skeletal Muscle Biogenesis

Shahragim Tajbakhsh[1]

1 Introduction

Skeletal muscle formation involves a complex interplay of cell movements, cell–cell signalling and the activation of key intracellular genes. In the mouse, the myogenic regulatory factors (MRFs) Myf5 and Myod, act as the gatekeepers into the skeletal muscle lineage. Since this entry point into myogenesis was described, more interest has focused on the immediate upstream and downstream events that control the establishment of the skeletal muscle programme. This review is centred on how certain key regulatory genes are involved in establishing skeletal muscles in the mouse embryo and some recent findings that have established skeletal myogenesis as an important paradigm for studying the restriction of cell fate during embryonic, fetal and postnatal development. Other aspects of myogenesis have been reviewed extensively in recent years, and the reader is invited to refer to them for more detailed information on the subject (Ordahl 2000 and references therein).

In vertebrates, somites are the first distinct sign of metamerization in the embryo, and they are responsible for the segmentation of several structures including the vertebrae, ribs and derivatives of the peripheral nervous system (see Christ and Ordahl 1995; Gossler and Hrabe de Angelis 1997). Cells in the epithelial somite are multipotent, and it is under the influence of signalling molecules emanating from the surrounding environment that somite cells gradually acquire their fate (see Borycki and Emerson 2000). During gastrulation, somite precursor mesenchymal cells in the presomitic mesoderm (PSM; also referred to as unsegmented paraxial mesoderm in mice, or segmental plate in avians) condense to form the somite unit, which can be morphologically distinguished from the PSM as an epithelial sphere surrounding a mesenchymal core. Somites are produced in pairs until mid-gestation; they flank the neuraxis and mature in a sequential rostrocaudal developmental gradient. In the mouse, the first somites form during gastrulation between embryonic day (E) 7.5–8 and by mid-gestation they can no longer be distinguished as an epithelial entity. These transitory structures will give rise to multiple cell types including cartilage, skeletal muscle, dermis, endothelial, smooth muscle and

[1] Department of Developmental Stem Cells & Development Biology, Pasteur Institute, 25 rue du Dr. Roux, 75724 Paris, Cedex 15, France

Results and Problems in Cell Differentiation, Vol. 38
B. Brand-Saberi (Ed.): Vertebrate Myogenesis

connective tissue (Christ and Ordahl 1995; Gossler and Hrabe de Angelis 1997). All skeletal muscles of the body proper (trunk, tail and limbs) and some head muscles originate from the somites. The majority of head muscles originate from paraxial head mesoderm (positioned anterior to the somites), and most anteriorly from prechordal mesoderm (see Wachtler and Christ 1992; Christ and Ordahl 1995). This review will be concerned primarily with skeletal muscles of the body proper.

Immediately after their formation, the ventral portion of the somite (facing the floor plate of the neural tube, and the notochord) disperses into mesenchymal sclerotomal cells, which will contribute to the cartilage of the vertebrae and ribs. The portion underlying the surface ectoderm remains epithelial for several days and is called the dermomyotome (see Tajbakhsh and Buckingham 2000). This structure expands by symmetric cell divisions along the dorsoventral and, to a lesser extent, the craniocaudal axis. In addition to the dorsal dermis, the dermomyotome gives rise to the first skeletal muscle, the underlying myotome, as well as skeletal muscles located outside the somite such as the limbs, diaphragm and tongue via long range migrations of muscle progenitor cells (MPCs).

2 The Restriction of Cell Fate and Views on Cell Determination

As a working definition, the terminology depicted in Fig. 1 has been adopted to define the stem to differentiated cell transitions (see Slack 1983). For example, for the skeletal muscle lineage, specification involves the establishment of a muscle progenitor cell. Subsequently, the acquisition of muscle iden-

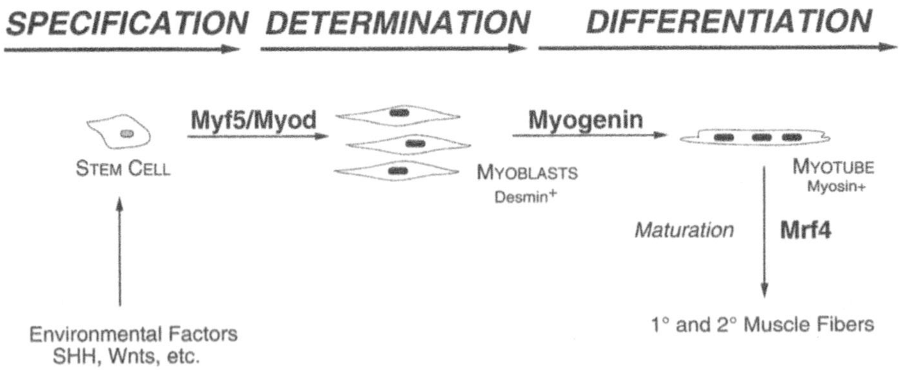

Fig. 1. Steps defining the stem to differentiated cell transitions for skeletal muscle. Specification involves the establishment of a muscle progenitor cell from a stem cell, under the influence of environmental signals. Determination (or commitment) is viewed as the step where myogenic identity is acquired, under the action of Myf5 and Myod. During the differentiation step, the muscle phenotype is elaborated by the expression of muscle-specific genes

tity from this MPC is referred to as the determination or commitment step and, finally, the elaboration of this identity into a multinucleate fibre defines the differentiation step. An important consideration for tissues and organs, particularly those comprised of post-mitotic cells such as skeletal muscles and neurons, is that reserve stem cells permit continued growth throughout development and adult life. Specification, determination and differentiation can be envisaged as multistep stages, each defined by a hierarchy of molecular and cellular events. It is generally accepted that multipotency is lost once a cell becomes determined (committed) to a particular lineage. As we will see below, the activation of the *Myf5* gene could be considered as a first step in determination to a muscle cell; however, in the absence of the Myf5 protein, these MPCs remain multipotent and determination is aborted. Consistent with this notion, experiments in avians have shown that threshold levels of Myod are necessary for cells to be engaged in the myogenic lineage, and precocious removal of critical axial signals can lead to the downregulation of this gene and failure to establish skeletal muscle (Pownall et al. 1996). Another example of the importance of threshold levels is provided by knock-in experiments in the Myogenin locus (Vivian et al. 1999). Here, it was elegantly demonstrated that the threshold requirements for this differentiation factor are remarkably sharp thereby providing one explanation why different phenotypes are observed sometimes in different genetic backgrounds of mice, probably because of an interplay of modifier genes resulting in haplo-insufficiency.

As far as determination is concerned, the concept of reversibility of the final cell fate and the resistance to change once this final cell fate has been established has classically been used as a guideline to define the determined state. This, in many cases, has been informative in embryological terms to distinguish different stages of cell commitment, and the respective signalling requirements (see Ordahl et al. 2000). For example, tissues secreting effector signalling molecules, such as SHH produced from the notochord, have been used to assess cell fate requirements for these signalling molecules (Brand-Saberi et al. 1993; Pourquié et al. 1993; Borycki and Emerson 2000). These studies have provided a working framework for identifying key components of the myogenic programme. It is unfortunate, however, that the notion of irreversibility has in some cases been the criterion for defining whether or not a cell is "determined". There are evident problems with this aspect of the definition. Certainly, incubating a cell in the presence of one signalling molecule or another, overexpressing a potent transcription factor within the cell, or transplanting a cell into a foreign environment where specific signalling molecules are secreted, will differentially influence the fate of the cell. It is difficult to quantitate these environmental influences. Indeed, there are no neutral sites in the embryo where autonomous cell fate decisions could be examined, e.g. the posterior vs anterior limb differentially expresses Shh. Therefore, it is problematic to consider "irreversibility" of cell fate as a criterion for determination because this would be clearly dependent on the site of transplantation. As one classic example, overexpression of MRF's in many cell types such as neurons or keratinocytes, provokes skeletal myogenesis (Weintraub et al. 1991). As another example, fully differentiated neurons expressing *Myf5* in vivo, when explanted in culture as monodisperse cells, are converted to skeletal muscle cells (Tajbakhsh et al. 1994). These manipulations, like the transplantation of a cell, or other in vivo or ex vivo challenging experiments, represent an insult to the cell (e.g. Bianco et al. 1999; Odelberg et al. 2000). If irreversibility were to be a criterion then, under these conditions, one would have to assume that the phenotype of the initial cell was not determined - this is clearly not the case. Therefore, it would be more appropriate to omit the notion of "irreversibility" from the definition of cell determination and, instead, use other more reproducible criteria such as cell morphology or the expression of lineage-restricted genes.

3 The Somite Is a Source of Multiple Cell Types

During embryonic and foetal development and during adult life, cellular identity is established from stem cells. This notion has been articulated more frequently recently as the interest in stem cells has been heightened because of the availability of cellular and molecular criteria that identify them. The timed release of different cell types from the somite as the embryo grows begs the following questions: what factors favour this selective differentiation of distinct cell types from a common ancestor in the somite? and, how is the stem cell pool maintained in the future tissues that are derived from the somite? The somite is a structure where multiple cell types are produced during the course of embryonic growth. For the muscle lineage, these stem cells are located in the dermomyotome. However, their precise location within this epithelium is unknown, and controversy still remains regarding the route of displacement of MPCs from the dermomyotome epithelium to form the myotome (Denetclaw et al. 1997; Kahane et al. 1998; Tajbakhsh and Buckingham 2000).

All skeletal muscle cells of the body proper and some located in the head, as well as dermal progenitors overlying the back, originate from the dermomyotome (see Tajbakhsh and Buckingham 2000). Subpopulations of somite progenitor cells also undergo long-range migrations to form muscle masses in the limbs, diaphragm and tongue. In addition, endothelial cell progenitors migrate from the somite into the limbs to contribute to vascular tissue (Christ and Ordahl 1995). It would be interesting to determine when endothelial progenitors leave the somite and migrate to the limb, and which genes guide this displacement with respect to those regulating the skeletal muscle pathway. Within the somite itself, subdivisions can be made according to the expression patterns of specific genes and cross-vertebrate evolutionary comparisons (Spörle 2001). Epaxial (dorsal) and hypaxial (ventral) body regions can be distinguished most clearly at interlimb levels from E 9.75 by the expression of genes such as *Myf5* and *Myod*. The epaxial myotome, arising from MPCs

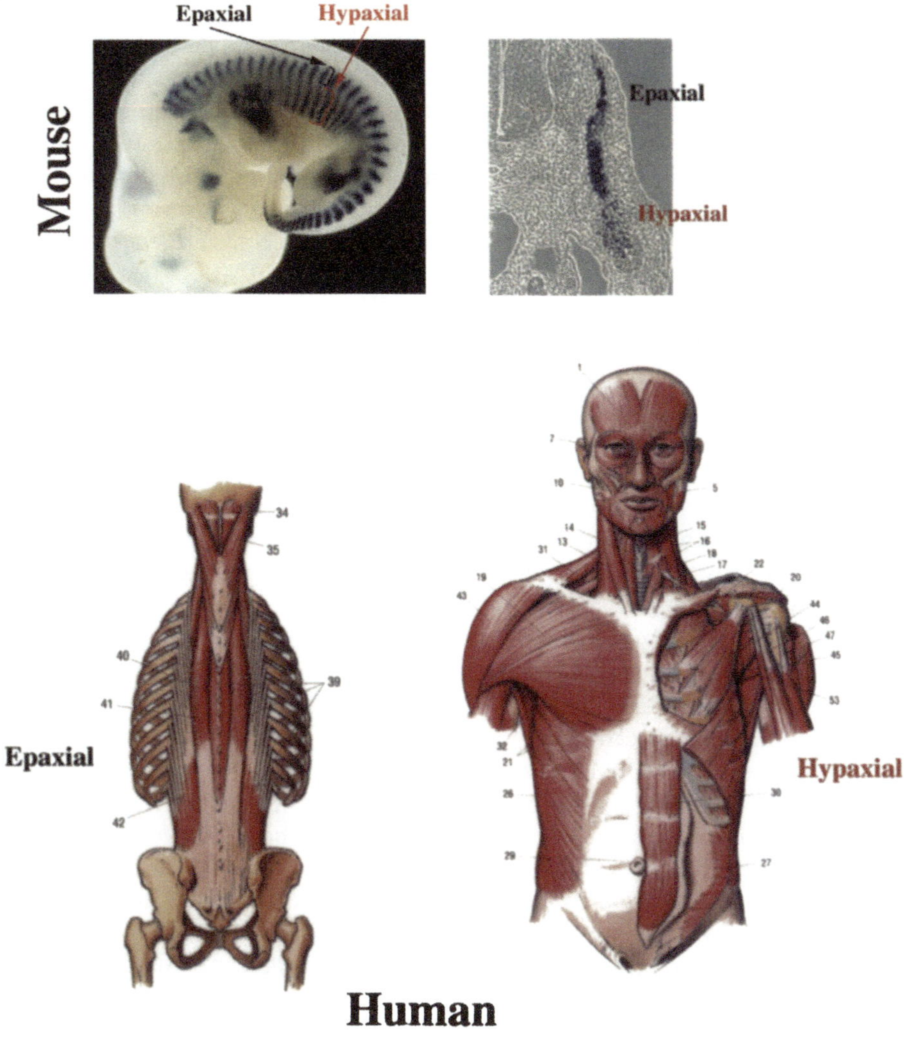

Fig. 2. Epaxial and hypaxial muscles in mouse and man. In the mouse, epaxial and hypaxial regions are indicated for an E11.5 *Myf5*$^{nlacZ/+}$ embryo, on whole mount and section. In the mouse, human and other higher vertebrates, it is clear that epaxial muscles comprise a minor fraction of the total muscle mass compared with hypaxial muscles

located in the epaxial-most or dorso-medial dermomyotome lip, subsequently contributes to the deep back muscles of the body proper. The hypaxial myotome, arising from MPCs located in the ventrolateral dermomyotome lip, contributes to the intercostal muscles at thoracic levels (see Fig. 2). Other hypaxial muscles not entering directly into the myotome, migrate laterally away from the dermomyotome to form body wall, diaphragm and tongue muscles, or at limb levels, the appendicular muscles (see Tajbakhsh and Buckingham 2000). Within the epaxial region, a so-called central intercalated epaxial region has been defined, which comprises the ventral region of the epaxial subdivision. In lower vertebrates such as fish, such regionalizations have also been described and it is intriguing to note that genes such as *Engrailed-1* mark the intercalated region in both lower and higher vertebrates. Further, it is possible that Shh plays a conserved role in governing the expression of *Engrailed-1* in both organisms (Tajbakhsh and Buckingham 2000; Spörle 2001).

The dermomyotome itself undergoes multiple rounds of symmetric cell divisions, enlarging in size, most notably along the dorsal–ventral body axis. The dermomyotome is structured like a "fitted bed sheet" where all lips curl underneath this epithelium and are thought to generate mesenchymal cells, albeit not all to the same extent. During this growth period, the central region of the dermomyotome loses its epithelial structure whereas the dorsal and ventral regions remain epithelial and continue to furnish muscle and dermal cells (see Tajbakhsh and Buckingham 2000). Subsequently, the location of these stem cells and their immediate progeny, which in the muscle lineage retain their replicative capacity to produce multiple waves of differentiating myoblasts, remains unknown. The lack of specific markers for these stem cells and some of their progeny has hampered progress in the field. Indeed, as indicated earlier, the precise location of skeletal muscle stem cells in the dermomyotome remains unclear.

Therefore, at least three major cell types are derived from the somite: cartilage, skeletal muscle and dermis. The former two have been well characterized, where specific markers can clearly distinguish these lineages and can also profoundly influence their fate. The skeletal muscle lineage, which is the best studied, will be discussed in more detail below. Pax1 and Pax9 are among the earliest markers for sclerotomal cells, and the Pax1/Pax9 double null mutants exhibit vertebral column and proximal rib defects (Peters et al. 1999). A heterogeneity in gene expression is also observed in the sclerotome, which may reflect a regionalization of this compartment (Spörle 2001). Although Pax1 and Pax9 are expressed early in the sclerotome, transplantation studies in the chick have shown that cell fate in the sclerotome remains flexible and does not become fixed to the cartilage lineage until very late stages, suggesting that multipotent cells for this lineage reside in this portion of the somite (Ordahl et al. 2000). If Pax1 and Pax9 can be considered to be the functional counterparts of Pax3 and Pax7 in the skeletal muscle lineage (see below), then the functional equivalent(s) of the Myf5 and Myod determination genes, if they exist, have yet to be identified. The dermal lineage is less well characterized, and some relatively later markers include Msx1 (Houzelstein et al. 2000) and Dermo1 (Li et al. 1995). Finally, our knowledge of the origins of endothelial, smooth muscle and connective tissue progenitors derived from the somite remains rudimentary.

4 The Acquisition of Cell Fate in the Somite: Myf5 and Myod Confer Skeletal Muscle Identity

How do skeletal muscle cells acquire their fate in the embryo? Myf5 and the other members of the myogenic regulatory factor (MRF) family, Myod, Myogenin and Mrf4, are key regulatory transcription factors of the basic-helix-loop-helix (bHLH) superfamily (Yun and Wold 1996; Tajbakhsh and Buckingham 2000). These factors act on the promoters of muscle genes as heterodimers with the products of ubiquitously expressed genes such as E2A and can activate their transcription. Notably, the overexpression of any of these factors in many non-muscle cell types results in a change in cell fate to the skeletal muscle phenotype, concomitant with the expression of muscle specific genes, hence their designation as master regulatory muscle determination genes (Weintraub et al. 1991).

Over the years, single, double and triple mutants, and some knock-ins have been performed for the *MRFs*. These studies have led to the general classification of the MRFs into two groups: the "determination" factors Myf5 and Myod, and the "differentiation" factors Myogenin and Mrf4 (Tajbakhsh and Buckingham 2000). Their complex expression patterns along the anterior–posterior axis of the developing embryo reflects the spatio-temporal control of the skeletal muscle programme (Tajbakhsh and Buckingham 2000). There are also clear differences between these MRFs with respect to their potency (Kablar et al. 1997; Valdez et al. 2000) or cell cycle regulation (Kitzmann et al. 1998; Lindon et al. 1998) and, indeed, it is these differences that will be of interest in the future. Whereas no single mutation in one of these MRFs totally abolishes skeletal muscle, significantly, *Myf5:Myod* double mutant fetuses are born totally lacking skeletal muscle (Rudnicki et al. 1993). Direct evidence that these are determination genes came from observations with a $Myf5^{nlacZ}$ knock-in allele where, in the absence of Myf5, β-galactosidase^{+} muscle progenitor cells were aberrantly patterned (Tajbakhsh et al. 1996a). Some became localized to ectopic sites in the embryo and adopted non-muscle fates such as dermis and cartilage. Subsequently, *Myod* is activated within these developmentally arrested MPCs and the majority of them are recruited into the myogenic lineage (S.T. and M. Buckingham, unpubl. data). These findings demonstrated that in the absence of Myf5 protein, muscle progenitor cells are multipotent and are capable of changing fate according to their local environment. It is still unknown how many steps are required for a skeletal muscle stem cell to become determined, and at what point during this restriction of cell fate the *Myf5* gene is activated. These studies now provide the framework for identifying the cellular and molecular mechanisms that direct these multipotent cells to the skeletal muscle or other lineages.

Another key regulatory gene in the somite is the homeobox/paired domain transcription factor *Pax3*, belonging to the *Pax* gene family, homologues of the *Drosophila paired* gene (Tremblay and Gruss 1994). Continuous signalling, mainly from surface ectoderm, is required to maintain the dermomyotome as an epithelium (Sosic et al. 1997), and to prolong *Pax3* expression, potentially via Wnt signalling (Borycki et al. 2000). Indeed *Pax3* and its paralogue *Pax7*, which is also expressed in the dermomyotome, play integral roles in establishing this epithelium. Embryos lacking Pax3 exhibit a reduction in the size of the dermomyotome, and in *Pax3:Pax7* double mutants the dermomyotome epithelium is almost entirely lacking (A. Mansouri and P. Gruss, pers. comm.; see Tajbakhsh and Buckingham 2000). These perturbations have clear repercussions on downstream cell differentiation events in multiple cell lineages.

The role of Pax3 in the somite has best been studied in the skeletal muscle lineage. For example, *Splotch* embryos, which carry a spontaneous mutation in Pax3, exhibit perturbations of skeletal muscles in the limbs, diaphragm and tongue (Bober et al. 1994; Goulding et al. 1994). The epistatic relationship between Pax3, Myf5 and Myod was established by examining *Sp:Myf5*^{nlacZ} double mutants (Tajbakhsh et al. 1997). As previously mentioned, Myf5 and Myod are the primary entry points into the myogenic lineage and they impose myogenic identity on multipotent progenitor cells. Unexpectedly, in *Sp:Myf5*^{nlacZ} double homozygous mutant embryos, skeletal muscles in the body proper are missing, whereas head muscles are present, indicating that in the absence of Pax3 and Myf5, Myod is not sufficient to rescue the myogenic programme. These studies demonstrated that *Pax3* and *Myf5* act in parallel pathways upstream of *Myod*. Consistent with this genetic hierarchy, overexpression of *Pax3* in non-muscle cells also resulted in the ectopic activation of *Myod* (Maroto et al. 1997). Unlike *Myf5*, *Pax3* is expressed throughout the dermomyotome and at higher levels in the dorsal and ventral dermomyotome lips. Curiously, some *Myod* activation does take place in the body of double mutants and consequently a few skeletal muscle fibres form. We suggested that this may be because either the spontaneous *splotch* mutant allele used is not a true *Pax3* null, or because some MPCs are specified and determined independently of Pax3 and Myf5 (Tajbakhsh et al. 1997). This and the precise role that Pax3 plays in the other dermomyotome lineages remains to be examined.

5 Subpopulations of Stem Cells Migrate from the Somite to the Limb

Skeletal muscle formation in the limb is an interesting paradigm because incoming progenitors from the somite enter the limb field under the control of specific genes. The remainder of the limb is derived primarily from the lateral mesoderm. These migrations depend essentially on the activities of Pax3, the tyrosine kinase receptor c-met, and its ligand, scatter factor/hepatocyte growth factor (SF/HGF), all of which act in the same genetic network (see Tajbakhsh and Buckingham 2000). c-met is expressed on the surface of cells undergoing long-range migrations from the dermomyotome, and this migration is directed by the expression of SF/HGF from cells located along the migratory route (Scaal et al. 1999; Birchmeier and Brohmann 2000).

Interestingly, muscle progenitor cell migrations from the dermomyotome into the (fore) limb occur within a 30-h period (E 9.25–10.5), at which time the limb is believed to inherit the totality of this progenitor cell pool to construct all of the skeletal muscle masses for the lifetime of the organism. Evidence that cells leaving the dermomyotome are heterogeneous comes from studies with *Lbx1* mutant embryos. In the *Lbx1* null, a subset of progenitors do not find their target correctly and accumulate outside of the limb bud, whereas other *Lbx1* expressing cells enter the limb (Schafer and Braun 1999; Brohmann et al. 2000; Gross et al. 2000). This finding indicates that a heterogeneity already exists in the stem cell pool, raising the possibility that other markers would be able to distinguish between these incoming cell populations. Differential expression of cell surface guidance molecules would probably distinguish this heterogeneous population of cells. Mislocated progenitors were also described in $Myf5^{nlacZ/nlacZ}$ embryos and, here too, it was postulated that cell surface molecules not expressed in the mutant in the absence of Myf5 protein were implicated in their correct localization. Some estimates suggest that only several hundred cells leave the dermomyotome and enter the limb (Christ and Ordahl 1995). The impressive increase in mass from such a small initial pool of progenitors suggests that these cells have a considerable capacity to self-renew. This observation, coupled with the fact that skeletal muscle is a post-mitotic tissue, strongly argues in favour of the notion that these are stem cells for the skeletal muscle lineage. In keeping with this notion, experiments with *Myf5:Myod* double mutant embryos have demonstrated that somite-derived muscle progenitor cells that enter the limb can change their fate to cartilage in the absence of Myf5 and Myod (Kablar et al. 1999; S.T. and M. Rudnicki, unpubl. observ.). Further, cell transplantation experiments using the chick/quail chimera system have shown that somite-derived cells that enter the limb can be fated to cartilage (Ordahl et al. 2000). Taken together, these studies indicate that at least the skeletal muscle and perhaps also endothelial stem cells entering the limb are bi- or multipotent.

Given the small initial input of cells into the limb from the somite, a stem to progenitor cell model must be operating where these stem cells progressively produce embryonic and fetal myoblasts, and post-natal satellite cells throughout development and adult life. According to the current stem cell paradigm, these cells subsequently undergo a limited number of cell divisions before differentiating. How are these multiple waves of embryonic, foetal and satellite cell populations generated? A differential response to a signalling molecule or to different signalling molecules may be one possible mechanism for the timed release of these distinct populations. Indeed, there is evidence for the former because embryonic and foetal myoblasts were shown to be differentially sensitive to TGFβ (Cusella-De Angelis et al. 1994).

Satellite cells, which can be readily identified morphologically underneath the basal laminae of muscle fibres, appear at late foetal and post-natal stages (Seale and Rudnicki 2000). These cells are necessary for postnatal muscle growth and regeneration of muscles during tissue damage and disease. Their true developmental capacity is under active investigation. Furthermore, it is not yet clear whether all of the stem cells that give rise to satellite cells originate from the somite exclusively because rare stem cells located in vascular tissue and the embryonic dorsal aorta have also been shown to contribute to skeletal muscle (De Angelis et al. 1999). It was recently shown that Pax7 plays a key role in governing satellite cell fate because in *Pax7*-null mice satellite cells are absent at least in the limb (Seale et al. 2000). Where in the stem cell hierarchy Pax7 intervenes, remains to be determined. In this context, it is noteworthy that *Myf5* is expressed in quiescent satellite cells and that the *Myf5*^{nlacZ} mouse is an important genetic tool for identifying and examining the developmental status of satellite cells (Beauchamp et al. 2000; Heslop et al. 2001). In the limbs of *Myod* null mice prior to E 12.5, as in satellite cells, *Myf5* expression is not sufficient to drive myogenesis, whereas in the somite *Myf5* expression is rapidly followed by *Desmin* expression and differentiation takes place in the absence of Myod. Therefore, the developmental context is clearly an important element when evaluating when a progenitor can be directed to differentiate.

6 Extrinsic Factors Direct Cell Identity in the Somite

In general, stem cells are postulated to reside in optimal microenvironments or "niches". It is thought that when a stem cell divides, one daughter remains in the niche whereas the other differentiates. This concept was first formulated for hematopoiesis where secreted factors selectively prevent the death of lineage committed progenitors (Weissman 2000). It is not clear, however, to what extent the hematopoesis model is applicable to the somite because cell signalling in the latter appears also to play an instructive role. Another well-studied example of a stem cell niche is in the *Drosophila* ovary (see Watt and Hogan 2000). An asymmetry in structure allows each stem cell in the *Drosophila* ovariole to receive differential activities of hedgehog and wingless signals compared with its daughter cell. Further, the BMP2/4 homologue dpp in *Drosophila* is required to maintain female germline stem cells and promote their division. Mutations in dpp accelerate stem cell loss whereas overexpression of dpp blocks the differentiation of these stem cells (Xie and Spradling 1998).

The two examples mentioned above demonstrate clearly the importance of environmental signals in establishing stem cells and regulating their proliferation and fate. Cells in the immature somite are also multipotent and it is now clear that signalling molecules emanating from the surrounding tissues can profoundly influence their fate. Experiments with avian and mouse explants have confirmed that positive and negative signals converge to precisely govern where and when sclerotome and skeletal muscle identity are initiated (Borycki and Emerson 2000; Tajbakhsh and Buckingham 2000). The known signalling molecules that mediate this induction include the Wnts and Sonic hedgehog.

In addition, *TGFβ* is expressed in the dermomyotome (Grass et al. 1996). Although Sonic hedgehog is clearly important for sclerotome differentiation, it is also implicated in the activation of *Myf5* and the initiation of axial myogenesis. Indeed Shh will, at different concentrations, promote either sclerotome (high) or muscle (low) cell fates (see Borycki and Emerson 2000). Many of these signalling molecules belong to families comprising several members and, in many cases, with overlapping expression patterns. Shh and Ihh are produced by the notochord and endoderm respectively. In the double *Shh:Ihh* null, *Pax1* expression is totally compromised in the sclerotome, whereas in the *Shh* null, some *Pax1* expression remains (Zhang et al. 2001). *Myf5* expression was also reported to be present, although noticeably reduced, in the somite of single *Shh* and double *Shh:Ihh* null embryos (Kruger et al. 2001; Zhang et al. 2001). It remains to be seen if this residual expression in the *Shh* null is mediated by the Wnts, and whether this represents an overall loss of expression in all cells expressing *Myf5* in the dermomyotome and myotome, or only a subset of Shh-dependent cells. *Myf5* expression was previously shown to be compromised independently of cell survival/death effects, strongly suggesting that Shh can play an instructive role in the muscle lineage (Borycki et al. 1999) as well as a cell survival role (Teillet et al. 1998). These signalling molecules can act either alone or in a combinatorial fashion to promote cell differentiation in the somite. Mouse mutants for the Wnts are also available, and here perhaps the most informative has been the *Wnt1:Wnt3a* double mutant; the dorsal dermomyotome is perturbed and the expression of *Myf5* is much reduced (Ikeya and Takada 1998). However, the cells of the epaxial-most dermomyotome are not present and, therefore, the role of Wnts in activating *Myf5* cannot be clearly evaluated. It is probable that these signalling molecules play dual roles, both as permissive and instructive agents.

Although the initiation of cell differentiation in the somite can be thought of as a response to positive signals, de-repression by molecules such as BMP4 add another level of precision for the spatiotemporal activation of key genes (Pourquié et al. 1996). BMP effects can be enhanced or mediated by other molecules. For example, in the presence of Shh, sclerotome cell fate is acquired in cultured explants of chick presomitic mesoderm and BMP signals can promote chondrogenesis, whereas in the absence of a Shh signal, BMP signals promote the activation of lateral plate mesoderm (non-somitic) markers suggesting that cell fate changes had occurred (Murtaugh et al. 1999). BMP4 can prevent the acquisition of muscle cell identity, and precise dosing of BMP concentrations on explant cultures have demonstrated that Pax3 and Myod gene expression can be uncoupled in response to different concentrations of this signalling molecule (Reshef et al. 1998).

Another potent inhibitor of cell differentiation is the Notch signalling pathway (Artavanis-Tsakonas et al. 1999). Notch signalling is involved in at least three fundamental processes: lateral specification, lineage decisions and boundary decisions. One of the roles of Notch signalling is to maintain stem and progenitor cells in a non-differentiated state. Although a role for Notch in more downstream events has been proposed (Nofziger et al. 1999; Delfini et al. 2000; Hirsinger et al. 2001), in *Drosophila*, *Notch* mutants affect early steps in myogenesis (Cossu et al. 1996; Baylies et al. 1998) and it remains to be seen if this will also be the case in vertebrates.

Deciphering which signalling molecules and their interactors in the somite environment promote different cell fates appears to be complex; however, some progress has been made in the field. For example, different Wnts have been shown to have differential effects on *Myf5* and *Myod* activation in explants of presomitic mesoderm, probably via β-catenin dependent and independent pathways (Tajbakhsh et al. 1998). Downstream intracellular components of these signalling pathways can now be examined in the somite. In particular for the Wnts, β-catenin and TCF have been implicated in governing stem cell fate. β-catenin, which binds to the cytoplasmic region of cadherins and plays a role in cell shape, is released from the membrane after Wnt signalling, and binds to HMG box transcription factors such as the Lef/TCF family members to activate the transcription of target genes. Some of the intracellular mediators of Wnt signalling are expressed in the somite prior to cell differentiation. Notably, β-catenin and Lef1 exhibit a highly regionalized expression pattern in the dermomyotome, and subsequently become restricted to the skeletal muscle lineage (Schmidt et al. 2000). Interestingly, a mutation in Tcf4, another component of Wnt signalling, revealed a role for this protein in maintaining the epithelial stem cell compartment of the small intestine (Korinek et al. 1998). However, stem cell compartments in other parts of the intestinal tract were not affected in this mutant, suggesting that specific mediators of signalling pathways may be effectors in different stem cell lineages. This may also turn out to be the case for stem cell lineages in the somite.

Signalling molecules have also been implicated in directing cell fate decisions via asymmetric cell divisions. Studies in *Drosophila* and *C. elegans* have shown that Wnts and their downstream effectors, as well as the putative receptor frizzled, all play a role in this process (see Hawkins and Garriga 1998). This raises the possibility that Wnt signalling may also be performing similar functions in selecting cell fates in somite-derived cells.

7 Manipulations of the Myf5 Locus and Distal Rib Phenotypes: the Complexities of Gene Regulation

One of the striking features of the Myf5 mutant has been the almost total absence of distal ribs (Braun et al. 1992; Tajbakhsh et al. 1996a). This phenotype was observed when either a *PGK-Neo-PGKpA* selection cassette (Braun et al. 1992) or an *nlacZ/RNApolII-Neo-HPRTpA* cassette was introduced into the first exon of the *Myf5* gene (Tajbakhsh et al. 1996a). Because in the Myf5 mutant, the early myotome is absent for 2–3 days, and Myf5 is not expressed in the sclerotome, it has been postulated that cell-to-cell signalling between the myotome and early rib precursors in the sclerotome is required for distal rib formation (Grass et al. 1996; Tallquist et al. 2000). The growth factors implicated include FGF4, FGF6 and PDGFA among others. Moreover, the respective receptors for these molecules are expressed in the sclerotome – for example, FGFR3 and PDGFRα (Soriano 1997; Tribioli and Lufkin 1999).

Although it has been generally accepted that ribs arise from the sclerotome and that the dermomyotome does not contribute to their formation (Christ and Ordahl 1995), this view has been recently challenged. Indeed, it has been suggested that progenitors of the distal and sternal ribs arise from the dermomyotome, whereas proximal ribs and the vertebral column are derived from the sclerotome (Kato and Aoyama 1998). This issue remains to be resolved because the sclerotomal origin or ribs was reaffirmed (Huang et al. 2000). Given that *Myf5* is activated in the dermomyotome and that MPCs are developmentally arrested in the *Myf5* null, these could have implications for distal rib formation if their progenitors are located in the dermomyotome, although our preliminary experiments suggest that the cellular perturbations in the *Myf5* null are not responsible for the distal rib phenotype (S.T. and M. Buckingham, unpubl. observ.).

As far as the distal rib phenotype in the *Myf5* null is concerned, it is now apparent that this is caused by manipulations of the *Myf5* locus itself. New alleles generated at this locus (Kaul et al. 2000; S.T. and M. Buckingham, unpublished observations) demonstrate that the distal rib phenotype, as well as fusions of some vertebrae, are caused by perturbations of the locus and that the muscle phenotype can be uncoupled from this effect. Indeed, the early myotome is absent for about 2 days in the new *Myf5* mutants and muscle is apparently absent in *Myf5:Myod* double mutants (Kaul et al. 2000). In one new *Myf5* allele, 3 kbp encompassing the promoter and 1st exon were deleted and this was considered to be null for Myf5 function (Kaul et al. 2000). In another allele where the C-terminus was deleted and the *Neomycin* gene was also removed using a cre-lox strategy, the distal ribs were also present and these new *Myf5* null mice are alive (S.T. and M. Buckingham, unpubl. observ.). Therefore, it is clear that manipulations of the *Myf5* locus can yield distal rib perturbations by provoking perturbations in *cis*. How can this be explained? One possibility is that a gene involved in rib formation is located in the vicinity of the *Myf5* locus, and manipulations of the latter perturbs regulatory or coding sequences of this candidate gene. This is the model favoured by Kaul and colleagues (2000). Another possibility is that perturbations of the *Myf5* locus down-regulated the expression of the nearby *Mrf4* gene whose transcriptional start sequences are located ca. 9 kbp upstream from that of *Myf5*. As indicated above, until presently, *Mrf4* has been described as a "late" expressing myogenic factor gene that becomes activated only when overt muscle differentiation is observed. Recently, however, we observed that during the first wave of *Mrf4* expression in the somite (ca. E9-E11), *Mrf4* is expressed prior to muscle differentiation in the hypaxial somitic bud (S.T. and M. Buckingham, unpubl. observ.). It is precisely this region of the somite that is implicated in distal rib formation. Therefore, it is possible that manipulations of the *Myf5* locus perturb *Mrf4* expression and thereby lower the myogenic factor threshold concentration required for distal rib formation, probably in a cell autonomous manner. In agreement with this notion, perturbations in the *Mrf4* locus provoke distal rib phenotypes; all three *Mrf4* alleles described to date, where the *Neomycin* selection cassette was retained in all cases, affect *Myf5* expression (Olson et al. 1996; Yoon et al. 1997). However, the simple removal of the selection cassette is not sufficient to explain the absence of distal ribs, at least in the case of the *Myf5* gene (Tallquist et al. 2000; S.T. and M. Buckingham, unpubl. observ.). Tallquist and colleagues showed that the introduction of *cre* sequences alone was sufficient to provoke distal rib phenotypes (Tallquist et al. 2000). A fine analysis of the locus for potential regulatory sequences responsible for these effects is necessary. In the context of the new *Myf5* null alleles where muscle and rib phenotypes have been uncoupled, the redundancy argument of Myogenin substituting for Myf5 for distal rib formation has to be reconsidered (Wang and Jaenisch 1997). Rib defects have also been reported for *Myogenin* null and *Mrf4:Myogenin* null mice; however, this apparently results from secondary effects because of severe skeletal muscle deficiencies, which are apparent at later stages with these mutants (Hasty et al. 1993; Nabeshima et al. 1993; Rawls et al. 1998). As far as the perturbation of a nearby gene is concerned, creating simple and double mutant mice carrying subtle mutations in both *Myf5* and *Mrf4* should clarify which model is operating in distal rib formation.

In light of the discussion above, it is noteworthy to consider that the regulation of the *Myf5/Mrf4* locus has proven to be complex (Hadchouel et al. 2000; Summerbell et al. 2000; Carvajal et al. 2001). In additional to proximal regulatory sequences for *Myf5* expression in the 6-kbp intergenic region, and within the *Myf5* gene itself (Patapoutian et al. 1993; Tajbakhsh et al. 1996b; Summerbell et al. 2000), regulatory sequences that direct *Myf5* expression in the limbs, brain and myotome are located between 48 and 58 kbp upstream of the *Myf5* gene (Hadchouel et al. 2000). Strikingly, these sequences are located within another gene, PTP-RQ (Wright et al. 1998), which extends over 100 kbp, and terminates upstream of *Mrf4* (Carvajal et al. 2001). This gene is expressed in mesangial cells of the kidney and, therefore, is presently not a likely candidate gene responsible for the distal rib phenotype. To add to this complexity, other sequences located greater than 90 kbp upstream of the *Myf5* gene are responsible for maintaining *Myf5* expression at later stages (Hadchouel et al. 2000; Carvajal et al. 2001). One of the key questions is determining which regulatory sequences in the locus are *Myf5* or *Mrf4* specific, or common to both. Where overlapping expression is not observed, such as in the brain, early limb bud and early arches, it is clear that these regulatory sequences are *Myf5* specific. As more regulatory sequences in this locus are identified, the orchestrated expression of *Myf5* and *Mrf4* could be examined in the context of chromatin configuration and juxtapositions of multiple enhancers, and this would provide another paradigm of how enhancers "choose" when, where and with whom they will interact during development.

8 Conclusions

In recent years, many more researchers in the field have turned to the embryo to answer questions that could not be adequately addressed with cultured cells lines. However, in the organism, we are confronted with the complexities of spatio-temporal control of gene expression for establishing the skeletal muscle programme. This has become particularly important as regulatory sequences of muscle genes have been used to direct reporter gene expression to specific sites in transgenic mice. Further, to understand how signalling molecules mediate their effects, and from where, a detailed understanding of the anatomy of the organism has been crucial. This knowledge is a necessary requisite to precisely localize stem cells for the skeletal muscle or other somite-derived lineages. Further, we need to identify new genetic tools that will permit us to monitor stem to differentiated cell transitions, particularly at later stages, when the somite is no longer a structural landmark.

Acknowledgements. I would like to thank Dr. M. Buckingham for her support. I would also like to thank the Human Frontiers Science Programme for their financial support.

References

Artavanis-Tsakonas S, Rand MD, Lake RJ (1999) Notch signaling: cell fate control and signal integration in development. Science 284:770–776

Baylies MK, Bate M, Ruiz Gomez M (1998) Myogenesis: a view from *Drosophila*. Cell 93:921–927

Beauchamp JR, Heslop L, Yu DS, Tajbakhsh S, Kelly RG, Wernig A, Buckingham ME, Partridge TA, Zammit PS (2000) Expression of CD34 and Myf5 defines the majority of quiescent adult skeletal muscle satellite cells. J Cell Biol 151:1221–1234

Bianco P, Cossu G, Odelberg SJ, Kollhoff A, Keating MT (1999) Uno, nessuno e centomila: searching for the identity of mesodermal progenitors. Myotubes can dedifferentiate when stimulated with the appropriate signals and that msx1 can contribute to the dedifferentiation process. Exp Cell Res 251:257–263

Birchmeier C, Brohmann H (2000) Genes that control the development of migrating muscle precursor cells. Curr Opin Cell Biol 12:725–730

Bober E, Franz T, Arnold HH, Gruss P, Tremblay P (1994) Pax-3 is required for the development of limb muscles: a possible role for the migration of dermomyotomal muscle progenitor cells. Development 120:603–612

Borycki AG, Emerson CP Jr (2000) Multiple tissue interactions and signal transduction pathways control somite myogenesis. Curr Top Dev Biol 48:165–224

Borycki AG, Brunk B, Tajbakhsh S, Buckingham M, Chiang C, Emerson CP Jr (1999) Sonic hedgehog controls epaxial muscle determination through Myf5 activation. Development 126:4053–4063

Borycki A, Brown AM, Emerson CP Jr (2000) Shh and Wnt signaling pathways converge to control Gli gene activation in avian somites. Development 127:2075–2087

Brand-Saberi B, Ebensperger C, Wilting J, Balling R, Christ B (1993) The ventralizing effect of the notochord on somite differentiation in chick embryos. Anat Embryol 188:239–245

Braun T, Rudnicki MA, Arnold HH, Jaenisch R (1992) Targeted inactivation of the muscle regulatory gene Myf-5 results in abnormal rib development and perinatal death. Cell 71:369–382

Brohmann H, Jagla K, Birchmeier C (2000) The role of Lbx1 in migration of muscle precursor cells. Development 127:437–445

Carvajal JJ, Cox D, Summerbell D, Rigby PW (2001) A BAC transgenic analysis of the Mrf4/Myf5 locus reveals interdigitated elements that control activation and maintenance of gene expression during muscle development. Development 128:1857–1868

Cossu G, Tajbakhsh S, Buckingham M (1996) How is myogenesis initiated in the embryo? Trends Genet 12:218–223

Christ B, Ordahl CP (1995) Early stages of chick somite development. Anat Embryol 191:381–396

Cusella-De Angelis MG, Molinari S, Donne AD, Coletta M, Vivarelli E, Bouche M, Molinaro M, Ferrari S, Cossu G (1994) Differential response of embryonic and fetal myoblasts to TGFβ: a possible regulatory mechanism of skeletal muscle histogenesis. Development 121:637–649

De Angelis L, Berghella L, Coletta M, Lattanzi L, Zanchi M, Cusella-De Angelis MG, Ponzetto C, Cossu G (1999) Skeletal myogenic progenitors originating from embryonic dorsal aorta coexpress endothelial and myogenic markers and contribute to postnatal muscle growth and regeneration. J Cell Biol 147:869–878

Delfini M, Hirsinger E, Pourquie O, Duprez D (2000) Delta1-activated Notch inhibits muscle differentiation without affecting Myf5 and Pax3 expression in chick limb myogenesis. Development 127:5213–5224

Denetclaw WF, Christ B, Ordahl CP (1997) Location and growth of epaxial myotome precursor cells. Development 124:1601–1610

Gossler A, Hrabe de Angelis M (1997) Somitogenesis. Curr Top Dev Biol 38:225–287

Goulding M, Lumsden A, Paquette AJ (1994) Regulation of Pax-3 expression in the dermomyotome and its role in muscle development. Development 120:957–971

Grass S, Arnold HH, Braun T (1996) Alterations in somite patterning of Myf-5 deficient mice: a possible role for FGF-4 and FGF-6. Development 122:141–150

Gross MK, Moran-Rivard L, Velasquez T, Nakatsu M. N, Jagla K, Goulding M (2000) Lbx1 is required for muscle precursor migration along a lateral pathway into the limb. Development 127:413–424

Hadchouel J, Tajbakhsh S, Primig M, Chang TH, Daubas P, Rocancourt D, Buckingham M (2000) Modular long-range regulation of myf5 reveals unexpected heterogeneity between skeletal muscles in the mouse embryo. Development 127:4455–4467

Hasty P, Bradley A, Morris JH, Edmondson DG, Venuti JM, Olson EN, Klein WH (1993) Muscle deficiency and neonatal death in mice with a targeted mutation in the myogenin gene. Nature 364:501–506

Hawkins N, Garriga G (1998) Asymmetric cell division: from A to Z. Genes Dev 12:3625–3638

Heslop L, Beauchamp JR, Tajbakhsh S, Buckingham ME, Partridge TA, Zammit PS (2001) Transplanted primary neonatal myoblasts can give rise to functional satellite cells as identified using the Myf5(nlacZl+) mouse. Gene Ther 8:778–783

Hirsinger E, Malapert P, Dubrulle J, Delfini MC, Duprez D, Henrique D, Ish-Horowicz D, Pourquie O (2001) Notch signalling acts in postmitotic avian myogenic cells to control MyoD activation. Development 128:107–116

Houzelstein D, Cheraud Y, Auda-Boucher G, Fontaine-Perus J, Robert B (2000) The expression of the homeobox gene Msx1 reveals two populations of dermal progenitor cells originating from the somites. Development 127:2155–2164

Huang R, Zhi Q, Schmidt C, Wilting J, Brand-Saberi B, Christ B (2000) Sclerotomal origin of the ribs. Development 127:527–532

Ikeya M, Takada S (1998) Wnt signaling from the dorsal neural tube is required for the formation of the medial dermomyotome. Development 125:4969–4976

Kablar B, Krastel K, Ying C, Asakura A, Tapscott SJ, Rudnicki MA (1997) MyoD and Myf-5 differentially regulate the development of limb versus trunk skeletal muscle. Development 124:4729–4738

Kablar B, Krastel K, Ying C, Tapscott SJ, Goldhamer DJ, Rudnicki MA (1999) Myogenic determination occurs independently in somites and limb buds. Dev Biol 206:219–231

Kahane N, Cinnamon Y, Kalcheim C (1998) The cellular mechanism by which the dermomyotome contributes to the second wave of myotome development. Development 125:4259–4271

Kato N, Aoyama H (1998) Dermomyotomal origin of the ribs as revealed by extirpation and transplantation experiments in chick and quail embryos. Development 125:3437–3443

Kaul A, Koster M, Neuhaus H, Braun T, Tribioli C, Lufkin T (2000) Myf-5 revisited: loss of early myotome formation does not lead to a rib phenotype in homozygous Myf-5 mutant mice. Cell 102:17–19

Kitzmann M, Carnac G, Vandromme M, Primig M, Lamb NJC, Fernandez A (1998) The muscle regulatory factors Myod and Myf-5 undergo distinct cell cycle-specific expression in muscle cells. J Cell Biol 142:1447–1459

Korinek V, Barker N, Moerer P, van Donselaar E, Huls G, Peters PJ, Clevers H (1998) Depletion of epithelial stem-cell compartments in the small intestine of mice lacking Tcf-4. Nat Genet 19:379–383

Kruger M, Mennerich D, Fees S, Schafer R, Mundlos S, Braun T (2001) Sonic hedgehog is a survival factor for hypaxial muscles during mouse development. Development 128:743–752

Li L, Cserjesi P, Olson EN (1995) Dermo-1: a novel twist-related bHLH protein expressed in the developing dermis. Dev Biol 172:280–292

Lindon C, Montarras D, Pinset C (1998) Cell cycle-regulated expression of the muscle determination factor Myf5 in proliferating myoblasts. J Cell Biol 140:111–118

Maroto M, Reshef R, Munsterberg AE, Koester S, Goulding M, Lassar AB (1997) Ectopic Pax-3 activates MyoD and Myf-5 expression in embryonic mesoderm and neural tissue. Cell 89:139–148

Murtaugh LC, Chyung JH, Lassar AB (1999) Sonic hedgehog promotes somitic chondrogenesis by altering the cellular response to BMP signaling. Genes Dev 13:225–237

Nabeshima Y, Hanaoka K, Hayasaka M, Esumi E, Li S, Nonaka I, Nabeshima Y (1993) Myogenin gene disruption results in perinatal lethality because of severe muscle defect. Nature 364: 532–535

Nofziger D, Miyamoto A, Lyons KM, Weinmaster G (1999) Notch signaling imposes two distinct blocks in the differentiation of C2C12 myoblasts. Development 126:1689–1702

Odelberg SJ, Kollhoff A, Keating MT (2000) Dedifferentiation of mammalian myotubes induced by msx1. Cell 103:1099–1109

Olson EN, Arnold H-H, Rigby PWJ, Wold BJ (1996) Know your neighbors: three phenotypes in null mutants of the myogenic bHLH gene *MRF4*. Cell 85:1–4

Ordahl CP (2000) Somitogenesis. Curr Top Dev Biol 48 (Pt 1 and 2)

Ordahl CP, Williams BA, Denetclaw W (2000) Determination and morphogenesis in myogenic progenitor cells: an experimental embryological approach. Curr Top Dev Biol 48:319–367

Patapoutian A, Miner JH, Lyons GE, Wold B (1993) Isolated sequences from the linked *myf*-5 and *MRF*4 genes drive distinct patterns of muscle-specific expression in transgenic mice. Development 118:61–69

Peters H, Wilm B, Sakai N, Imai K, Maas R, Balling R (1999) Pax1 and Pax9 synergistically regulate vertebral column development. Development 126:5399–5408

Pourquié O, Coltey M, Teillet M-A, Ordahl C, Le Douarin NM (1993) Control of dorsoventral patterning of somitic derivatives by notochord and floor plate. Proc Natl Acad Sci USA 90:5242–5246

Pourquié O, Fan C-M, Coltey M, Hirsinger E, Wanatabe Y, Bréant C, Francis-West P, Brickell P, Tessier-Lavigne M, Le Douarin N (1996) Lateral and axial signals involved in avian somite patterning: a role for BMP4. Cell 84:461–471

Pownall ME, Strunk KE, Emerson CP Jr (1996) Notochord signals control the transcriptional cascade of myogenic bHLH genes in somites of quail embryos. Development 122:1475–1488

Rawls A, Valdez MR, Zhang W, Richardson J, Klein WH, Olson EN (1998) Overlapping functions of the myogenic bHLH genes MRF4 and MyoD revealed in double mutant mice. Development 125:2349–2358

Reshef R, Maroto M, Lassar AB (1998) Regulation of dorsal somitic cell fates: BMPs and Noggin control the timing and pattern of myogenic regulator expression. Genes Dev 12:290–303

Rudnicki MA, Schneglesberg PNJ, Stead RH, Braun T, Arnold H-H, Jaenisch R (1993) MyoD or myf-5 is required for the formation of skeletal muscle. Cell 75:1351–1359

Scaal M, Bonafede A, Dathe V, Sachs M, Cann G, Christ B, Brand-Saberi B (1999) SF/HGF is a mediator between limb patterning and muscle development. Development 126:4885–4893
Schafer K, Braun T (1999) Early specification of limb muscle precursor cells by the homeobox gene Lbx1h. Nat Genet 23:213–216
Schmidt M, Tanaka M, Munsterberg A (2000) Expression of β-catenin in the developing chick myotome is regulated by myogenic signals. Development 127:4105–4113
Seale P, Rudnicki MA (2000) A new look at the origin, function, and "stem-cell" status of muscle satellite cells. Dev Biol 218:115–124
Seale P, Sabourin LA, Girgis-Gabardo A, Mansouri A, Gruss P, Rudnicki MA (2000) Pax7 is required for the specification of myogenic satellite cells. Cell 102:777–786
Slack JMW (1983) From egg to embryo: determinative events in early development. Cambridge University Press, Cambridge
Soriano P (1997) The PDGF alpha receptor is required for neural crest cell development and for normal patterning of the somites. Development 124:2691–2700
Sosic D, Brand-Saberi B, Schmidt C, Christ B, Olson EN (1997) Regulation of paraxis expression and somite formation by ectoderm- and neural tube-derived signals. Dev Biol 185:229–243
Spörle R (2001) Epaxial–adaxial–hypaxial regionalisation of the vertebrate somite: evidence for a somite organiser and a mirror-image duplication. Dev Genes Evol 211:198–217
Summerbell D, Ashby PR, Coutelle O, Cox D, Yee S, Rigby PW (2000) The expression of Myf5 in the developing mouse embryo is controlled by discrete and dispersed enhancers specific for particular populations of skeletal muscle precursors. Development 127:3745–3757
Tajbakhsh S, Buckingham M (2000) The birth of muscle progenitor cells in the mouse: spatiotemporal considerations. Curr Top Dev Biol 48:225–268
Tajbakhsh S, Spörle R (1998) Somite development: constructing the vertebrate body. Cell 92:9–16
Tajbakhsh S, Vivarelli E, Cusella-De-Angelis G, Rocancourt D, Buckingham M, Cossu G (1994) A population of myogenic cells derived from the mouse neural tube. Neuron 13:813–821
Tajbakhsh S, Rocancourt D, Buckingham M (1996a) Muscle progenitor cells failing to respond to positional cues adopt non-myogenic fates in *myf-5* null mice. Nature 384:266–270
Tajbakhsh S, Bober E, Babinet C, Pournin S, Arnold H, Buckingham M (1996b) Gene targeting the myf-5 locus with LacZ reveals expression of this myogenic factor in mature skeletal muscle fibres as well as early embryonic muscle. Dev Dyn 206:291–300
Tajbakhsh S, Rocancourt D, Cossu G, Buckingham M (1997) Redefining the genetic hierarchies controlling skeletal myogenesis: *Pax-3* and *Myf-5* act upstream of *MyoD*. Cell 89:127–138
Tajbakhsh S, Vivarelli E, Kelly R, Papkoff J, Duprez D, Buckingham M, Cossu G (1998) Differential activation of Myf5 and MyoD by different Wnts in explants of mouse paraxial mesoderm and the later activation of myogenesis in the absence of Myf5. Development 125:4155–4162
Tallquist MD, Weismann KE, Hellstrom M, Soriano P (2000) Early myotome specification regulates PDGFA expression and axial skeleton development. Development 127:5059–5070
Teillet MA, Watanabe Y, Jeffs P, Duprez D, Lapointe F, Ledouarin NM (1998) Sonic Hedgehog is required for survival of both myogenic and chondrogenic somitic lineages. Development 125: 2019–2030
Tremblay P, Gruss P (1994) Pax: genes for mice and men. Pharmacol Therapeutics 61:205–226
Tribioli C, Lufkin T (1999) The murine Bapx1 homeobox gene plays a critical role in embryonic development of the axial skeleton and spleen. Development 126:5699–5711
Valdez MR, Richardson JA, Klein WH, Olson EN (2000) Failure of Myf5 to support myogenic differentiation without Myogenin, MyoD, and MRF4. Dev Biol 219:287–298
Vivian JL, Gan L, Olson EN, Klein WH (1999) A hypomorphic myogenin allele reveals distinct myogenin expression levels required for viability, skeletal muscle development, and sternum formation. Dev Biol 208:44–55
Wachtler F, Christ B (1992) The basic embryology of skeletal muscle formation in vertebrates: the avian model. Semin Dev Biol 3:217–227
Wang Y, Jaenisch R (1997) Myogenin can substitute for Myf5 in promoting myogenesis but less efficiently. Development 124:2507–2513
Watt FM, Hogan BL (2000) Out of Eden: stem cells and their niches. Science 287:1427–1430

Weintraub H, Davis R, Tapscott S, Thayer M, Krause M, Benezra R, Blackwell TK, Turner D, Rupp R, Hollenberg S et al. (1991) The MyoD gene family: nodal point during specification of the muscle cell lineage. Science 251:761–766

Weissman IL (2000) Translating stem and progenitor cell biology to the clinic: barriers and opportunities. Science 287:1442–1446

Wright MB, Hugo C, Seifert R, Disteche CM, Bowen-Pope DF (1998) Proliferating and migrating mesangial cells responding to injury express a novel receptor protein-tyrosine phosphatase in experimental mesangial proliferative glomerulonephritis. J Biol Chem 273:23929–23937

Xie T, Spradling AC (1998) decapentaplegic is essential for the maintenance and division of germline stem cells in the *Drosophila* ovary. Cell 94:251–260

Yoon JK, Olson EN, Arnold HH, Wold BJ (1997) Different MRF4 knockout alleles differentially disrupt Myf-5 expression: *cis*-regulatory interactions at the MRF4/Myf-5 locus. Dev Biol 188: 349–362

Yun KS, Wold B (1996) Skeletal muscle determination and differentiation – story of a core regulatory network and its context. Curr Opin Cell Biol 8:877–889

Zhang XM, Ramalho-Santos M, McMahon AP (2001) Smoothened mutants reveal redundant roles for Shh and Ihh signaling including regulation of L/R asymmetry by the mouse node. Cell 105:781–792

Somite Patterning: a Few More Pieces of the Puzzle

Christophe Marcelle, Cynthia Lesbros, and Claudia Linker[1]

1 Introduction

A number of excellent reviews have been written in recent years on the various aspects of somite differentiation (Christ and Ordahl 1995; Brand-Saberi et al. 1996; Lassar and Münsterberg 1996; Gossler and Hrabe de Angelis 1998; Dietrich 1999; Hirsinger et al. 2000; Stockdale et al. 2000; Tajbakhsh and Buckingham 2000, and a compilation of reviews on somitogenesis in Ordahl 2000, to cite only a few). Thus, our purpose is not to provide yet another in-depth review of the ever growing literature on that subject, but, rather, to discuss some recent data and alternative views that might hopefully shed new light on some specific aspects of somite differentiation. We will mostly focus on dermomyotome and skeletal muscle differentiation. Because a considerable amount of information on this subject has been gathered in chick, most of our discussion will refer to this work. However, differences between mouse and chick somitogenesis will be pointed out when appropriate. In addition, we have integrated the data acquired from genetic studies performed in mouse, which have been crucial to our understanding of somite differentiation. Given the immense literature pertaining to this subject, it is likely that we have not cited everyone: we would like to apologize here to any of our colleagues that might feel we have overlooked their data.

2 Segmental Plate Morphology

Somites are mesodermal structures that are characteristic of vertebrate embryos. During gastrulation, mesoderm is formed by the recruitment and ingression of cells from the epiblast through the primitive streak and later through the tail bud. Paraxial and lateral mesoderm is laid down on either side of the axial structures (notochord and neural tube) as two wide strips of loose mesenchyme (Fig. 1; see also Fig. 3). As it matures, and thereby becomes displaced rostrally, the mesoderm which lies closest to the axial structures (called the segmental plate mesoderm in birds, and pre-somitic mesoderm in mouse)

[1] University Aix-Marseille II, LGPD, CNRS UMR 6545, Developmental Biology Institute of Marseille, Campus de Luminy, Case 907, 13288 Marseille cedex, France

Results and Problems in Cell Differentiation, Vol. 38
B. Brand-Saberi (Ed.): Vertebrate Myogenesis

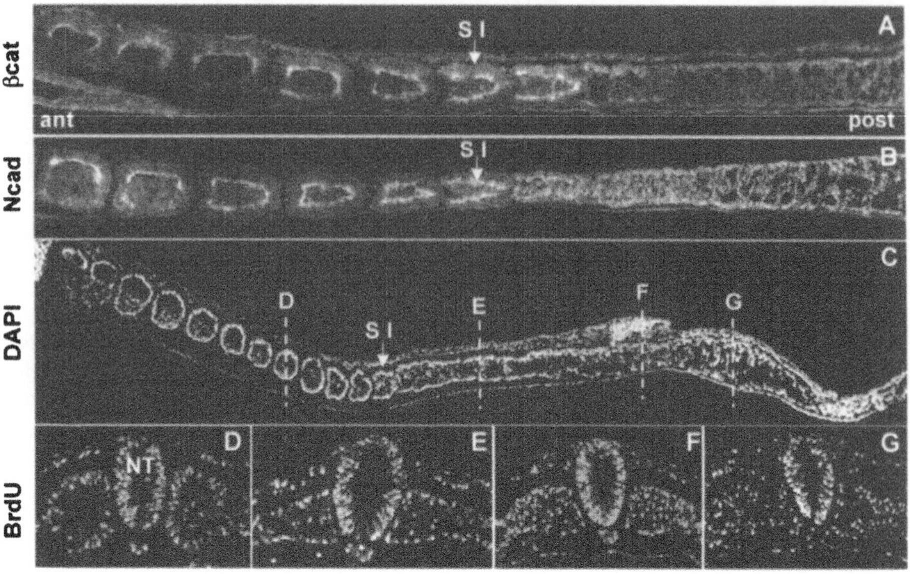

Fig. 1A–G. Morphology of the segmental plate and somites in chick. Sagittal sections of stage 12HH (**C**) and 11HH (**A, B**) chick embryos were labelled with mouse monoclonal antibodies directed against *β*catenin (Transduction Laboratories, **A**) and N-Cadherin (Sigma, **B**) or stained with DAPI (**C**). **D–G** Transversal sections of a stage 11HH chick embryo labelled with a mouse monoclonal antibody directed against BrdU (Becton-Dickinson) In **C**, *dotted lines* indicate the level at which transversal sections presented in **D–G** were performed. *SI* indicates the newly formed somite (i.e. stage I somite according to Christ and Ordahl 1995)

becomes separated from the lateral mesoderm by a groove, where intermediate mesoderm (i.e. primitive kidney) forms. As a consequence of segmental plate maturation, epithelial units, called somites, bud off from the rostral end of the segmental plate at regular time intervals. In chick, one somite pair is formed roughly every 90 min, in mouse every 120 min. The periodic segmentation of paraxial mesoderm into somites is called somitogenesis. Recent evidence has indicated that a network of molecules (to which members of the Notch pathway have been shown to belong) act as readouts and/or key components of a molecular clock that regulates this process (for review see Maroto and Pourquié 2001 and references therein). However, the molecular mechanism(s) that regulates somite number, a species-specific parameter that varies considerably between species (reviewed in Richardson et al. 1998), is totally unknown.

The structure of the segmental plate is not homogenous throughout its antero-posterior axis. Two distinct regions are easily recognizable: at its caudal end, the morphology of this tissue is that of a loose mesenchyme (Fig. 1C, F, G); from the mid-segmental plate level anteriorward, cells become organized into two cell populations (easily recognizable after performing nuclei stainings such as DAPI or after BrdU staining; Fig. 1C, E): whereas centrally located cells retain an unorganized mesenchymal structure, the nuclei of the outer cell layer become aligned and regularly spaced, and these cells seem to come into close contact. This cellular arrangement is reminiscent of an epithelium, although the presence of a basal lamina (which characterizes epithelia) should be confirmed by electron microscopy or immunological examination. Changes in cell morphology are paralleled by molecular modifications because cells present at the periphery of the mid-segmental plate initiate the production of cell adhesion molecules such as N-cadherin, N-CAM and β-catenin (Duband et al. 1987; our observations; see Fig. 1A, B). Therefore, the paraxial mesoderm becomes organized into two morphologically and molecularly distinct cell populations before somites are formed.

3 Somite Differentiation

3.1 Muscle Formation

3.1.1 Epaxial and Hypaxial Muscle Derivatives

Somites give rise to all the skeletal muscles of the body, with the exclusion of head muscles (ocular, mandibular and superficial facial expression muscles), which derive from the unsegmented head mesoderm and pre-chordal mesoderm (reviewed in Christ and Ordahl 1995; Brand-Saberi and Christ 2000). Trunk and neck muscles are subdivided into epaxial muscles, which derive from the medial portion of the somite, and hypaxial muscles, which arise from the lateral half somite (Ordahl and Le Douarin 1992).

Because skeletal muscles regulate all aspects of animal movements, it is not surprising that considerable variations (in muscle number, length, insertion points . . .) have appeared during evolution. In particular, the transition from an aquatic organism (where lateral movements regulate swimming behaviour) to a terrestrial tetrapod (where the body weight is supported by the limb and girdle muscles) supposes a complete reorganization of many muscles of the body. Thus, rather than focusing in detail on one or another organism, we will very briefly discuss the general patterns pertinent to muscle formation as observed amongst vertebrates. In adults, epaxial muscles are located dorsal to the ribs in the upper trunk region, and dorsal to the transverse processes of the vertebrae in the rest of the body (Fig. 2). They collectively function to both extend the vertebral column and to allow lateral flexion of the body. Within the group of epaxial muscles, intervertebrals (which connect successive vertebral processes) are the deepest muscles and they are the only ones that retain the original metamerism inherited from the somites. Longitudinal epaxial muscles are well developed in amniotes; they are located dorsal to the vertebral column; they are not segmented and they extend over multiple segments. Hypaxial muscles exert a number of functions in the adult: subvertebral muscles, which lie ventral to the vertebral column, form a longitudinal band

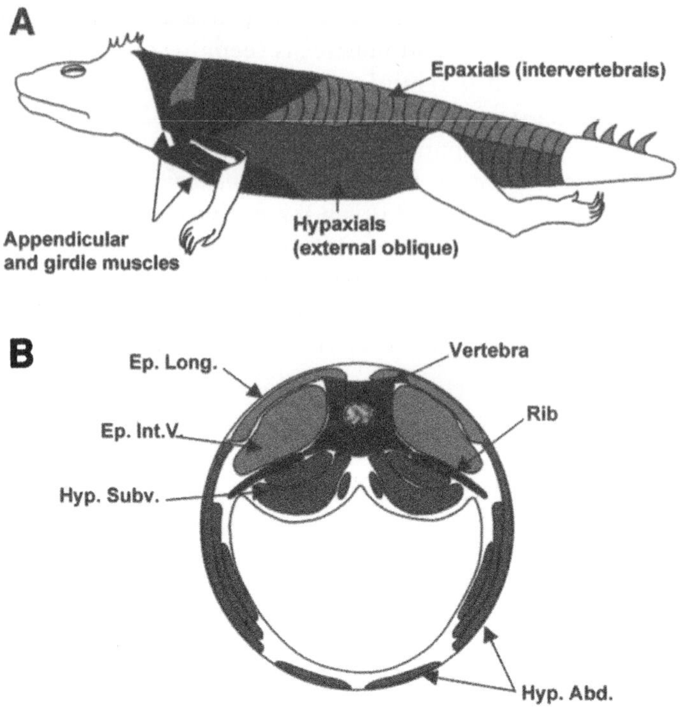

Fig. 2A,B. Disposition of epaxial and hypaxial muscles in vertebrates. **A** Schematic representation of epaxial and hypaxial muscles in a primitive vertebrate (*Sphenodon*). In this animal, epaxial muscles (intervertebrals) are clearly metameric, whereas hypaxial muscles consist of broad sheets of muscles (the outer layer being the external oblique) that extend over a number of segments and mask the original segmentation pattern. In higher vertebrates, intervertebrals are reduced and covered by longitudinal epaxial muscles, thereby hiding any external sign of skeletal muscle metamerism (**B**). **B** Schematic illustration of the components of the axial musculature in the abdominal wall of a mammal. *Ep. Long.* Epaxial longitudinal muscles covering the *Ep. Int. V* (epaxial intervertebral) muscles. *Hyp. Subv.* Hypaxial subvertebral muscles; *Hyp. Abd.* hypaxial abdominal muscles

of flexors whose action is opposite to that of longitudinal epaxials. Oblique and transverse muscles form a multi-layered sheet of muscles that surrounds the abdominal cavity; in the rib cage region, they also play a major role in respiration. In mammals, breathing movements from the action of these muscles are largely superseded by that of the diaphragm (which belongs to the hypaxial muscle group as well). Finally, longitudinal muscles of the abdomen, which lie parallel to its midline, assist in flexing the trunk and supporting abdominal viscera.

Muscles of the limb and girdles are also hypaxial, as they arise during embryogenesis from the lateral portion of the somite, but are often anatomically classified within an appendicular muscle group. In higher vertebrates (birds and mammals), girdle muscles have become extremely well developed, inserting as far dorsal as the spinous processes (thus occupying a region spanning the neck and the trunk down to the lumbar region) and as far ventral as the sternum; as a consequence, they cover a large portion of the trunk epaxial and hypaxial muscles.

3.1.2 A Distinct Embryonic Origin for Epaxial and Hypaxial Muscles?

Lineage studies performed in chick and mouse (Spratt 1955; Nicolet 1971; Ooi et al. 1986; Selleck and Stern 1991; Schoenwolf et al. 1992; Psychoyos and Stern 1996; Wilson and Beddington 1996) and retrospective clonal analyses carried out in mouse (Nicolas et al. 1996; Eloy-Trinquet et al. 2000) have shown that the origin of myotomal cells can be traced back to the primitive streak. A remarkable finding was the observation that a self-renewing muscle stem cell population (estimated to be 100–150 cells in the mouse) resides within the streak and apparently provides all muscle cells (i.e. epaxial and hypaxial) of the myotome. These data indicate that the myotome is specified long before any sign of muscle differentiation has occurred in the embryo, i.e. as early as in the primitive streak. In addition, some degree of regionalization among these muscle stem cells can be observed within the streak because fate mapping studies have shown that cells contributing to epaxial muscles tend to originate from more rostral regions of the primitive streak than those that will give rise to hypaxial muscles (Selleck and Stern 1991; Psychoyos and Stern 1996). It is unclear whether the specification of an epaxial and a hypaxial domains within the streak is important for, or because of, the differential expression of genes regulating the correct patterning of these muscles or whether it results from physical constraints imposed by migratory movements during gastrulation.

3.1.3 Epaxial Muscle Formation

To define a reliable way to characterize somites at various stages of their differentiation, Bodo Christ and Charles Ordahl (1995) have proposed to name somites according to their age, starting at the time of their formation; the newly formed somite would be a stage I somite, the second youngest would be a stage II, etc. Clearly, this terminology is valid only if somites of similar stages are identical regardless of the overall embryonic age. In fact, this is true only within certain time windows of development. Here, we will use Ordahl's somite staging, keeping in mind that the description that will follow is true for chick embryos at 2 to 3.5 days of development (roughly 15 to 35 somite embryos).

The stage I somite, which forms at the rostral end of the segmental plate, comprises an epithelial ball containing an inner cavity, the somitocoel, which is filled with a loose mesenchyme (Fig. 3C); the outer epithelium is termed the cortex. The cellular arrangement of the early somites, therefore, is fundamentally similar to that of the rostral half of the segmental plate. An important

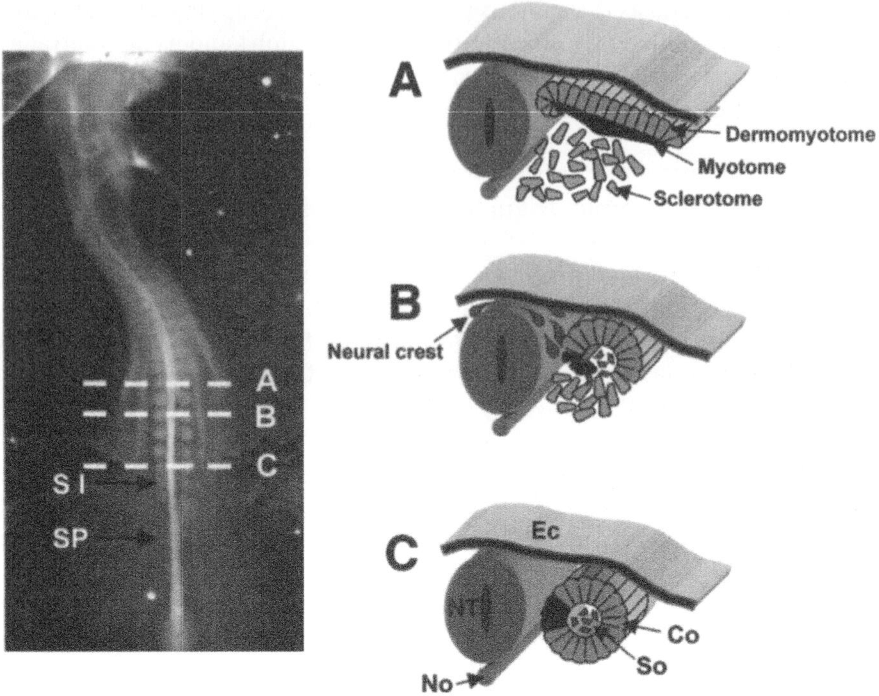

Fig. 3A–C. Schematic representation of the different steps of somite differentiation. **C** Stage I–III somites display an epithelial structure. Cells of the cortex (*Co*) surround the somitocoel (*So*), filled with a loose mesenchyme. Cells of the medial wall of the somite initiate the expression of MyoD (*dark shaded area*). **B** At the level of stage IV–V somite, the ventral portion of the epithelial somite disaggregates into a mesenchyme. Together with cells of the somitocoel, these cells form the sclerotome. Cells of the medial wall fold under the dorsal epithelium, named the dermomyotome. **A** Myotome growth is initiated at the level of somite stage V–VI as cells of the medial portion of the dermomyotome translocate in the myotome. *NT* Neural tube; *No* notochord; *Ec* ectoderm; *SI* stage I somite; *SP* segmental plate

modification resides in the cortex cells, which elongate radially as somites segment. Cortex cells are bottle-shaped and polarized (apical facing the somitocoel, basal to the outside); their nuclei are not perfectly aligned, giving the impression of a multi-layered structure (see for instance Fig. 1D): this arrangement is typical of a pseudo-stratified epithelium, which is quite similar, in fact, to the structure of the neural tube. Multiple cellular extensions reaching towards the ectoderm are observed at the basal end of the cortex cells (Sosic et al. 1997, our observations). At the somitic level, mitotic figures are adjacent to the somitocoel: 75% are oriented tangentially to the somitocoel, the remaining 25% are oriented radially (Williams 1910; Langman and Nelson 1968; Stern and Bellairs 1984). In contrast, within the segmental plate, the orientation of mitotic figures is random.

A strong expression of myogenic differentiation factors is first observed in cells of the medial wall of the stage I–II somite (Ott et al. 1991; Pownall and

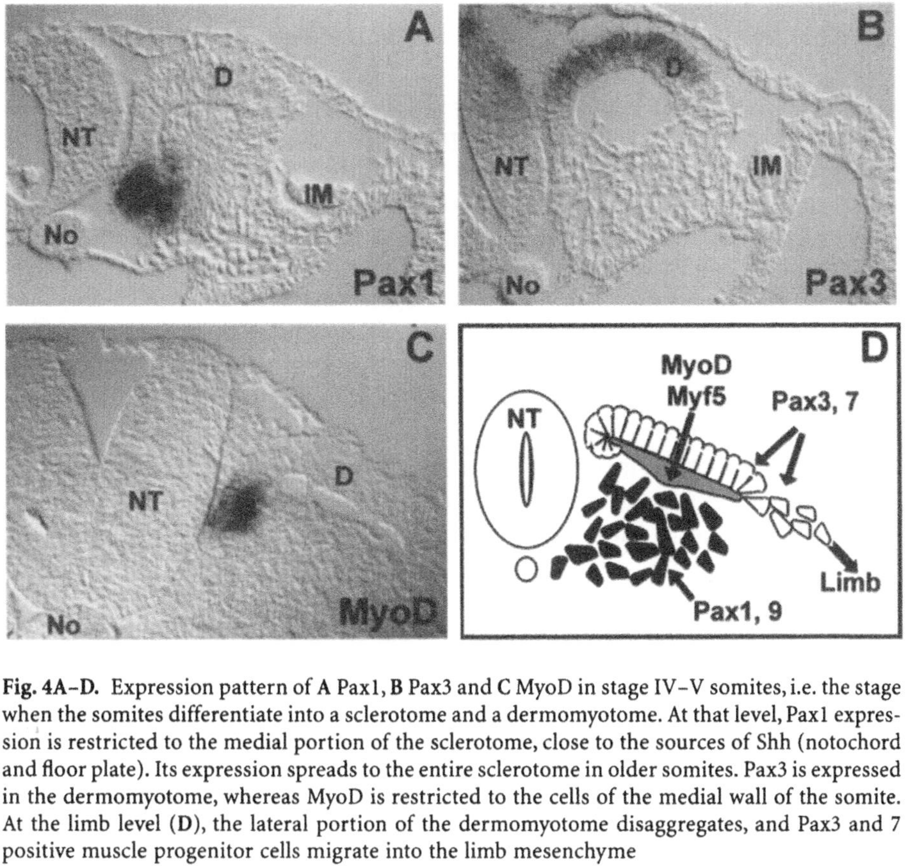

Fig. 4A–D. Expression pattern of **A** Pax1, **B** Pax3 and **C** MyoD in stage IV–V somites, i.e. the stage when the somites differentiate into a sclerotome and a dermomyotome. At that level, Pax1 expression is restricted to the medial portion of the sclerotome, close to the sources of Shh (notochord and floor plate). Its expression spreads to the entire sclerotome in older somites. Pax3 is expressed in the dermomyotome, whereas MyoD is restricted to the cells of the medial wall of the somite. At the limb level (**D**), the lateral portion of the dermomyotome disaggregates, and Pax3 and 7 positive muscle progenitor cells migrate into the limb mesenchyme

Emerson 1992; Borycki et al. 1997; see Fig. 3C). Therefore, although these cells display a typical pseudo-epithelial morphology and are still proliferating, they can be considered as the earliest myotomal cells. Recent lineages studies using the lipophilic fluorescent markers DiI and DiO have resulted in the emergence of a coherent model for epaxial as well as hypaxial myotome formation (Denetclaw et al. 1997, 2001; Kahane et al. 1998; Cinnamon et al. 1999; Denetclaw and Ordahl 2000; Ordahl et al. 2001).

At somite stage IV–V of development, the ventral portion of the cortex epithelium disaggregates into a mesenchyme; together with the cells of the somitocoel; these mesenchymal cells form the sclerotome (recognized by the onset of expression of the pair-rule transcription factors Pax1 and Pax9 at this developmental stage somite; Fig. 4A). These sclerotomal cells will differentiate into axial cartilage and bones. The medial wall of the somite, directly apposed to the neural tube and the dorsal part of the somite, remains epithelial and continues to express MyoD (Fig. 4C). Subsequently, the medial somitic wall detaches from the neural tube and folds under the dorsal epithelial portion of the somite, named the dermomyotome, which expresses Pax3 and Pax7 (Goulding et al. 1991, 1994; Fig. 4B). Concomitant with the separation of the medial wall of the somite from the neural tube, neural crest cells embark on their ventral migratory route between the neural tube and the somites (Fig. 3B). An intriguing possibility is that proteases secreted by neural crest cells (e.g. plasminogen activator), in order to modify their extracellular environment to facilitate their migration (Valinsky and Le Douarin 1985), might also activate the separation of the medial wall of the somite from the neural tube.

At somite stage V–VI, epaxial myotome formation is initiated as cells located at the medial portion of the dermomyotome (termed the medial lip) start translocating beneath the dermomyotome and finally come to lie under it in a compartment termed the myotome (Denetclaw et al. 1997, 2001; Ordahl et al. 2001). Myotome growth is the result of a continuous passage of medial lip cells into the myotomal compartment. Quail-chick grafting experiments have demonstrated that the medial lip contains most, if not all, progenitor cells that contribute to early epaxial myotome growth (Huang and Christ 2000; Ordahl et al. 2001). Once in the myotomal compartment of the somite, the muscle progenitors, termed myocytes, become post-mitotic (Williams 1910; Langman and Nelson 1968), and elongate both rostrally and caudally in order to reach the borders of the dermomyotome (Fig. 5F).

3.1.4 Hypaxial Muscle Formation

After an initial lag phase of about 24 h, hypaxial muscles of the trunk region form in very much the same way that epaxial muscles do, with the exception that they translocate into the myotomal compartment through the lateral lip of the dermomyotome (Denetclaw and Ordahl 2000). Epaxial and hypaxial progenitors stop proliferating once they have entered the myotomal compartment of the somite. In the fore- and hindlimb regions, muscle progenitors located in the lateral dermomyotome undergo an epithelio-mesenchymal transition and migrate as single cells into the limb mesenchyme where they later differentiate into hypaxial appendicular muscles (Chevallier et al. 1977; Christ et al. 1977; Fig. 4D). Early microscopic studies have demonstrated that muscle progenitor emigration from the lateral somite into the limb mesenchyme occurs as early as stage 14 HH (22 somite embryos, Jacob et al. 1978, reviewed in Wachtler and Christ 1992). More recent molecular studies, using Pax3 as a molecular marker for migrating hypaxial progenitors, have identified the first signs of emigration at stage 18 HH (30–35 somite embryos), i.e. almost a day later, at the moment that the forelimb bud starts bulging (Goulding et al. 1994; Williams and Ordahl 1994, our observation). Although this apparent discrepancy might indicate that a first wave of muscle progenitors that do not express any identified molecular marker might leave the lateral dermomyotome prior to the second, Pax3-positive wave, an alternative hypothesis is that the cells that were shown, by electron microscopy, to leave the somite before limb bud

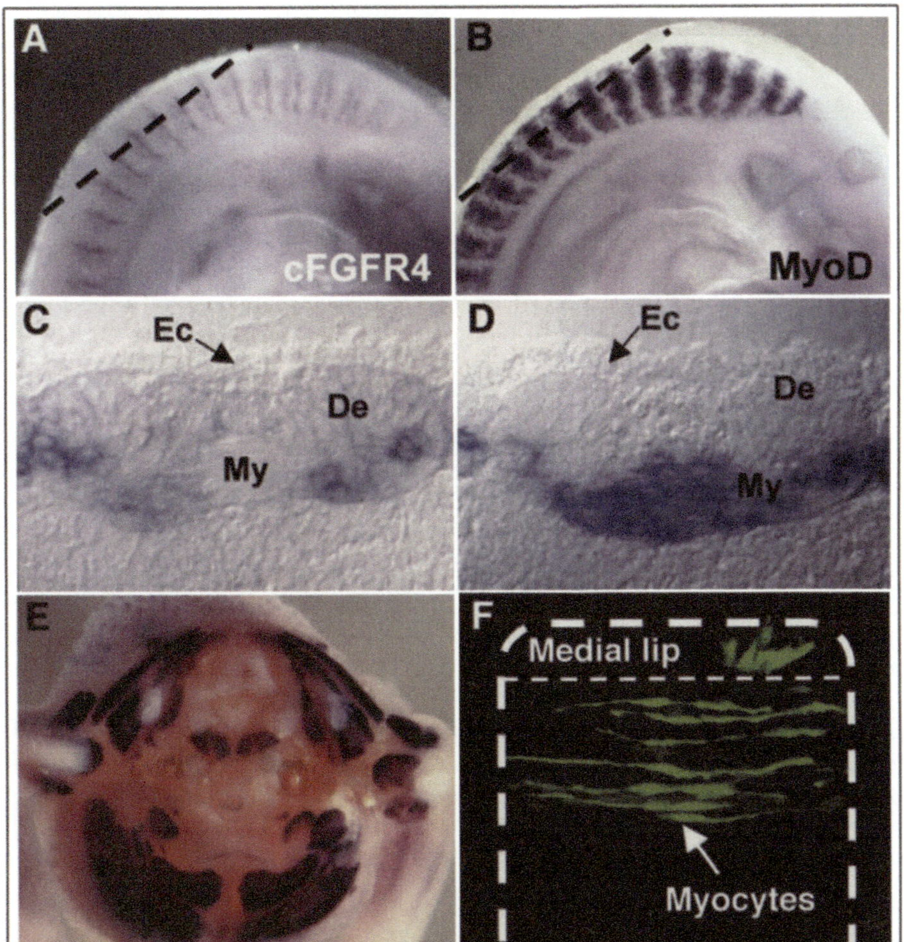

Fig. 5A–F. Comparison of the expression patterns of cFGFR4/FREK and MyoD in 3-day-old chick embryos. In **A** and **B**, the *dotted line* shows the angle of sections in **C** and **D**. **C** and **D** are, therefore, frontal sections through cervical somites. In whole-mount, cFGFR4 expression is observed at the anterior and posterior borders of the somites, whereas MyoD expression is restricted to the medial region of each somite. On sections, cFGFR4 expression is mostly observed at the anterior and posterior borders of the dermomyotome (*De*), as well as in a few cells scattered within the myotome (*My*) In contrast, MyoD expression is mostly observed in the medial region of the myotome. *Ec* Ectoderm. **E** In 9-day-old quail, as myoblast proliferation reaches its peak, FGFR4 is highly expressed in every skeletal muscle of the body. **F** Organization of myocytes in the early myotome. The medial portion of a newly formed somite was electroporated with an expression plasmid coding for the Green Fluorescent Protein (GFP) cloned under chick β-actin promoter/CMV enhancer. One day later, the electroporated cells and their progeny are observed with a confocal microscope. A few cells of the medial lip are labelled, as well as a number of myocytes that extend from the anterior to the posterior border of the somite. The *thick dotted line* delineates the borders of the somite; the *thin dotted line* indicates the position of the medial lip of the dermomyotome formation are not, in fact, muscle progenitors. In support of this second hypothesis is the observation that a portion of the endothelial precursors of the limbs and the dorsal aorta originate in the somite (Solursh et al. 1987; Noden 1990; Wilting et al. 1995). Thus, it is possible that these progenitors leave the somite prior to muscle progenitors. Furthermore, VEGFR2, which is a marker for endothelial progenitors, has been shown to be temporarily expressed in the lateral portion of somites apposed to the wing field prior to the emigration of Pax3-positive myogenic progenitors into the limb bud mesenchyme (Eichmann et al. 1993). In light of these data, it would be interesting to follow the migration of endothelial precursors into the limb bud in relation to that of myogenic precursors.

3.1.5 A Second Wave of Proliferative Muscle Progenitors

The demonstration that myocytes of the myotome are post-mitotic and that the dermomyotome is the sole source of muscle cells (Christ et al. 1977) raises a paradox because the dermomyotome (including the medial and lateral lips) as a morphologically distinguishable structure disappears between days 4 and 5 of development (unpubl. observ.), whereas muscles grow continuously during embryogenesis and even after birth. Therefore, one must assume that muscle progenitor cells capable of proliferation and self-renewal (i.e. muscle stem cells or myoblasts) have to enter the myotomal compartment of the somite before the dermomyotome disappears. Indeed, this appears to be the case: at around day 3 of chick embryogenesis, mitotically competent muscle progenitor cells that express the FGF receptor FREK/cFGFR4, appear in the myotome (Marcelle et al. 1994, 1995; Fig. 5A, C). The migration of cFGFR4-expressing cells into the myotome, therefore, represents a second wave of myotomal precursors that arises about 15 h after the first wave of MyoD-positive, post-mitotic cells that can be observed emanating from the medial lip of the dermomyotome (Fig. 5B, D). Because FGFR4 initiates its expression at the anterior and posterior borders of the dermomyotome (Kahane et al. 1998, Fig. 5A, C), it is likely that FGFR4-positive myotomal precursors originate from these regions of the dermomyotome. This idea recently gained support by the observations that mitotically competent myotomal precursors, distinct from those arising from the medial lip, migrate into the myotome from the anterior and posterior borders of the dermomyotome during early chick development (Kahane et al. 1998, 2001; Huang and Christ 2000). During development, the early myotome is therefore formed by at least two waves of myogenic cells: a first wave of cells ingressing from the medial and lateral lips of the dermomyotome that never express FGFR4, and which rapidly begins to express myogenic differentiation markers while becoming post-mitotic. This is followed by a second wave of FGFR4-positive cells that proliferate in the myotome before undergoing terminal myogenic differentiation (Marcelle et al. 1995). As development proceeds, myoblasts divide rapidly to build up the fast-growing muscle masses. At the peak of myoblast proliferation (E9–E10 in chick) most, if not all, proliferating cells in the skeletal muscle masses of the body strongly express FGFR4 (Fig. 5E). Interestingly, FGFR4 is not only expressed in embryonic myoblasts, but also in satellite cells of the adult muscle of chick (Halevy et al. 1994; Marcelle et al. 1994) and mouse (Kastner et al. 2000), which are the only cells of the adult muscle that are capable of proliferation and self-renewal to replace damaged muscle fibres.

In contrast to the situation described above for somites, the formation of appendicular hypaxial muscle seems to result from only one wave of myogenic precursors because all Pax3-positive cells emigrating from the lateral dermomyotome and that enter the limb bud mesenchyme initiate FGFR4 expression and proliferate before expressing MyoD and undergoing terminal myogenic differentiation (Marcelle et al. 1995).

3.2 Dermis Formation

Because of the lack of morphological or molecular criteria to follow the differentiation of dermal cells, dermis formation is far less well understood than myotomal or sclerotomal differentiation. Between the medial and the lateral lips of the dermomyotome lies a region that is fated to become the dermis of the back; the dermis of the limbs is derived from the somatopleura, and the dermis of the face derives from cranial neural crest (Couly and Le Douarin 1988; Christ and Ordahl 1995; Olivera-Martinez et al. 2000). To form the dermis, cells within the dermomyotome undergo an epithelio-mesenchymal transition and migrate away from the dermomyotome towards the ectoderm (Brill et al. 1995). In chick, de-epithelialization of the dermomyotome is first observed in its lateral portion (i.e. close to the lateral lip) in the cervical region of a stage 18 HH embryo (30–35 somite embryos). With time, this de-epithelialization process progresses both along the antero-posterior axis of the embryo and medially towards the medial lip of the dermomyotome. For instance, at stage 21 HH (43–46 somite embryo), delamination is initiated in the lateral portion of the dermomyotome at the level of the hindlimb, whereas at the cervical level of the same embryo, the entire dermomyotome (but for the medial and lateral lips) has undergone de-epithelialization (K. Tosney, pers. comm., our observations). As cells that have delaminated from the dermomyotome condense to form the dermis subjacent to the epidermis, they initiate the expression of the bHLH molecule Dermo-1 (Li et al. 1995; Scaal et al. 2001).

4 Tissue and Molecular Regulation of Somite Differentiation

At the time that somites form, paraxial mesoderm is not committed to a definitive fate: when challenged in a novel environment, rostral segmental plate and early somitic tissue can adopt very different fates, such as endodermal or lateral plate identities (Veini and Bellairs 1991). Moreover, if somites and rostral segmental plate are rotated dorso-ventrally or switched medio-laterally, cells fated to differentiate into sclerotome, epaxial or hypaxial muscles will develop according to their new location (Aoyama and Asamoto 1988; Ordahl and Le Douarin 1992; Dockter and Ordahl 2000). These observations have led to the fundamental notion that early somites are comprised of largely naive cells that adopt their various fates solely through the inductive influences of environmental cues. In the search for these cues, a recurrent and complicated issue that arises is whether the signals emitted by the environment of the somites are inductive (i.e. that they instruct a naive tissue to adopt a particular fate) or permissive (i.e. that they promote an intrinsic or pre-existing potential for a specific differentiation program). For half a century, a number of studies performed in vitro and in vivo have shown that tissues surrounding the somite (i.e. notochord, neural tube, ectoderm and lateral plate mesoderm) indeed display various regulatory roles towards somite development. More recently, this has led to the identification of a number of molecules that can mediate these activities. Although the present review focuses predominantly on muscle and dermomyotome differentiation, we will also mention sclerotomal development because many of the differentiation mechanisms (if not molecules) identified to date are thought to pattern all compartments of the somite in a similar manner.

4.1 The Notochord and Floor Plate Exert a Ventralizing Activity on the Somite: a Role for Sonic Hedgehog?

Following the pioneering work of Hall (1977), several groups have demonstrated the importance of notochord- and floor-plate-derived signals in terms of sclerotome differentiation. Supernumerary notochords or floor plates transplanted dorsally to the somites greatly expand the sclerotomal compartment of the somite (as judged by the expression of Pax-1 and Pax-9) at the expense of the dermomyotome and myotome (Brand-Saberi et al. 1993; Pourquié et al. 1993; Bober et al. 1994; Goulding et al. 1994; Borycki et al. 1997; Dietrich et al. 1997; Teillet et al. 1998). In vitro culture experiments have led to the same conclusion (Fan and Tessier-Lavigne 1994; Ebensperger et al. 1995; Muller et al. 1996). The effect of the notochord does not seem to promote the overgrowth of the sclerotomal compartment; rather, quail-chick grafting experiments have demonstrated that the notochord is actually capable of changing the fate of dermomyotome into sclerotome (Williams and Ordahl 1997). Conversely, early ablation of the notochord leads to an absence of sclerotomal markers and to a ventral expansion of dorsal dermomyotomal markers, such as Pax3 and Pax7 (Goulding et al. 1994; Pownall et al. 1996; Dietrich et al. 1997; Teillet et al. 1998). These data suggest that notochord signalling exerts a ventralizing activity on somites that is necessary and sufficient for sclerotome induction.

Sonic Hedgehog (Shh) has been thought to mediate these notochord and floor plate activities (Fan and Tessier-Lavigne 1994; Fan et al. 1995; Münsterberg et al. 1995; Maroto et al. 1997; Reshef et al. 1998). Shh is initially expressed in the notochord; it induces its own expression in the ventral neural tube, where it promotes the differentiation of the floor plate (Roelink et al. 1995). In vitro and in vivo, Shh is capable of antagonizing the induction of dermomyotomal markers by the dorsal neural tube (Fan and Tessier-Lavigne 1994; Johnson et al. 1994; Maroto et al. 1997). In vitro approaches have shown that Shh can induce Pax-1 expression in uncommitted paraxial mesoderm explants (Fan and Tessier-Lavigne 1994; Fan et al. 1995; Münsterberg et al. 1995). In vivo, Shh is able to activate Pax-1 expression regardless of the presence (Johnson et al. 1994; Borycki et al. 1998) or absence (Marcelle et al. 1999) of axial structures. Further differentiation of the sclerotome into cartilage is directly promoted by Shh, as this molecule augments the competence of ventral somite cells to respond to chondrogenic molecules, such as BMP (Murtaugh et al. 1999). Shh activity is mediated by its putative receptors Patched and Smoothened (Stone et al. 1996) and their downstream effectors, the Gli transcription factors (Sasaki et al. 1997). Shh is able to activate the expression of its putative receptor Patched in a dose-dependent manner; which leads to an amplification of Shh activity in the close vicinity of its source of secretion (Tabata and Kornberg 1994; Goodrich et al. 1996). In chick, Gli1, 2 and 3 are differentially expressed in the developing somite (Borycki et al. 1998, 2000); Gli1 expression is up-regulated in response to Shh, whereas Gli2 and 3 are positively regulated by dorsal signals from the neural tube and/or ectoderm and repressed by the notochord (Borycki et al. 2000). Although the precise roles played by the Gli molecules during somite patterning remain to be demonstrated, these observations indicate a complex regulation of Shh activity in tissues exposed to its action and suggest that Shh might mediate various functions through different Gli effectors within different somitic cell populations. In addition to this positive autoregulation of its own signalling pathway, Shh signalling might actively oppose the dorsalizing activity of Wnt molecules (see below), by inducing in the ventral compartment of somites the expression of SFRP2, a putative antagonist of Wnt signalling (Lee et al. 2000).

The experiments described above all point towards a central role for Shh in the induction of sclerotome. If Shh exerts its effect upon a naive tissue, the segmental plate, one would predict blocking this activity would result in the absence of sclerotomal fate. Thus, it came as a surprise that Shh-null mice display a close to normal, although transient expression of Pax1 in somites (Chiang et al. 1996). Because no other member of the Hedgehog family is expressed at the appropriate time and place to compensate for the loss of Shh activity (Bitgood and McMahon 1995), this observation is difficult to reconcile with a central role for Shh in the induction of sclerotome; this suggests that Shh may rather play a maintenance role, thereby indicating a permissive rather than an instructive function in this process. This hypothesis is further supported by the observation that Shh is a potent mitogen for somitic cells

(Fan et al. 1995; Teillet et al. 1998; Marcelle et al. 1999). Taken together, these observations indicate that Shh might promote and/or amplify an intrinsic or pre-existing potential for sclerotomal differentiation. A tempting hypothesis, therefore, is that other molecules might play a role in the initial induction of Pax1 expression. In vitro, small, but reproducible induction of Pax1 is achieved by addition of the BMP-inhibiting molecule noggin to segmental plate explants. This activation is independent of Shh signalling because this induction is not compromised by the addition of Shh blocking antibody or forskolin, which both antagonize Shh signalling (McMahon et al. 1998). In contrast to the Shh-null mice, noggin-null mutant mice display a severely reduced Pax1-expression domain. This effect is seen despite the presence of Shh expression in the axial structures, which is normal in these mutant animals (McMahon et al. 1998). These results suggest that premature Pax1 expression in the segmental plate and early somites may be prevented by a BMP signal acting on these tissues. This hypothesis implies that sclerotomal fate could either be a default pathway for paraxial mesoderm observed once BMP signalling is inhibited or, alternatively, it could be induced much earlier than expected (by a yet unknown molecule), but would be repressed until after somites have formed. The identity and origin of the BMP signal is unclear: putative candidates are BMP4, which is expressed in the caudal segmental plate and lateral plate mesoderm (Pourquié et al. 1996), BMP7 (expressed in the notochord and in the ectoderm overlying the segmental plate mesoderm) and BMP3, which is expressed in the notochord (Lyons et al. 1995; Dudley and Robertson 1997). Because BMP molecules have also been shown to inhibit muscle and dermomyotome differentiation (see below), a simple model holds that BMP signalling generally inhibits premature somite differentiation. This model implies a tight temporal and spatial control over the release of BMP inhibition, along the antero-posterior axis of the embryo. One possible way to achieve this would be to stop the production of BMP protein. An alternative means would be to oppose its effect by activating the expression of BMP-inhibiting molecules. Interestingly, noggin expression is initiated in paraxial mesoderm prior to somitogenesis (Hirsinger et al. 1997; Marcelle et al. 1997; Reshef et al. 1998). This expression pattern is compatible with a function for noggin in alleviating the BMP inhibitory activity and indicates that, in vivo, noggin might actively participate in the activation of the sclerotomal fate.

4.2 Dorsalizing Activity of Wnt Molecules in the Dorsal Ectoderm and Neural Tube

Embryonic manipulations aimed at determining the role of dorsal structures during somitogenesis have shown that both the ectoderm (via contact-mediated factor) and the dorsal neural tube (via a soluble factor) can promote the formation of the dermomyotome (Gallera 1966; Fan and Tessier-Lavigne 1994; Spence et al. 1996; Dietrich et al. 1997, 1998; Maroto et al. 1997; Sosic et al. 1997; Reshef et al. 1998). In addition, it is possible that there exists a synergy between the dorsal neural tube and ectoderm because the removal of both structures (but not simply one) leads to the ventralization of somites (Dietrich et al. 1997). Interestingly, apart from its putative role in dermomyotome formation, the ectoderm is also required for epithelialization of the PSM, and is required to maintain somite segmentation (Sosic et al. 1997). Taken together, these results indicate that the ectoderm and the dorsal neural tube have a necessary and possibly redundant dorsalizing activity on somites. The dorsal signals can be overridden by ventralizing signals because grafting a notochord dorsal to the segmental plate leads to a complete absence of dermis and muscles in the vicinity of the graft (Pourquié et al. 1993). Although ectoderm and dorsal neural tube signals are likely to play an important role in the differentiation and maintenance of a dorsal compartment of the somite, the dorsal neural tube might play an additional role in the later differentiation of the dermis because it was shown to be required for the epithelial-mesenchyme conversion of the dermis precursors (Brill et al. 1995).

The most likely candidates to elicit this effect of somite dorsalization are the Wnt molecules. In vivo, the ectopic expression of Wnt1, 3a and 4 in somites result in the expansion of the dorsal compartment of the somite and a concomitant reduction of the ventral compartment (Capdevila et al. 1998; Wagner et al. 2000). In vitro experiments support the model that a number of Wnt proteins are able to mediate the dorsalizing activity of the dorsal neural tube and ectoderm because Wnt1 and 3a mediate a diffusible dorsalizing activity, whereas Wnt6 and 4 need close contact with the paraxial mesoderm to mediate their activity (Fan et al. 1997; Maroto et al. 1997). Experiments performed in vitro elegantly demonstrated the antagonistic and dose-dependent actions of Shh and Wnt molecules on the dorso-ventral patterning of somites. When uncommitted mesoderm is positioned between two sources of Shh and Wnt proteins, the extent of the Pax1 (i.e. sclerotome) and the Pax3 (i.e. dermomyotome) domains vary with respect to the amount of Shh and Wnt protein to which they are exposed (Fan et al. 1997). These data indicate that Wnt and Shh signalling pattern the paraxial mesoderm dorso-ventrally in a mutually antagonistic manner, suggestive of a concentration-dependent mechanism. Key questions that arise from this work are the exact Wnt molecules involved and their tissue source of origin. A number of Wnt molecules are expressed in the dorsal neural tube and ectoderm and they have all been shown to play a role in this process. An additional point of confusion in the literature is that some of the Wnt candidate molecules are differentially expressed when comparing chick and mouse: although Wnt1, 3a and 4 are expressed in the dorsal neural tube of both species (Parr et al. 1993; Parr and McMahon 1994; Hollyday et al. 1995; Marcelle et al. 1997), Wnt4 and Wnt7a are expressed in the ectoderm overlying the mouse paraxial mesoderm whereas, in chick, only Wnt6 is expressed in the ectoderm (Parr et al. 1993; Marcelle et al. 1997; Tajbakhsh et al. 1998; and pers. observ.; see Fig. 6A, B). Embryonic manipulations performed in chick clarify the respective roles played by the ectoderm

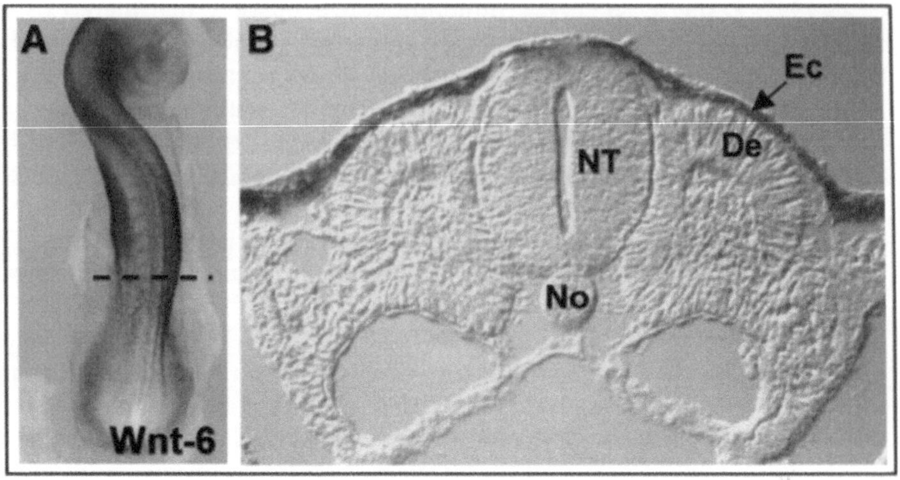

Fig. 6. **A.** Expression pattern of Wnt6 in chick. In-situ hybridization of a stage 17 HH chick embryo, using a chick Wnt6 probe (generous gift of Andy McMahon). Wnt6 is expressed in the ectoderm overlying paraxial and lateral plate mesoderm. The *dotted line* indicates the level of the section presented in **B.** *NT* Neural tube, *No* notochord; *De* dermomyotome and *Ec* ectoderm

and neural tube in the dorso-ventral polarity of somites. Surgical separation of the segmental plate mesoderm from the neural tube prior to the initiation of Pax3 and paraxis expression (used as dermomyotomal markers) leads to (1) the maturation of somites into a ventral mesenchyme (albeit Pax1-negative) and a dorsal epithelial dermomyotome and (2) the normal (i.e. dorsal) expression of Pax3 (Marcelle et al. 1997) and paraxis (unpubl. observ.). This observation argues against an involvement of axial structures (and thus the Wnts expressed in the dorsal neural tube) in the formation of the dorsal compartment of the somite, and imply that, in vivo, ectodermally expressed Wnt molecules might play a more important role in this process. Thus, the activities of Wnt1, 3a and 4 (expressed in the dorsal neural tube), which have been observed in vitro and in vivo, may only mimic that of the ectodermal Wnt. Whether ectoderm plays a role in the induction or maintenance of the demomyotome is unclear at the moment. Unlike the situation described above for Shh, knockout studies in mice do not seem to clarify this point; this is partly because of the large number of Wnt molecules expressed at the time of somite differentiation, with redundant and/or compensatory functions occurring during this process. Further work will be needed to analyse the exact function of each of the Wnt molecules during dermomyotome formation.

At around somite stage V–VI, the dermomyotome is patterned into a medial compartment (the "medial lip"), which contains epaxial muscle progenitors, a central region that gives rise to the dermis of the back, and a lateral compartment where hypaxial muscle progenitors originate. The secreted factor Wnt-11, the transcription factor Engrailed-1 and the basic helix–loop–helix transcription factor *Sim1* are specific markers of the medial, central and lateral compartments of the dermomyotome, respectively (Gardner and Barald 1992; Marcelle et al. 1997; Pourquié et al. 1996). It was shown that BMP in the dorsal neural tube indirectly induces the formation of a Wnt11-positive medial lip by up-regulating Wnt1 and Wnt3a expression in the neural tube (Marcelle et al. 1997); likewise, BMP4, secreted by the lateral plate mesoderm, regulates Sim1 expression, thereby specifying the hypaxial muscle lineage (Pourquié et al. 1996). It is possible that neural tube specifies the formation of the dermis as well (Brill et al. 1995). Therefore, although axial and lateral structures are not likely to play a role in the initial formation of the dermomyotome, they display important functions in the specification of its medio-lateral axis.

4.3 Tissue and Molecular Regulation of Myogenesis: an Instructive or Permissive Process?

The means by which myogenesis is initiated in somites has been a highly controversial question over the years. The intense search for muscle-inducing tissues has led to confusing, if not opposing results. Nevertheless, all of these studies have postulated that myogenesis is initiated as somites form. The neural tube has been repeatedly demonstrated to play an important role in myogenesis (Watterson et al. 1954; Teillet and Le Douarin 1983; Vivarelli and Cossu 1986; Kenny-Mobbs and Thorogood 1987; Christ et al. 1992; Rong et al. 1992; Borman and Yorde 1994; Buffinger and Stockdale 1994; Stern et al. 1995). Yet, data from several studies also indicate that myogenic-promoting activity is also elicited by other tissues, including the surface ectoderm (Christ et al. 1972; Kuratani et al. 1994; Cossu et al. 1996), and the notochord/ventral neural tube (Buffinger and Stockdale 1994; Stern and Hauschka 1995). Moreover, combinatorial signals emanating from the notochord/floor plate and the dorsal neural tube have also been proposed to be required for muscle induction (Münsterberg and Lassar 1995; Münsterberg et al. 1995). Together, these observations indicate an inductive role for tissues surrounding the somite. On the other hand, the fact that epaxial muscles degenerate when axial structures are ablated, implies that axial structures might merely perform a maintenance and/or proliferative role during muscle development rather than an inductive one. An additional level of complexity results from the observation that hypaxial muscles are less affected than epaxial muscles by the removal of axial structures, indicating a differential role of axial structures on the differentiation of various populations of skeletal muscles (Teillet and Le Douarin 1983; Rong et al. 1992; Kruger et al. 2001).

To identify the molecular signals that regulate myogenesis in early somites, co-culture of unsegmented mesoderm or caudal somites in the presence of Shh and Wnt have been performed. Although Shh alone is not able to activate muscle marker expression (Fan and Tessier-Lavigne 1994; Münsterberg et al. 1995; Kos et al. 1998; Tajbakhsh et al. 1998), segmental plate mesoderm cultured in the presence of Wnt protein (Wnt1, 4, 5a, 6 or 7a) results in a low-level induction of myogenic precursors (Stern et al. 1995; Tajbakhsh et al. 1998). In contrast, when both signalling pathways are applied simultaneously to unsegmented mesoderm, a robust upregulation of both MyoD and Myf5 is observed (Münsterberg et al. 1995; Maroto et al. 1997; Reshef et al. 1998; Tajbakhsh et al. 1998). In light of these data, it was proposed that myogenesis induction in segmental plate mesoderm requires the instructive combinatorial action of Wnt and Shh (Münsterberg et al. 1995). Because these experiments were performed in parallel to address sclerotome, muscle and dermomyotome induction, an extension of this model predicts that high concentrations of Shh and low levels of Wnt induce sclerotome, intermediate levels of both signals result in muscle induction, whereas low levels of Shh and high levels of Wnt favour dermomyotome differentiation. Although this is a very simple and (therefore) attractive model to broadly explain somite patterning, a number of observations question the validity of this model. As was mentioned above for the case of sclerotome induction, the instructive role played by Shh in muscle induction is doubtful with regard to the Shh mouse knock-out where Myf5 (the earliest myogenic differentiation factor expressed in mice) expression in somites is present, albeit temporarily and at a reduced level in early somites (Chiang et al. 1996). Re-examination of the Shh-null mice further shows that epaxial muscles are more sensitive to the absence of Shh signalling than hypaxial muscles, confirming the notochord extirpation experiments mentioned above (Kruger et al. 2001). These results indicate that Shh acts to maintain and/or to expand the population of Myf5-positive cells, thereby performing a predominantly permissive function. This finding was confirmed in chick by the observation that Shh maintains, but cannot induce MyoD expression, and exerts a proliferative role on somitic tissues as well as other tissues in vitro and in vivo. In the absence of a source of Shh, somite cells rapidly stop proliferating and eventually die (Fan et al. 1995; Duprez et al. 1998; Teillet et al. 1998; Ahlgren and Bronner-Fraser 1999; Marcelle et al. 1999). In light of such a strong requirement for Shh for somite survival, one understands that it is extremely complicated to test whether this molecule plays any additional patterning role on these tissues.

If the understanding of Shh function in myogenesis is still far from clear, that of Wnt signalling in this process is even more obscure. The observation that Wnts can induce reproducible expression of myogenic markers when added to uncommitted segmental plate mesoderm indicates that a Wnt protein might represent the initial inducer of myogenesis. Wnt molecules have been reported to be expressed in the dorsal neural tube and the ectoderm roughly at the time that myogenesis is initiated in early somites; likely candidate molecules are Wnt1, Wnt3a and Wnt4 in the dorsal neural tube, and Wnt6 (chick) or Wnt4 (mouse) in the ectoderm (see above). The observation that myogenesis is initiated in the medial part of somites directly apposed to the neural tube has prompted the search for inducing molecules in the neural tube.

Indeed, all Wnts expressed in the neural tube display some level of myogenic inducing activity (see above). However, removal of the dorsal neural tube before myogenesis is initiated does not impede the initiation of muscle markers (Marcelle et al. 1997), indicating that the dorsal neural tube is not necessary for the initial expression of muscle markers. On the other hand, although ectodermal Wnts (6 and 4) have the ability to activate myogenesis, it is unlikely that they could restrict muscle marker expression to the medial somite because they are uniformly expressed in the ectoderm. As mentioned above in the case of sclerotome formation, mouse mutants for Wnt are not informative as to the specific role of distinct Wnt molecules during muscle induction. It is likely that careful analyses will be needed to uncover the specific role of each of the Wnt molecules expressed at the moment that myogenesis is initiated.

Is induction the only plausible way to explain the initiation of myogenesis in somites? An alternative hypothesis is that the myogenic fate is a default pathway that develops in the absence of signals from the environment. A number of observations are in support of this idea. Classical experiments (Ellison et al. 1969; O'Hare 1972; Holtzer and Holtzer 1976; Wachtler et al. 1982; Choi et al. 1989) have shown that somitic tissue are able to differentiate into muscle fibres in vitro. Certainly, it is possible that the medium in which grafts were cultured in these studies is not neutral and the tissues might have found, in these conditions, similar muscle-inducing activities that they normally encounter. To address this question, George-Weinstein and collaborators have assayed the capacity of several embryonic tissues to activate a muscle phenotype in defined culture conditions. They have shown that myogenesis represents the preferred pathway of the epiblast, segmental plate and somitic tissues because over 95% of these cells express muscle markers when they are dissociated to produce a single cell suspension, and are subsequently cultured in serum-free medium (George-Weinstein et al. 1996, 1997). These surprising observations led the authors to hypothesize that myogenesis is a default pathway that is repressed in the mesoderm until needed. The cell dissociation step, which is crucial for myogenic differentiation, is reminiscent of the situation observed in *Xenopus* ectoderm, where dissociation of animal pole cells promotes the formation of neural tissue (Godsave and Slack 1989; Grunz and Tacke 1989; Sato and Sargent 1989). In *Xenopus*, cell dissociation is thought to impede the paracrine action of a neural inhibitor molecule secreted by ectodermal cells themselves. The neural-inhibiting molecule has been identified as a member of the BMP family of growth factors (Hemmati-Brivanlou and Melton 1997; reviewed in Hemmati-Brivanlou and Melton 1992). Remarkably, BMP was demonstrated to be a strong inhibitor of myogenic differentiation and members of this family are expressed around the segmental plate, in the posterior lateral plate mesoderm and possibly in the ectoderm that overlays this structure (Lyons et al. 1995; Pourquié et al. 1996; Dudley and Robertson 1997; Solloway and Robertson 1999). In accordance with these data, the noggin-null mice display a strong reduction of myogenesis, indicating an important role for BMP signalling in the repression of the myogenic programme (McMahon et al. 1998).

Where then would myogenesis be initiated? Although George-Weinstein's observations hypothesize that most cells of gastrulating embryos have a myogenic potential, retrotracing lineage studies have determined the emergence of a restricted muscle stem cell population, distinct from other mesodermal lineages, within the primitive streak (Nicolas et al. 1996). The two observations are not necessarily incompatible because George-Weinstein proposes that a small population of randomly distributed cells stably committed towards the myogenic pathway, recruit uncommitted cells to become myogenic through cell–cell interactions. This "cell community" effect, which has been shown to be important for muscle differentiation in both *Xenopus* and mouse (Gurdon 1988; Cossu et al. 1995), has been experimentally demonstrated by the observation that high density, but not low density cultures of epiblast or segmental plate cells, undergo efficient myogenesis (George-Weinstein et al. 1998). Although the existence of a muscle stem cell population, either within the epiblast or the primitive streak, has as yet no supporting molecular evidence, two observations reinforce the notion that myogenesis might be under way well before somites form: (1) onset of MyoD expression, which is believed to be a strong indication that myogenesis is under way, although never detected by in-situ hybridization in the chick segmental plate, is detectable by PCR in this tissue (George-Weinstein et al. 1996); (2) improved in-situ hybridization techniques and the choice of new probes have enabled the detection of Myf5 in the rostral portion of the segmental plate (Delfini et al. 2000; Kiefer and Hauschka 2001). The expression of Myf5 in this region of the segmental plate, which is fated to become not only muscle, but also dermis, endothelial cells . . ., does not commit these cells to the myogenic lineage, but might indicate a possible bias or competence of the entire rostral paraxial mesoderm towards the myogenic pathway prior to somitogenesis. It is plausible that a signal(s) intrinsic or extrinsic to the rostral segmental plate initiates Myf5 expression in this tissue. An appealing hypothesis is that Wnts, which are expressed at various levels along the segmental plate (Parr et al. 1993; Hollyday et al. 1995, our observations) and that display myogenic potentials, participate in this process. Whether or not low level expression of Myf5 confers myogenic competence to segmental plate cells, it is important for correct body patterning that full-blown myogenesis be initiated only after somitogenesis has taken place. Indeed, what would happen if muscle fibres were to arise within the segmental plate? Without somite borders, it is likely that fibres would be completely disorganized, leading to important defects in motility. The activation of noggin expression in the rostral segmental plate of the chick is likely to play an important role in the timely release of the BMP-mediated repression of myogenesis (Hirsinger et al. 1997; Marcelle et al. 1997; Reshef et al. 1998). Because BMP was shown to inhibit sclerotome formation as well, this molecule therefore acts as a general inhibitor of somite differentiation.

5 Conclusion

The experiments and observations described here point towards the same conclusion: at the time that somitogenesis is taking place, paraxial mesoderm is not entirely naive, myogenic precursors are specified and myogenesis (together with sclerotome formation) might in fact be initiated, but be actively repressed by BMP signalling (Fig. 7). Myotome, sclerotome or dermomyotome formation, which have been examined in the past as one-step events, are likely to involve a number of successive steps that occur as mesodermal cells progress along the antero-posterior axis of the embryo. In the case of muscle formation, a number of steps have been experimentally demonstrated: (1) specification of muscle stem cells in the primitive streak, (2) Myf5 expression in the rostral segmental plate, (3) MyoD expression in the medial somite and (4) myotome growth. It is likely that in this cascade of events, each step stabilizes the previous one in a process that we could name "maturation". An important challenge will now reside in the identification of the signals that regulate these maturation steps. Shh and Wnt emanating from the axial structures have been thought to induce the formation of the different somitic compartments. In the light of the data presented here, it is possible that these molecules participate in "late" events of somite differentiation, which involve maintenance or amplification of pre-existing potentials.

Over the past few years, it has become clear that an ever growing number of BMP and Wnt molecules are expressed at a time and in a location compat-

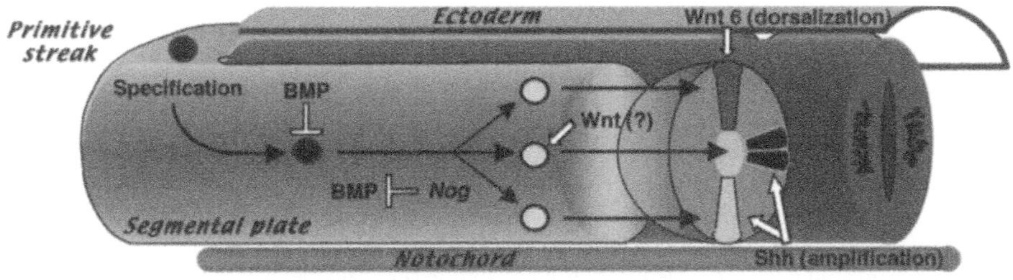

Fig. 7. A model describing somite patterning. As early as in the primitive streak, muscle (and maybe sclerotome) progenitors are specified. Along the segmental plate, their differentiation into muscle and sclerotomal cells is inhibited by BMP signalling. It is possible that additional steps of maturation are required during their migration along the segmental plate. The first signs of dermomyotome differentiation (Pax3 or paraxis expression), which takes place in the rostral half of the segmental plate are independent of axial structures. As mesodermal cells reach the anterior third of the segmental plate, noggin expression releases the BMP-mediated inhibition of sclerotome and myotomal fate. Wnt in the paraxial mesoderm might activate the myogenic potential of these cells, leading to the activation of MyoD in the medial portion of the newly formed somite (in *black*) Early somites are under the influence of Shh (emanating from the notochord and the floor plate), which acts as a potent maintenance and/or proliferative factor for sclerotomal and myotomal cells. It is likely that ectoderm (via Wnt6) plays a role in the maintenance of the dorsal compartment of the somite (dermomyotome)

ible with their playing a possible role in the various steps of somite differentiation. A rather personal and subjective opinion is that it is unlikely that embryos would usurp energy to activate the strong and specific expression of a number of factors and their signalling pathways, if each of those ultimately only play redundant roles in somite differentiation. It is more likely that each of these molecular players has a well-defined role to play during this process. Early somite differentiation involves a large number of distinct steps that have to occur in only a few hours: segmentation, somite border formation, organization of the somite into a pseudo-stratified epithelium, initiation of myogenesis, all these occur almost simultaneously. A few hours later, the ventral epithelium disaggregates, Pax1 expression is induced, while myotome growth and organization are initiated. It is plausible that a large number of molecules are needed to tightly regulate each of these important steps of somite differentiation and it is quite possible that most of the signalling molecules have now been identified. Now is the time to assemble the puzzle. In the future, therefore, it will be extremely important to (1) re-evaluate the temporal and spatial expression patterns of each of these molecules in the light of recent discoveries and (2) determine not so much whether these molecules can activate this or that pathway or tissue fate, but whether they actually do it in vivo. Given the number of candidate molecules involved, it is likely that there lie in front of us a number of years of hard work!

Acknowledgements. We would like to thank Dr. Kim Dale, Dr. Martin Scaal for critical reading of the manuscript and David Liotta for his help in the artwork. Work in our laboratory is supported by grants from the Actions Concertées Incitatives (ACI), the Association Française contre les Myopathies (AFM) and the Association pour le Recherche sur le Cancer (ARC). Claudia Linker is a fellow from the Boehringer Ingelheim Fonds and Cynthia Lesbros is a Fellow from the Ministère de l'Education Nationale, de la Recherche et des Technologies (MENRT).

References

Ahlgren SC, Bronner-Fraser M (1999) Inhibition of sonic hedgehog signaling in vivo results in craniofacial neural crest cell death. Curr Biol 9:1304–1314

Aoyama H, Asamoto K (1988) Determination of somite cells: independence of cell differentiation and morphogenesis. Development 104:15–28

Bitgood MJ, McMahon AP (1995) Hedgehog and Bmp genes are coexpressed at many diverse sites of cell-cell interaction in the mouse embryo. Dev Biol 172:126–138

Bober E, Brand-Saberi B, Ebensperger C, Wilting J, Balling R, Paterson BM, Arnold HH, Christ B (1994) Initial steps of myogenesis in somites are independent of influence from axial structures. Development 120:3073–3082

Borman WH, Yorde DE (1994) Barrier inhibition of a temporal neuraxial influence on early chick somitic myogenesis. Dev Dyn 200:68–78

Borycki AG, Strunk KE, Savary R, Emerson CP Jr (1997) Distinct signal/response mechanisms regulate pax1 and QmyoD activation in sclerotomal and myotomal lineages of quail somites. Dev Biol 185:185–200

Borycki AG, Mendham L, Emerson CP Jr (1998) Control of somite patterning by Sonic hedgehog and its downstream signal response genes. Development 125:777–790

Borycki A, Brown AM, Emerson CP Jr (2000) Shh and Wnt signaling pathways converge to control Gli gene activation in avian somites. Development 127:2075–2087
Brand-Saberi B, Christ B (2000) Evolution and development of distinct cell lineages derived from somites. Curr Top Dev Biol 48:1–42
Brand-Saberi B, Ebensperger C, Wilting J, Balling R, Christ B (1993) The ventralizing effect of the notochord on somite differentiation in chick embryos. Anat Embryol (Berl) 188:239–245
Brand-Saberi B, Wilting J, Ebensperger C, Christ B (1996) The formation of somite compartments in the avian embryo. Int J Dev Biol 40:411–420
Brill G, Kahane N, Carmeli C, von Schack D, Barde YA, Kalcheim C (1995) Epithelial-mesenchymal conversion of dermatome progenitors requires neural tube-derived signals: characterization of the role of Neurotrophin-3 Development 121:2583–2594
Buffinger N, Stockdale FE (1994) Myogenic specification in somites: induction by axial structures. Development 120:1443–1452
Capdevila J, Tabin C, Johnson RL (1998) Control of dorsoventral somite patterning by Wnt-1 and beta-catenin. Dev Biol 193:182–194
Chevallier A, Kieny M, Mauger A (1977) Limb-somite relationship: origin of the limb musculature. J Embryol Exp Morphol 41:245–258
Chiang C, Litingtung Y, Lee E, Young KE, Corden JL, Westphal H, Beachy PA (1996) Cyclopia and defective axial patterning in mice lacking Sonic hedgehog gene function. Nature 383:407–413
Choi J, Schultheiss T, Lu M, Wachtler F, Kuruc N, Franke WW, Bader D, Fischman DA, Holtzer H (1989) Cellular and molecular biology of muscle development. Alan R. Liss, pp 27–36
Christ B, Ordahl CP (1995) Early stages of chick somite development. Anat Embryol (Berl) 191: 381–396
Christ B, Jacob HJ, Jacob M (1972) Experimentelle Untersuchungen zur Somitenentwicklung bein HΩhnerembryo. Z Anat Entwicklungsges 138:82–97
Christ B, Jacob HJ, Jacob M (1977) Experimental analysis of the origin of the wing musculature in avian embryos. Anat Embryol (Berl) 150:171–186
Christ B, Brand-Saberi B, Grim M, Wilting J (1992) Local signalling in dermomyotomal cell type specification. Anat Embryol (Berl) 186:505–510
Cinnamon Y, Kahane N, Kalcheim C (1999) Characterization of the early development of specific hypaxial muscles from the ventrolateral myotome. Development 126:4305–4315
Cossu G, Kelly R, Di Donna S, Vivarelli E, Buckingham M (1995) Myoblast differentiation during mammalian somitogenesis is dependent upon a community effect. Proc Natl Acad Sci USA 92:2254–2258
Cossu G, Kelly R, Tajbakhsh S, Di Donna S, Vivarelli E, Buckingham M (1996) Activation of different myogenic pathways: myf-5 is induced by the neural tube and MyoD by the dorsal ectoderm in mouse paraxial mesoderm. Development 122:429–437
Couly G, Le Douarin NM (1988) The fate map of the cephalic neural primordium at the presomitic to the 3-somite stage in the avian embryo. Development 103:101–113
Delfini M, Hirsinger E, Pourquié O, Duprez D (2000) Delta 1-activated notch inhibits muscle differentiation without affecting Myf5 and Pax3 expression in chick limb myogenesis. Development 127:5213–5224
Denetclaw WF Jr, Christ B, Ordahl CP (1997) Location and growth of epaxial myotome precursor cells. Development 124:1601–1610
Denetclaw WF, Ordahl CP (2000) The growth of the dermomyotome and formation of early myotome lineages in thoracolumbar somites of chicken embryos. Development 127:893–905
Denetclaw WF, Berdougo E, Venters SJ, Ordahl CP (2001) Morphogenetic cell movements in the middle region of the dermomyotome dorsomedial lip associated with patterning and growth of the primary epaxial myotome. Development 128:1745–1755
Dietrich S (1999) Regulation of hypaxial muscle development. Cell Tissue Res 296:175–182
Dietrich S, Schubert FR, Lumsden A (1997) Control of dorsoventral pattern in the chick paraxial mesoderm. Development 124:3895–3908
Dietrich S, Schubert FR, Healy C, Sharpe PT, Lumsden A (1998) Specification of the hypaxial musculature. Development 125:2235–2249

Dockter J, Ordahl CP (2000) Dorsoventral axis determination in the somite: a re-examination. Development 127:2201–2206
Duband JL, Dufour S, Hatta K, Takeichi M, Edelman GM, Thiery JP (1987) Adhesion molecules during somitogenesis in the avian embryo. J Cell Biol 104:1361–1374
Dudley AT, Robertson EJ (1997) Overlapping expression domains of bone morphogenetic protein family members potentially account for limited tissue defects in BMP7 deficient embryos. Dev Dyn 208:349–362
Duprez D, Fournier-Thibault C, Le Douarin N (1998) Sonic Hedgehog induces proliferation of committed skeletal muscle cells in the chick limb. Development 125:495–505
Ebensperger C, Wilting J, Brand-Saberi B, Mizutani Y, Christ B, Balling R, Koseki H (1995) Pax-1, a regulator of sclerotome development is induced by notochord and floor plate signals in avian embryos. Anat Embryol (Berl) 191:297–310
Eichmann A, Marcelle C, Breant C, Le Douarin NM (1993) Two molecules related to the VEGF receptor are expressed in early endothelial cells during avian embryonic development. Mech Dev 42:33–48
Ellison ML, Ambrose EJ, Easty GC (1969) Myogenesis in chick embryo somites in vitro. J Embryol Exp Morph 21:341–346
Eloy-Trinquet S, Mathis L, Nicolas JF (2000) Retrospective tracing of the developmental lineage of the mouse myotome. Curr Top Dev Biol 47:33–80
Fan CM, Tessier-Lavigne M (1994) Patterning of mammalian somites by surface ectoderm and notochord: evidence for sclerotome induction by a hedgehog homolog. Cell 79:1175–1186
Fan CM, Porter JA, Chiang C, Chang DT, Beachy PA, Tessier-Lavigne M (1995) Long-range sclerotome induction by Sonic Hedgehog: direct role of the amino-terminal cleavage product and modulation by the cyclic AMP signaling pathway. Cell 81:457–465
Fan CM, Lee CS, Tessier-Lavigne M (1997) A role for WNT proteins in induction of dermo-myotome. Dev Biol 191:160–165
Gallera J (1966) Mise en évidence du r™le de l'ectoblaste dans la différenciation des somites chez les oiseaux (Evidence for the role of the ectoblast in the differentiation of somites in birds). Rev Suisse Zool 73:492–503
Gardner CA, Barald KF (1992) Expression patterns of engrailed-like proteins in the chick embryo. Dev Dyn 193:370–388
George-Weinstein M, Gerhart J, Reed R, Flynn J, Callihan B, Mattiacci M, Miehle C, Foti G, Lash JW, Weintraub H (1996) Skeletal myogenesis: the preferred pathway of chick embryo epiblast cells in vitro. Dev Biol 173:279–291
George-Weinstein M, Gerhart J, Blitz J, Simak E, Knudsen KA (1997) N-cadherin promotes the commitment and differentiation of skeletal muscle precursor cells. Dev Biol 185:14–24
George-Weinstein M, Gerhart J, Mattiacci-Paessler M, Simak E, Blitz J, Reed R, Knudsen K (1998) The role of stably committed and uncommitted cells in establishing tissues of the somite. Ann NY Acad Sci 842:16–27
Godsave SF, Slack JM (1989) Clonal analysis of mesoderm induction in *Xenopus laevis*. Dev Biol 134:486–490
Goodrich LV, Johnson RL, Milenkovic L, McMahon JA, Scott MP (1996) Conservation of the Hedgehog/Patched signaling pathway from flies to mice: induction of a mouse patched gene by Hedgehog. Genes Dev 10:301–312
Gossler A, Hrabe de Angelis M (1998) Somitogenesis. Curr Top Dev Biol 38:225–287
Goulding MD, Chalepakis G, Deutsch U, Erselius JR, Gruss P (1991) Pax-3, a novel murine DNA binding protein expressed during early neurogenesis. EMBO J 10:1135–1147
Goulding M, Lumsden A, Paquette AJ (1994) Regulation of Pax-3 expression in the dermo-myotome and its role in muscle development. Development 120:957–971
Grunz H, Tacke L (1989) Neural differentiation of *Xenopus laevis* ectoderm takes place after disaggregation and delayed reaggregation without inducer. Cell Differ Dev 28:211–217
Gurdon JB (1988) A community effect in animal development. Nature 336:772–774

Halevy O, Monsonego E, Marcelle C, Hodik V, Mett A, Pines M (1994) A new avian fibroblast growth factor receptor in myogenic and chondrogenic cell differentiation. Exp Cell Res 212: 278–284

Hall BK (1977) Chondrogenesis of the somitic mesoderm. Advan Anat Embryol Cell Biol 53:3–47

Hemmati-Brivanlou A, Melton DA (1992) A truncated activin receptor inhibits mesoderm induction and formation of axial structures in *Xenopus* embryos. Nature 359:609–614

Hemmati-Brivanlou A, Melton D (1997) Vertebrate neural induction. Annu Rev Neurosci 20: 43–60

Hirsinger E, Duprez D, Jouve C, Malapert P, Cooke J, Pourquié O (1997) Noggin acts downstream of Wnt and Sonic Hedgehog to antagonize BMP4 in avian somite patterning. Development 124:4605–4614

Hirsinger E, Jouve C, Dubrulle J, Pourquié O (2000) Somite formation and patterning. Int Rev Cytol 198:1–65

Hollyday M, McMahon JA, McMahon AP (1995) Wnt expression patterns in chick embryo nervous system. Mech Dev 52:9–25

Holtzer H, Holtzer S (1976) Lineages, quantal cell cycles, and cell diversification. In: Muller-Berat H (ed) Progress in differentiation research. North-Holland, Amsterdam

Huang R, Christ B (2000) Origin of the epaxial and hypaxial myotome in avian embryos. Anat Embryol (Berl) 202:369–374

Jacob M, Christ B, Jacob HJ (1978) On the migration of myogenic stem cells into the prospective wing region of chick embryos. A scanning and transmission electron microscope study. Anat Embryol (Berl) 153:179–193

Johnson RL, Laufer E, Riddle RD, Tabin C (1994) Ectopic expression of Sonic hedgehog alters dorsal-ventral patterning of somites. Cell 79:1165–1173

Kahane N, Cinnamon Y, Kalcheim C (1998) The cellular mechanism by which the dermomyotome contributes to the second wave of myotome development. Development 125:4259–4271

Kahane N, Cinnamon Y, Bachelet I, Kalcheim C (2001) The third wave of myotome colonization by mitotically competent progenitors: regulating the balance between differentiation and proliferation during muscle development. Development 128:2187–2198

Kastner S, Elias MC, Rivera AJ, Yablonka-Reuveni Z (2000) Gene expression patterns of the fibroblast growth factors and their receptors during myogenesis of rat satellite cells. J Histochem Cytochem 48:1079–1096

Kenny-Mobbs T, Thorogood P (1987) Autonomy of differentiation in avian branchial somites and the influence of adjacent tissues. Development 100:449–462

Kiefer JC, Hauschka SD (2001) Myf-5 is transiently expressed in nonmuscle mesoderm and exhibits dynamic regional changes within the presegmented mesoderm and somites I–IV. Dev Biol 232:77–90

Kos L, Chiang C, Mahon KA (1998) Mediolateral patterning of somites: multiple axial signals, including Sonic hedgehog, regulate Nkx-31 expression. Mech Dev 70:25–34

Kruger M, Mennerich D, Fees S, Schafer R, Mundlos S, Braun T (2001) Sonic hedgehog is a survival factor for hypaxial muscles during mouse development. Development 128:743–752

Kuratani S, Martin JF, Wawersik S, Lilly B, Eichele G, Olson EN (1994) The expression pattern of the chick homeobox gene gMHox suggests a role in patterning of the limbs and face and in compartmentalization of somites. Dev Biol 161:357–369

Langman J, Nelson GR (1968) A radioautographic study of the development of the somite in the chick embryo. J Embryol Exp Morphol 19:217–226

Lassar AB, Münsterberg AE (1996) The role of positive and negative signals in somite patterning. Curr Opin Neurobiol 6:57–63

Lee CS, Buttitta LA, May NR, Kispert A, Fan CM (2000) SHH-N upregulates Sfrp2 to mediate its competitive interaction with WNT1 and WNT4 in the somitic mesoderm. Development 127: 109–118

Li L, Cserjesi P, Olson EN (1995) Dermo-1: a novel twist-related bHLH protein expressed in the developing dermis. Dev Biol 172:280–292

Lyons KM, Hogan BL, Robertson EJ (1995) Colocalization of BMP 7 and BMP 2 RNAs suggests that these factors cooperatively mediate tissue interactions during murine development. Mech Dev 50:71–83

Marcelle C, Eichmann A, Halevy O, Breant C, Le Douarin NM (1994) Distinct developmental expression of a new avian fibroblast growth factor receptor. Development 120:683–694

Marcelle C, Wolf J, Bronner-Fraser M (1995) The in vivo expression of the FGF receptor FREK mRNA in avian myoblasts suggests a role in muscle growth and differentiation. Dev Biol 172:100–114

Marcelle C, Stark MR, Bronner-Fraser M (1997) Coordinate actions of BMPs, Wnts, Shh and noggin mediate patterning of the dorsal somite. Development 124:3955–3963

Marcelle C, Ahlgren S, Bronner-Fraser M (1999) In vivo regulation of somite differentiation and proliferation by Sonic Hedgehog. Dev Biol 214:277–287

Maroto M, Pourquié O (2001) A molecular clock involved in somite segmentation. Curr Top Dev Biol 51:221–248

Maroto M, Reshef R, Münsterberg AE, Koester S, Goulding M, Lassar AB (1997) Ectopic Pax-3 activates MyoD and Myf-5 expression in embryonic mesoderm and neural tissue. Cell 89: 139–148

McMahon JA, Takada S, Zimmerman LB, Fan CM, Harland RM, McMahon AP (1998) Noggin-mediated antagonism of BMP signaling is required for growth and patterning of the neural tube and somite. Genes Dev 12:1438–1452

Muller TS, Ebensperger C, Neubuser A, Koseki H, Balling R, Christ B, Wilting J (1996) Expression of avian Pax1 and Pax9 is intrinsically regulated in the pharyngeal endoderm, but depends on environmental influences in the paraxial mesoderm. Dev Biol 178:403–417

Münsterberg AE, Lassar AB (1995) Combinatorial signals from the neural tube, floor plate and notochord induce myogenic bHLH gene expression in the somite. Development 121: 651–660

Münsterberg AE, Kitajewski J, Bumcrot DA, McMahon AP, Lassar AB (1995) Combinatorial signaling by Sonic hedgehog and Wnt family members induces myogenic bHLH gene expression in the somite. Genes Dev 9:2911–2922

Murtaugh LC, Chyung JH, Lassar AB (1999) Sonic hedgehog promotes somitic chondrogenesis by altering the cellular response to BMP signaling. Genes Dev 13:225–237

Nicolas JF, Mathis L, Bonnerot C, Saurin W (1996) Evidence in the mouse for self-renewing stem cells in the formation of a segmented longitudinal structure, the myotome. Development 122: 2933–2946

Nicolet G (1971) Avian gastrulation. Advan Morphogen 9:231–262

Nishita M, Hashimoto MK, Ogata S, Laurent MN, Ueno N, Shibuya H, Cho KW (2000) Interaction between Wnt and TGF-beta signalling pathways during formation of Spemann's organizer. Nature 403:781–785

Noden DM (1990) Origins and assembly of avian embryonic blood vessels. Ann NY Acad Sci 588: 236–249

O'Hare JM (1972) Differentiation of chick embryo somites in chorioallantoic culture. J Embryol Exp Morph 27:215–228

Olivera-Martinez I, Coltey M, Dhouailly D, Pourquié O (2000) Mediolateral somitic origin of ribs and dermis determined by quail-chick chimeras. Development 127:4611–4617

Ooi VE, Sanders EJ, Bellairs R (1986) The contribution of the primitive streak to the somites in the avian embryo. J Embryol Exp Morphol 92:193–206

Ordahl CP (2000) Somitogenesis (parts 1 and 2) In: Current topics in developmental biology, vol 47. Academic Press, London

Ordahl CP, Le Douarin NM (1992) Two myogenic lineages within the developing somite. Development 114:339–353

Ordahl CP, Berdougo E, Venters SJ, Denetclaw WF (2001) The dermomyotome dorsomedial lip drives growth and morphogenesis of both the primary myotome and dermomyotome epithelium. Development 128:1731–1744

Ott MO, Bober E, Lyons G, Arnold H, Buckingham M (1991) Early expression of the myogenic regulatory gene, myf-5, in precursor cells of skeletal muscle in the mouse embryo. Development 111:1097–1107

Parr BA, McMahon AP (1994) Wnt genes and vertebrate development. Curr Opin Genet Dev 4:523–528

Parr BA, Shea MJ, Vassileva G, McMahon AP (1993) Mouse Wnt genes exhibit discrete domains of expression in the early embryonic CNS and limb buds. Development 119:247–261

Pourquié O, Coltey M, Teillet MA, Ordahl C, Le Douarin NM (1993) Control of dorsoventral patterning of somitic derivatives by notochord and floor plate. Proc Natl Acad Sci USA 90:5242–5246

Pourquié O, Fan CM, Coltey M, Hirsinger E, Watanabe Y, Breant C, Francis-West P, Brickell P, Tessier-Lavigne M, Le Douarin NM (1996) Lateral and axial signals involved in avian somite patterning: a role for BMP4 Cell 84:461–471

Pownall ME, Emerson CP Jr (1992) Sequential activation of three myogenic regulatory genes during somite morphogenesis in quail embryos. Dev Biol 151:67–79

Pownall ME, Strunk KE, Emerson CP Jr (1996) Notochord signals control the transcriptional cascade of myogenic bHLH genes in somites of quail embryos. Development 122:1475–1488

Psychoyos D, Stern CD (1996) Fates and migratory routes of primitive streak cells in the chick embryo. Development 122:1523–1534

Reshef R, Maroto M, Lassar AB (1998) Regulation of dorsal somitic cell fates: BMPs and Noggin control the timing and pattern of myogenic regulator expression. Gene Dev 12:290–303

Richardson MK, Allen SP, Wright GM, Raynaud A, Hanken J (1998) Somite number and vertebrate evolution. Development 125:151–160

Roelink H, Porter JA, Chiang C, Tanabe Y, Chang DT, Beachy PA, Jessell TM (1995) Floor plate and motor neuron induction by different concentrations of the amino-terminal cleavage product of sonic hedgehog autoproteolysis. Cell 81:445–455

Rong PM, Teillet MA, Ziller C, Le Douarin NM (1992) The neural tube/notochord complex is necessary for vertebral but not limb and body wall striated muscle differentiation. Development 115:657–672

Sasaki H, Hui C, Nakafuku M, Kondoh H (1997) A binding site for Gli proteins is essential for HNF-3beta floor plate enhancer activity in transgenics and can respond to Shh in vitro. Development 124:1313–1322

Sato SM, Sargent TD (1989) Development of neural inducing capacity in dissociated *Xenopus* embryos. Dev Biol 134:263–266

Scaal M, Fuchtbauer EM, Brand-Saberi B (2001) cDermo-1 expression indicates a role in avian skin development. Anat Embryol (Berl) 203:1–7

Schoenwolf GC, Garcia-Martinez V, Dias MS (1992) Mesoderm movement and fate during avian gastrulation and neurulation. Dev Dyn 193:235–248

Selleck MA, Stern CD (1991) Fate mapping and cell lineage analysis of Hensen's node in the chick embryo. Development 112:615–626

Solloway MJ, Robertson EJ (1999) Early embryonic lethality in Bmp5;Bmp7 double mutant mice suggests functional redundancy within the 60A subgroup. Development 126:1753–1768

Solursh M, Drake C, Meier S (1987) The migration of myogenic cells from the somites at the wing level in avian embryos. Dev Biol 121:389–396

Sosic D, Brand-Saberi B, Schmidt C, Christ B, Olson EN (1997) Regulation of paraxis expression and somite formation by ectoderm- and neural tube-derived signals. Dev Biol 185:229–243

Spence MS, Yip J, Erickson CA (1996) The dorsal neural tube organizes the dermamyotome and induces axial myocytes in the avian embryo. Development 122:231–241

Spratt NT (1955) Analysis of the organizer center in the early chick embryo. I. Localisation of the prospective notochord and somite cells. J Exp Zool 128:121–162

Stern CD, Bellairs R (1984) Mitotic activity during somite segmentation in the early chick embryo. Anat Embryol 169:97–102

Stern HM, Hauschka SD (1995) Neural tube and notochord promote in vitro myogenesis in single somite explants. Dev Biol 167:87–103

Stern HM, Brown AM, Hauschka SD (1995) Myogenesis in paraxial mesoderm: preferential induction by dorsal neural tube and by cells expressing Wnt-1 Development 121:3675–3686

Stockdale FE, Nikovits W Jr, Christ B (2000) Molecular and cellular biology of avian somite development. Dev Dyn 219:304–321

Stone DM, Hynes M, Armanini M, Swanson TA, Gu Q, Johnson RL, Scott MP, Pennica D, Goddard A, Phillips H et al. (1996) The tumour-suppressor gene patched encodes a candidate receptor for Sonic hedgehog. Nature 384:129–134

Tabata T, Kornberg TB (1994) Hedgehog is a signaling protein with a key role in patterning *Drosophila* imaginal discs. Cell 76:89–102

Tajbakhsh S, Buckingham M (2000) The birth of muscle progenitor cells in the mouse: spatiotemporal considerations. Curr Top Dev Biol 48:225–268

Tajbakhsh S, Borello U, Vivarelli E, Kelly R, Papkoff J, Duprez D, Buckingham M, Cossu G (1998) Differential activation of Myf5 and MyoD by different Wnts in explants of mouse paraxial mesoderm and the later activation of myogenesis in the absence of Myf5 Development 125: 4155–4162

Teillet MA, Le Douarin NM (1983) Consequences of neural tube and notochord excision on the development of the peripheral nervous system in the chick embryo. Dev Biol 98:192–211

Teillet M, Watanabe Y, Jeffs P, Duprez D, Lapointe F, Le Douarin NM (1998) Sonic hedgehog is required for survival of both myogenic and chondrogenic somitic lineages. Development 125: 2019–2030

Valinsky JE, Le Douarin NM (1985) Production of plasminogen activator by migrating cephalic neural crest cells. EMBO J 4:1403–1406

Veini M, Bellairs R (1991) Early mesoderm differentiation in the chick embryo. Anat Embryol 183:143–149

Vivarelli E, Cossu G (1986) Neural control of early myogenic differentiation in cultures of mouse somites. Dev Biol 117:319–325

Wachtler F, Christ B (1992) The basic embryology of skeletal muscle formation in vertebrates: the avian model. Semin Dev Biol 3:217–227

Wachtler F, Christ B, Jacob HJ (1982) Grafting experiments on determination and migratory behaviour of presomitic, somitic and somatopleural cells in avian embryos. Anat Embryol 164:369–378

Wagner J, Schmidt C, Nikowits W Jr, Christ B (2000) Compartmentalization of the somite and myogenesis in chick embryos are influenced by wnt expression. Dev Biol 228:86–94

Watterson R, Fowler I, Fowler BJ (1954) The role of the neural tube and notochord in the development of the axial skeleton of the chick. Am J Anat 95:337–399

Williams LW (1910) The somites of the chick. Am J Anat 11:55–100

Williams BA, Ordahl CP (1994) Pax-3 expression in segmental mesoderm marks early stages in myogenic cell specification. Development 120:785–796

Williams BA, Ordahl CP (1997) Emergence of determined myotome precursor cells in the somite. Development 124:4983–4997

Wilson V, Beddington RS (1996) Cell fate and morphogenetic movement in the late mouse primitive streak. Mech Dev 55:79–89

Wilting J, Brand-Saberi B, Huang R, Zhi Q, Kontges G, Ordahl CP, Christ B (1995) Angiogenic potential of the avian somite. Dev Dyn 202:165–171

Transcription Factors in Skeletal Myogenesis of Vertebrates

Petra Neuhaus and Thomas Braun[1]

1 Myogenesis

All skeletal muscles of the vertebrate body and some head muscles are derived from the somites. Somites are transient metameric structures that are laid down in a rostral to caudal direction on each side of the neural tube and notochord during embryogenesis. Initially, somites bud off from the unsegmented paraxial mesoderm and form epithelial, ball-shaped structures with a centrally located somitocoel. During further development the somites differentiate and eventually will give rise to the sclerotome and the dermomyotome (Christ and Ordahl 1995; Brand-Saberi and Christ 2000). Because of inductive and repressive instructions from surrounding tissues, the somite acquires a dorsal–ventral, anterior–posterior and medial–lateral polarity. Cells in the ventral part of the somite de-epithelialise and become mesenchymal cells of the sclerotome forming the axial skeleton. Cells in the dorsal portion of the somite retain their epithelial character and form the dermomyotome, which will give rise to the dermis of the skin and most of the skeletal muscle of the body. The dermomyotome is further separated in a medial and a lateral population of cells, giving rise to the epaxial muscles (deep muscles of the back) and the hypaxial muscles (appendicular musculature, abdominal muscles, diaphragm, hypoglossal chord), respectively (Ordahl and Le Douarin 1992; Christ and Ordahl 1995).

Muscle precursor cells are initially maintained in a proliferating, undifferentiated status. Some cells, such as the muscle precursor cells forming the epaxial musculature of the back, do not have to migrate long distances to reach their target location, whereas the muscle precursor cells giving rise to the appendicular and tongue musculature, as well as the diaphragm, have to undergo long-range migration to reach their target area. During their migration, these cells maintain a proliferating, undifferentiated status. Only after muscle progenitor cells have reached their destination, they start to differentiate into myocytes and eventually form mature multinucleated myotubes.

A small number of cells (5%), which are located adjacent to mature myotubes and termed satellite cells, are not terminally differentiated, but

[1] Martin-Luther-Universität Halle-Wittenberg, Medizinische Fakultät, Institut für physiologische Chemie, Hollystr.1, 06097 Halle/Saale, Germany

Results and Problems in Cell Differentiation, Vol. 38
B. Brand-Saberi (Ed.): Vertebrate Myogenesis

represent a stem-cell pool that is activated upon injury and/or degeneration of mature fibres. Activated satellite cells then undergo a differentiation process that is similar to the process of embryonic myogenesis and fuse with or replace the injured myotubes (Seale and Rudnicki 2000).

2 Determination and Differentiation of Muscle Precursor Cells

Since the activation and maintenance of the myogenic differentiation programme might be rather similar during embryogenesis, postnatal muscle growth and regeneration processes, we will first focus on transcription factors involved in the general myogenic differentiation programm. Later, we will continue with transcription factors involved in the development of specific subsets of skeletal muscle or regeneration processes in adult life.

2.1 MRFs

The first family of transcription factors reported to be essential for muscle development is the *MyoD* family of muscle regulatory factors (MRFs). *MyoD* was the first member of the family isolated and shown to induce non-muscle cells (10T1/2 fibroblasts) to acquire a muscle phenotype in vitro (Davis et al. 1987). Subsequently, three other members of the MRF gene family were discovered: *Myf5*, *Myogenin* and *MRF4* (*Myf6*, *Herculin*; Braun et al. 1989a, 1989b; Edmondson and Olson 1989; Rhodes and Konieczny 1989; Wright et al. 1989; Miner and Wold 1990), and also shown to convert non-muscle cells to the myogenic lineage. MRFs are basic helix–loop–helix transcription factors that activate muscle-specific gene expression by binding to *cis*-acting DNA control elements, referred to as E-boxes, in the promoter of target genes (Olson 1990; Buckingham 1992; Olson and Klein 1994; Arnold and Braun 1996). The MRF–bHLH domains, which mediate DNA-binding and dimerisation of the MRFs with ubiquitously expressed bHLH proteins of the E2A family are highly homologous to each other. The less conserved N- and C-terminal domains contain transactivation domains, which are essential for transcriptional activation of target genes (Braun et al. 1990; Weintraub et al. 1991; Mak et al. 1992; Winter et al. 1992).

Expression analyses in vitro have shown that *Myf5* and *MyoD* are expressed in muscle precursor cells, whereas *MRF4* and *myogenin* are activated only after differentiation of the myoblasts to myocytes is already underway. This led to the hypothesis that *Myf5* and *MyoD* are required for the determination of myogenic precursor cells, whereas *myogenin* and *MRF4* are responsible for terminal differentiation of myoblasts. However, each of the MRFs is able to induce the phenotypic switch of non-muscle cells to a myogenic phenotype in vitro, suggesting overlapping functions during determination and differentiation processes. The distinction of specific functions of different MRFs in vitro is complicated by the extensive cross-regulation capacity of MRFs in numerous cell types. This might also be of importance for the maintenance of the muscle phenotype, once any single MRF was activated (Braun et al. 1989a; Thayer et al. 1989).

Expression analyses during mouse embryogenesis and in adult muscle revealed distinct as well as overlapping spatio-temporal expression patterns for MRF genes in muscle cells and their progenitors in somites, visceral arches and limb buds (Sassoon et al. 1989; Bober et al. 1991; Ott et al. 1991). In mice, *Myf5* is the first MRF expressed in the somites and limb buds. Expression starts at E 8 in the dorso-medial region of the epithelial somite and later spreads throughout the entire myotome and myoblasts of the body. *Myf5* expression is maintained in the dermomyotome until E 10.5 and then rapidly decreases in a craniocaudal direction (Ott et al. 1991; Tajbakhsh et al. 1996a). *Myf5* expression is followed by *myogenin*, which is expressed in the most rostral, and, therefore, most mature somites at E 8.5 in the mouse (Sassoon et al. 1989). *Myogenin* expression is followed by *MRF4* expression with a delay of approximately 12 h (Bober et al. 1991). However, similar to *Myf5*, *MRF4* expression in the somite is only transient and, after 2 days, expression ceases. In contrast to the other MRFs, *MRF4* expression is not detected in limb buds between E 10.5 and E 14.5 in the mouse, despite its expression in the somites. After E 16, *MRF4* expression strongly increases in all skeletal muscle and *MRF4* becomes the predominant MRF expressed in the late foetal and adult muscle of the mouse (Bober et al. 1991). *MyoD* expression starts at E 10.5 in the mature myotome and in the myoblasts of facial and limb bud musculature (Sassoon et al. 1989).

Although the in vitro data showed very similar biochemical and myogenic properties for the different MRF genes, the differential expression patterns in vivo suggested distinct functions for muscle development. Therefore, all MRF genes were investigated by targeted mutation in mice.

2.1.1 *Myf5*

The first targeted mutations of *Myf5* led to homozygous mutant mice in which the initial round of muscle cell formation failed although, later during life, no major defect in skeletal muscle development was detected (Braun et al. 1992; Tajbakhsh et al. 1996b). The delayed initiation of myogenesis coincides with the onset of *MyoD* expression, suggesting that *MyoD* compensates for the loss of *Myf5*, and that either *Myf5* or *MyoD* are required for the determination of muscle precursor cells in the somite and thus for induction of myogenesis. Surprisingly, homozygous *Myf5* mutant mice die perinatally because of a severe rib defect. In homozygous mutant animals, the ribs do not attach to the sternum, resulting in an unstable rib cage and respiratory failure of newborn animals (Braun et al. 1992; Tajbakhsh et al. 1996b). This unexpected phenotype was explained by a non-cell autonomous defect, caused by instructive interactions between the developing myotome and sclerotome of the somite. Variable rib phenotypes were observed by Tallquist et al., who generated different *Myf5* knock-in alleles to analyse the cause of the rib phenotype, and the potential involvement of PDGFA in myotome-sclerotome signalling (Tallquist et al. 2000). Interestingly, they observed a more or less severe rib phenotype depending on the presence or absence of the neomycin selection gene in the mutated allele. In addition, recently a conditional mutation of the *Myf5* locus (*Myf5*$^{loxP/loxP}$) has been described. Deletion of the floxed *Myf5* gene in the germ line resulted in healthy, homozygous mutant pups (*Myf5*$^{\Delta loxP/\Delta loxP}$) without any rib defect (Kaul et al. 2000). Because the only difference between the new *Myf5* alleles and the originally described mutations is the insertion of a transcription unit (selection marker gene) into the *Myf5* locus, the rib defect in the earlier *Myf5* mutations is most likely a result of long-range effects caused by the inserted promoters/genes on a yet unknown gene in the vicinity of *Myf5*. This hypothesis is supported by the observation of the variable rib phenotype depending on the presence or absence of the neomycin gene in *Myf5*$^{PDGFA/PDGFA}$ mice.

As in the original *Myf5* mutations, homozygous mutant *Myf5*$^{\Delta loxP/\Delta loxP}$ mice do not display an obvious defect in skeletal muscle development at birth, although - in accordance with the original *Myf5* mutations - myotome development is delayed, until expression of *MyoD* occurs (Fig. 1). Taken together, analysis of the different *Myf5* mutations led to the conclusion that although *Myf5* is essential for the early onset of myogenic determination, it is dispensable for skeletal muscle development during embryogenesis.

2.1.2 *MyoD*

MyoD mutant mice do not show any obvious muscle defects at birth, similar to *Myf5* mutant mice (Rudnicki et al. 1992). A weak phenotype was detected in the limbs of *MyoD* (–/–) mice characterised by a slight delay in limb muscle development compared with wild-type controls. Therefore, *MyoD* function is specifically required for limb muscle development, and loss of *MyoD* during early limb bud morphogenesis cannot be compensated for by other myogenic factors (Kablar et al. 1997).

Because neither the *Myf5* deletion, nor the *MyoD* deletion by itself cause a severe defect in the determination of muscle precursor cells, the mutant strains were intercrossed to analyse whether *Myf5* and *MyoD* display functional redundancies. Double homozygous animals (*Myf5–/–* ; *MyoD–/–*) developed no skeletal muscle at all. In those animals, not even mononucleated muscle precursor cells were detected as judged by the expression of myoblast marker genes, demonstrating that either *Myf5* or *MyoD* is required for the determination of mesenchymal cells to the myogenic lineage (Rudnicki et al. 1993). Interestingly, in compound mutants, a dose effect was observed. In *Myf5–/–*; *MyoD+/–* mice myotome formation and myogenesis takes place normally,

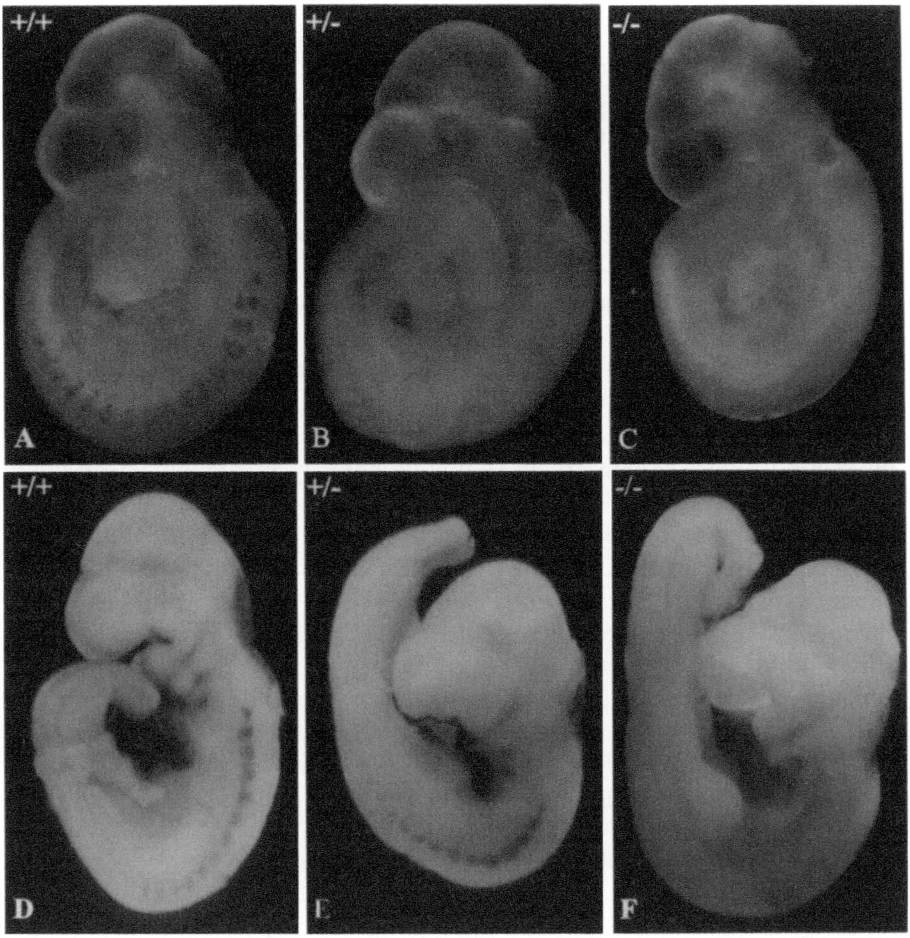

Fig. 1. Lack of myotome formation in homozygous Myf-$5^{\Delta loxP/\Delta loxP}$ embryos at E 9.75. Myf-$5^{loxP/loxP}$ (**A, D**), Myf-$5^{loxP/\Delta loxP}$ (**B, E**) and Myf-$5^{\Delta loxP/\Delta loxP}$ (**C, F**) embryos were subjected to whole-mount *in situ* hybridisation with a myogenin antisense riboprobe (**A–C**) and to whole-mount immunohistochemistry using a mAb against MyHC (MF20) The expression level of myogenin mRNA at E 9.75 correlated strictly with the dosage of functional Myf-5 alleles. Hence it is reduced in heterozygous Myf-$5^{loxP/\Delta loxP}$ and completely absent in Myf-$5^{\Delta loxP/\Delta loxP}$ embryos. Expression of MyHC is indistinguishable between Myf-$5^{loxP/loxP}$ and Myf-$5^{loxP/\Delta loxP}$ embryos but completely absent in Myf-$5^{\Delta loxP/\Delta loxP}$ embryos. Similar results were obtained with a number of other sarcomeric protein markers. Note the strong staining for MyHC in the hearts of all the embryos shown in **D–F**, which serves as an endogenous control

whereas *MyoD–/–*; *Myf5+/–* mice display severely reduced muscle masses. These data clearly indicate that although *MyoD* and *Myf5* share functional redundancies, they are not functionally equivalent *in vivo*, which might be because of differences in expression levels and cross-regulation between these factors.

2.1.3 *Myogenin*

In contrast to mice lacking *MyoD* and *Myf5*, the early steps of myogenesis occur normally in mice lacking *myogenin* (Venuti et al. 1995). However, secondary myogenesis, which leads to the formation of myofibres, does not take place in *myogenin* mutants, and homozygous mice die perinatally because of the complete absence of differentiated muscle. Instead of myotubes containing multinucleated myocytes, only mononucleated cells are present in muscle tissue derived from *myogenin -/-* mice (Hasty et al. 1993; Nabeshima et al. 1993). This defect also affects the diaphragm and renders it non-functional, leading to perinatal death.

Because of its partly overlapping expression pattern with *Myf5* and *MyoD* in somites, it was still a matter of debate if *myogenin* could play a role in muscle cell determination, although primary myogenesis occurred normally in *myogenin-/-* mice. Therefore, mice double homozygous either for *Myf5* and *myogenin* or *MyoD* and *myogenin* were analysed. These analyses indicated separate functions of those genes because the phenotypes of double mutants resembled the combined defects of single mutants, and not a drastically reduced primary myogenesis as would have been expected if *myogenin* played an essential role in muscle cell determination (Rawls et al. 1995). Thus, under normal conditions, *myogenin* acts in a genetic pathway downstream of *Myf5* and *MyoD*. However, if *myogenin* is placed under the transcriptional control of the *Myf5* promoter, replacing the *Myf5* coding sequence ($Myf5^{myog/myog}$), the delay of myotome formation in *Myf5* homozygous mutants is rescued by the myogenin expression (Wang et al. 1996). This suggests that *myogenin* can substitute for *Myf5* during early myogenesis, revealing an intrinsic redundancy of *Myf5* and *myogenin*. On the other hand, mice double homozygous for the $Myf5^{myog/myog}$ knock-in allele and the *MyoD* mutation die shortly after birth because of reduced skeletal muscle formation including the diaphragm, indicating that *myogenin* is not functionally equivalent to *Myf5* (Wang and Jaenisch 1997). Furthermore, *myogenin* expression under the control of the *Myf5* promoter cannot rescue the differentiation defect observed in *myogenin-/-* mice because mice double homozygous for the $Myf5^{myog/myog}$ knock-in allele and the *myogenin* mutation lack differentiated myocytes as do *myogenin-/-* mice. The latter effect may be explained by the downregulation of the *Myf5* locus in normal development, leading to reduced *myogenin* expression of the knock-in allele at stages when high *myogenin* expression is required for terminal differentiation of myocytes. In addition, the results from the $Myf5^{myog/myog}$ knock-in allele have to be reconsidered in the light of the $Myf5^{\Delta loxP/\Delta loxP}$ allele. One main conclusion of Wang and colleagues was that the rib phenotype of the original *Myf5* mutation was rescued by myogenin. The authors concluded that the formation of the myotome should occur normally in $Myf5^{myog/myog}$ mice and they paid only little attention to the onset of primary myogenesis. Because it is now clear that the rib phenotype is caused by the dis-regulation of another gene, it seems sensible to re-investigate $Myf5^{myog/myog}$ mice in terms of myotome development.

2.1.4 *MRF4*

Mutations of *MRF4*, the last member of the MRF family that is activated during myogenesis, were accomplished in three different laboratories (Braun and Arnold 1995; Patapoutian et al. 1995; Zhang et al. 1995). All three mouse strains display subtle defects in myogenesis, resulting in a slight reduction of the expression of muscle-specific genes, but did not generate any major problem for muscle development. However, rib development of mutant mice is affected more or less severely depending on the targeted allele. Phenotypes ranged from viable mice, with minor rib cage defects (Zhang et al. 1995), to perinatal lethality of mice displaying a rib cage defect similar to the original *Myf5* mutation (Braun and Arnold 1995). Later, it was shown that the mutation of the *MRF4* locus, which resides within 8 kb of the *Myf5* locus on mouse chromosome 10, affects the expression of the *Myf5* gene (Floss et al. 1996). The effect of the *MRF4* mutation on the *Myf5* gene correlated with the observed defect of rib development in each of the three mutant alleles. The stronger *Myf5* is down regulated, the more severe is the observed rib phenotype. Thus, it appeared likely that the rib phenotype observed in *MRF4* mutants was not because of the loss of *MRF4* function, but caused by the down regulation of *Myf5*. Given that the new *Myf5*$^{\Delta loxP/\Delta loxP}$ mutation has no rib cage defect (Kaul et al. 2000), it is reasonable to assume that the targeted inactivation of the *MRF4* gene not only leads to a downregulation of *Myf5*, but also affects the same unknown gene that causes the malformation of the ribs in the original *Myf5* mutation.

Summing up the results of the different mutations and double mutants, a genetic pathway of the function of the different MRFs is revealed. Some overlapping functions (particularly between *Myf5* and *MyoD*) as well as some unique functions are evident. However, it is not yet clear, whether these differences are because of inherent functional differences, or whether they are mainly caused by different expression patterns and/or transcriptional activation of MRFs. It has been proposed that the level of transcription factors expressed at a certain time is critical for normal myogenesis. This is supported by the fact that MRFs form a regulatory loop that activates and represses one another.

2.2 *MEF2* Transcription Factors

Several other genes have been shown to act synergistically with MRFs in activating muscle-specific genes and, thus, in inducing myogenic determination and differentiation. The *MEF2* family of MADS-box transcription factors contains four genes, *MEF2A–D* (Ludolph and Konieczny 1995). In vivo studies have revealed that *MEF2* genes are expressed in developing cardiac, skeletal and smooth muscle cells during embryogenesis and that expression in these cell types decreases after birth (Naya and Olson 1999). Because *MEF2* genes have been shown to activate several muscle-specific genes and to increase the efficiency of myogenic conversion of non-muscle cells in vitro if used in combination with MRFs (Kaushal et al. 1994; Molkentin et al. 1995), this expression pattern is consistent with their supposed function in myocyte differentiation. Specifically, *MEF2C* and *myogenin* were shown to reside in a positive feedback loop enhancing each others transcription, with *MEF2C* being essential for the correct spatio-temporal expression of myogenin (Ridgeway et al. 2000). None of the other MRFs is activated by *MEF2C*, and *MEF2C* is not strongly activated by any other MRF (Ridgeway et al. 2000). However, targeted mutation of *MEF2C* does not result in a phenocopy of the myogenin mutation, but in the loss of heart looping. No obvious defects in skeletal muscle development are observed in these mutants (Lin et al. 1997).

In *Drosophila*, only one *MEF2* gene was described, which is essential for the development of all muscle lineages in the fly, cardiac as well as somatic muscles (Lilly et al. 1995). So far, in vertebrates, besides the *MEF2C* deletion, no mutation for any of the other *MEF2* family members has been published. Thus, further proof for the involvement of MEF2 factors in vertebrate skeletal muscle development in vivo has to await mutational analysis of the different family members.

Whereas MRFs and MEF2 genes seem to be involved in the development of all skeletal muscles, some transcription factors have been shown to be essential only for specific aspects of myogenesis, as for example regeneration, or for the development of specific subsets of muscle groups.

3 Hypaxial Muscle Development

3.1 *Pax3*

Pax3 is a member of the paired box transcription factors that are involved in the regulation of various proliferation and differentiation processes during embryogenesis (for review see Mansouri et al. 1999). *Pax3* is initially expressed throughout the paraxial mesoderm before it becomes restricted to the dermomyotome. Cells of the lateral dermomyotomal lip that express *Pax3* delaminate from the dermomyotomal epithelium and, after a long range migration, form the skeletal muscles of the tongue, diaphragm and limbs (Williams and Ordahl 1994).

Splotch mutant mice, which lack a functional *Pax3* gene, display irregular shaped somite borders. In these mutants, limb musculature does not form, indicating an essential role for *Pax3* in hypaxial muscle development (Bober et al. 1994; Goulding et al. 1994; Daston et al. 1996). However, axial muscles as well as ventral body wall muscles seem to be largely unaffected by the loss of *Pax3*. Because ventral dermomyotomal cells lacking a functional *Pax3* gene are able to differentiate into myocytes after transplantation into chicken limb buds, the lack of hypaxial musculature cannot be because of a defect in muscle cell determination, but has to be caused by a differentiation and/or migratory defect of hypaxial muscle precursor cells (Daston et al. 1996).

Several genes known to be involved in either the delamination of hypaxial precursor cells from the dermomyotome, or in the long-range migration of hypaxial precursor cells were shown to be misexpressed in Splotch mutant mice, as for example the receptor tyrosine kinase *c-Met* (Epstein et al. 1996). The c-Met/HGF signalling system is essential for delamination of hypaxial precursor cells because in mice lacking either the receptor (*cMet*) or the ligand (*HGF*-hepatocyte growth factor), delamination from the dermomyotome and long-range migration of these cells does not take place (Bladt et al. 1995; Dietrich et al. 1999). Another gene that is down regulated by the loss of *Pax3* is the homeobox containing transcription factor *Lbx1*, which is a homologue of the *Drosophila labybird* gene (see below; Mennerich et al. 1998; Dietrich et al. 1999).

Besides the essential function of *Pax3* for migration of hypaxial muscle precursor cells, *Pax3* is also involved in the determination of all body muscle precursor cells via a pathway including *Myf5* and *MyoD*. Mice double homozygous for the *Pax3* mutation and *Myf5* mutation lack most of their trunk muscles, whereas head muscles are formed normally (Tajbakhsh et al. 1997). Closer analysis of these mutants revealed a loss of *MyoD* expression, resulting in mutants resembling *Myf5-/-* ; MyoD-/- mice, with respect to the trunk muscles. Therefore *Pax3* and *Myf5* are located upstream of *MyoD*, which is in accordance with the fact that ectopic expression of *MyoD* can be induced by *Pax3*, even in neural tissue (Maroto et al. 1997).

3.2 *Lbx1*

As mentioned above, *Lbx1* is a homeobox containing transcription factor, which is expressed in migrating hypaxial muscle precursor cells, indicating a potential role for *Lbx1* in the long-range migration of these cells (Jagla et al. 1995). *Lbx1* is co-expressed with *Pax3* and *cMet* in the lateral dermomyotomal lip and most likely acts downstream of *Pax3* (Mennerich et al. 1998). Targeted mutation of *Lbx1* leads to perinatal death and to drastically reduced hypaxial muscle formation in homozygous mutant animals (Fig. 2). Skeletal muscle is completely absent in the hindlimbs of homozygous mutant animals, whereas in forelimbs the ventral flexor muscles do form and only the dorsal extensor muscles are missing (Schäfer and Braun 1999; Brohmann et al. 2000; Gross et al. 2000). However, the tongue and diaphragm, which also belong to hypaxial muscles, are not affected by the mutation, demonstrating that the pathfinding of different muscle precursor cells is dependent on more than one factor. Hypaxial muscle precursor cells in *Lbx1* mutant mice delaminate from the dermomyotome and initiate migration. Muscle precursor cells of the

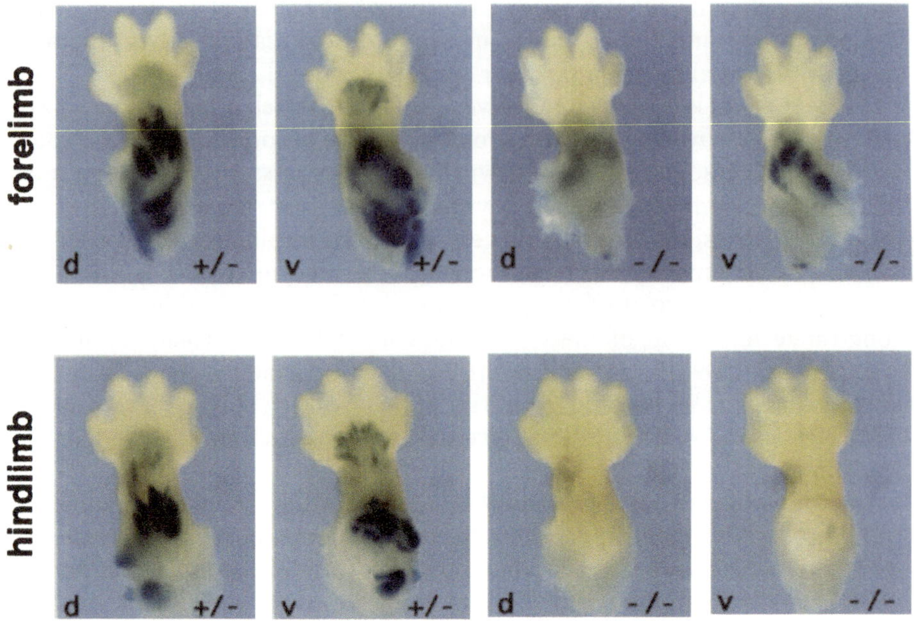

Fig. 2. Lbx1h-/- mice lack *β*-Gal+ muscle precursor cells in hindlimbs and on the dorsal side of the forelimbs. *β*Gal staining of fore- and hindlimbs dissected from Lbx1h-/- and Lbx1h+/- embryos at E 13.5 are shown in dorsal (*d*) and ventral (*v*) views. In the dorsal view of forelimbs from Lbx1h-/- mice staining of flexor muscles is visible through the limb. No *β*Gal-positive cells are detectable on the dorsal side. Hindlimbs of Lbx1h-/- mice lack any *β*Gal-positive muscle precursor cells, whereas all muscle anlagen form readily in Lbx1h+/- mice. The missing proximal muscle anlagen in the ventral view of the -/- forelimb were accidentally removed during preparation

hypoglossal chord migrate normally, although slightly delayed. In contrast, hypaxial precursor cells forming appendicular muscle appear to be unable to find their correct way and to enter the limb bud for muscle formation. Instead of invading the limb buds, *Lbx1* mutant cells accumulate at ectopic positions in the lateral plate mesoderm and later seem to adopt a new cell fate according to their ectopic position (Schäfer and Braun 1999).

Thus, the genetic analyses of *Pax3*, *cMet/HGF* and *Lbx1* function seem to suggest a genetic hierarchy of these four genes: *Pax3* is necessary for a normal morphological assembly of the dermomyotomal epithelium and acts upstream of *cMet*, probably by controlling its transcription. The tyrosine kinase receptor *cMet* and its ligand *HGF* are essential for the delamination of hypaxial muscle precursor cells from the dermomyotome. *Lbx1* marks hypaxial precursor cells that are about to leave the dermomyotome and is indispensable for the correct path-finding of a large subset of limb muscle precursor cells.

3.3 *Mox2*

Another gene expressed in migrating muscle precursor cells is *Mox2*, also a homeobox containing transcription factor. *Mox2*, and the closely related *Mox1* genes are initially co-expressed in the epithelial somite. Later, during differentiation of the somite, the sclerotome continues to co-express *Mox1* and *Mox2*, whereas dermomyotomal cells at this stage do not express *Mox2*, but *Mox1*. This picture is reversed in hypaxial cells that express *Mox2* in the ventrolateral tip of the dermomyotome and during migration into limb buds. Double-labelling techniques identified these cells as being *Pax3* positive, indicating that they are undifferentiated muscle precursor cells. Unlike *Pax3* and *Lbx1*, *Mox2* is not expressed solely in migrating hypaxial muscle precursor cells, but displays an additional expression domain in the limb bud mesenchyme (Mankoo et al. 1999). Targeted mutation of the *Mox2* gene led to the death of 80% of the homozygous mutants before 3 weeks of age. Surviving *Mox2* homozygous mutants display all characteristics of severe muscle dystrophy. Histological analyses revealed a severe reduction of appendicular muscles with a lack of distinct muscle groups in the forelimbs. In contrast to the *Lbx1* mutation, this loss of muscle does not affect either extensor or flexor muscles specifically, but affects both muscle groups evenly. In the hindlimbs of *Mox2* mutant mice, all muscle groups are present, but the overall muscle mass is reduced, and many fibres are myopathic (Mankoo et al. 1999). In addition to the observed histological defects, several genes known to affect limb muscle development are deregulated in *Mox2* mutant embryos. In particular the down-regulation of *Pax3* and *Myf5* in these mutants might explain the observed phenotype. Because mice lacking *Pax3* and *Myf5* do not form the bulk of the trunk and limb muscles, a reduction of *Pax3* and *Myf5* expression in *Mox2* mutants might result in a reduction of the most vulnerable muscle masses. Another explanation, supported by the mesenchymal expression of *Mox2* in the limb bud, might be faulty patterning of the limb connective tissue, which leads to disturbed muscle morphogenesis (Mankoo et al. 1999).

4 Regeneration of Skeletal Muscle

Postnatal growth, maintenance and regeneration of skeletal muscle is dependent on a distinct lineage of myogenic precursor cells, termed satellite cells (for review see Seale and Rudnicki 2000). In adult skeletal muscle, satellite cells are mitotically silent cells, which become activated by stress and/or injury. Upon activation they enter the cell cycle, differentiate along the myogenic differentiation pathway and eventually form new muscle fibres or fuse with existing myofibres and replenish them (Bischoff and Heintz 1994; Rosenblatt et al. 1994).

4.1 MRFs

The functions of MRF genes during determination and differentiation of skeletal muscle have been discussed above. In this section, we will focus on their function in the regeneration of skeletal muscle.

Because of the perinatal lethality in the original *Myf5* mutations, the role of *Myf5* for regeneration of skeletal muscle was difficult to assess. However, an initial analysis of *Myf5*$^{\Delta loxP/\Delta loxP}$ mutants demonstrates that *Myf5* function is essential for muscle regeneration. After experimental induction of muscle injuries, *Myf5*$^{\Delta loxP/\Delta loxP}$ mice show a reduced regeneration capacity compared with control littermates. This phenotype is most probably because of a drastically reduced number of satellite cells in *Myf5*$^{\Delta loxP/\Delta loxP}$ myofibres (Kaul et al., in prep.).

Another MRF that was shown to be essential for muscle regeneration in adult life is *MyoD*. As with *Myf5*$^{\Delta loxP/\Delta loxP}$ mice, homozygous *MyoD* mice display severe deficiencies in muscle regeneration after skeletal muscle damage (Megeney et al. 1996). Electron microscopy revealed that *MyoD* homozygous mutant myotubes display an increased number of satellite cells in comparison with myotubes from wild-type mice. In vitro studies performed on isolated satellite cells from *MyoD-/-* mice show that these cells are capable of proliferation, but are differentiation defective, which could explain the observed regeneration phenotype (Sabourin et al. 1999; Cornelison et al. 2000).

So far, neither *MRF4* nor *myogenin* have been conclusively shown to play a major role in muscle regeneration. *Myogenin* mutant mice die perinatally so that the analysis of its effect on regeneration processes has to await a conditional mutation.

4.2 *Pax7*

Pax7, a member of the paired box transcription factors (for review see: Mansouri et al. 1999), is expressed in neural and somitic tissue in a partly overlapping pattern to *Pax3*. Whereas *Pax3* is expressed early during somite development (see above), *Pax7* expression is not induced until somites mature into the different cell lineages. Within somites, *Pax7* is mainly expressed in the central and dorso-medial regions of the dermomyotome, giving rise to epaxial muscle (Jostes et al. 1990). During embryonic myogenesis, as well as in regenerating muscles, *Pax7* is selectively expressed in proliferating myoblasts and is downregulated upon terminal differentiation of myocytes (Seale et al. 2000). Surprisingly, despite the mitotically inactive state of quiescent satellite cells, it has been reported that *Pax7* is also expressed in quiescent satellite cells (Seale et al. 2000).

Pax7 homozygous mutant mice fail to thrive and die within the first 2 weeks of life, probably because of cephalic neural crest defects affecting facial structures such as the nose and maxilla (Mansouri et al. 1996). However, recent data also report a defect in skeletal muscle development, with homozygous mutant animals displaying severely reduced skeletal muscle mass in comparison to wild-type littermates. This reduction is because of a reduced diameter of myofibres whereas the number of myofibres is apparently normal. Several muscles are affected, including appendicular muscles and the diaphragm. Most interestingly, *Pax7* mutants completely lack satellite cells on their myofibres. This loss of satellite cells was considered to be the cause of the muscle mass reduction (Seale et al. 2000) because satellite cells are one of the major sources of postnatal skeletal muscle growth (Bischoff and Heintz 1994). Thus, *Pax7* might play an essential role in the ontogeny of satellite cells and, therefore, in postnatal skeletal muscle growth, maintenance and regeneration.

4.3 *MNF*

Two transcription factors shown to play an essential role during muscle cell regeneration are *MNF-α* (myocyte nuclear factor α) and *MNF-β* (myocyte nuclear factor *β*). Both factors are derived from the same gene by alternative splicing (Bassel-Duby et al. 1994; Yang et al. 1997) and belong to the forkhead/winged-helix transcription factor family (Lai et al. 1993). Although the two transcripts are derived from the same gene and are both expressed in myogenic cells, they exhibit different expression patterns. *MNF-β* is expressed transiently in the early stages of myogenesis in the myotome and limb buds, as well as in the heart tube. In differentiating myogenic cells, *MNF-β* expression is downregulated so that, in adult skeletal muscle, the only muscle cells still expressing *MNF-β* are quiescent satellite cells (Garry et al. 1997). The other splice isoform, *MNF-α*, is activated upon differentiation of the muscle precursor cells and, therefore, may play a different role than *MNF-β* (Yang et al. 1997). The alternating activation of the *MNF-α* and *-β* isoforms is also seen upon activation of satellite cells after injury (Garry et al. 2000). Whereas quiescent satellite cells express *MNF-β*, expression decreases after satellite cell differentiation and *MNF-α* expression is induced. When the satellite cells return to the quiescent stage, the expression pattern is reversed again. This dynamic expression pattern, together with data from in vitro binding essays using cell-cycle regulators suggested that the expression of either form of *MNF* might be a molecular switch, regulating induction of the cell cycle and, thus, activation of satellite cells (Garry et al. 2000). This hypothesis is supported by the fact that mice with a targeted mutation of the *MNF* gene show a drastically reduced capacity of muscle fibre regeneration after injury (Garry et al. 2000). In addition, the *MNF* mutation crossed onto an mdx background (Duchenne muscular dystrophy; Deconinck et al. 1997) caused severe muscular dystrophy, resulting in lethality of the affected double mutants at 2 weeks of age (Garry et al. 2000). These data clearly demonstrate the functional importance of *MNF*s in satellite cell function. Yet, it is still an open and interesting question of whether it is the distinct activation of the two *MNF* isoforms that regulates the cell cycle entry of satellite cells. Because the knock out of the *MNF* gene affected both isoforms, it is presently difficult to distinguish between their biological activities.

5 Perspectives

Our knowledge concerning the determination and differentiation programm of skeletal myogenesis has been expanded significantly by a number of new mutations in transcription factors that regulate different aspects of skeletal myogenesis in vivo. In order to fully understand the processes involved in cell cycle regulation and, thus, in determination and differentiation of different muscle precursor cells, we now need to identify downstream targets, which are regulated by different transcription factors. Transcription factors are on top of the hierarchy-controlling mediators that lead to the activation of genes responsible for the phenotypic change of a pluripotent precursor cell into a determined cell fulfilling its developmental or regenerational potential. Besides the elucidation of the regulatory network in which myogenic transcription factors exert their function during myogenesis, the next challenge will be to achieve a better understanding of the cellular apparatus that transmits a transcriptional signal into a cells response. It appears likely that this avenue will eventually lead to an improved accessibility of the muscle regeneration process for therapeutic manipulations.

Acknowledgements. We thank Herbert Neuhaus for critically reading the manuscript and Herbert Neuhaus and Konstanze Schäfer for providing the figures. Work in the authors laboratory was supported by grants from the DFG, the FCI, the Volkswagen Foundation and the Boehringer Ingelheim Fonds.

References

Arnold HH, Braun T (1996) Targeted inactivation of myogenic factor genes reveals their role during mouse myogenesis: a review. Int J Dev Biol 40:345–353

Bassel-Duby R, Hernandez MD, Yang Q, Rochelle JM, Seldin MF, Williams RS (1994) Myocyte nuclear factor, a novel winged-helix transcription factor under both developmental and neural regulation in striated myocytes. Mol Cell Biol 14:4596–4605

Bischoff R, Heintz C (1994) Enhancement of skeletal muscle regeneration. Dev Dyn 201:41–54

Bladt F, Riethmacher D, Isenmann, S, Aguzzi A, Birchmeier C (1995) Essential role for the c-met receptor in the migration of myogenic precursor cells into the limb bud (see comments). Nature 376:768–771

Bober E, Lyons GE, Braun T, Cossu G, Buckingham M, Arnold HH (1991) The muscle regulatory gene, Myf-6, has a biphasic pattern of expression during early mouse development. J Cell Biol 113:1255–1265

Bober E, Franz T, Arnold HH, Gruss P, Tremblay P (1994) Pax-3 is required for the development of limb muscles: a possible role for the migration of dermomyotomal muscle progenitor cells. Development 120:603–612

Brand-Saberi B, Christ B (2000) Evolution and development of distinct cell lineages derived from somites. Curr Top Dev Biol 48:1–42
Braun T, Arnold HH (1995) Inactivation of Myf-6 and Myf-5 genes in mice leads to alterations in skeletal muscle development. EMBO J 14:1176–1186
Braun T, Bober E, Buschhausen-Denker G, Kohtz S, Grzeschik KH, Arnold HH, Kotz S (1989a) Differential expression of myogenic determination genes in muscle cells: possible autoactivation by the Myf gene products [published erratum appears in EMBO J 1989 Dec; 8(13):4358]. EMBO J 8:3617–3625
Braun T, Buschhausen-Denker G, Bober E, Tannich E, Arnold HH (1989b) A novel human muscle factor related to but distinct from MyoD1 induces myogenic conversion in 10T1/2 fibroblasts. EMBO J 8:701–709
Braun T, Winter B, Bober E, Arnold HH (1990) Transcriptional activation domain of the muscle-specific gene-regulatory protein myf5. Nature 346:663–665
Braun T, Rudnicki MA, Arnold HH, Jaenisch R (1992) Targeted inactivation of the muscle regulatory gene Myf-5 results in abnormal rib development and perinatal death. Cell 71:369–382
Brohmann H, Jagla K, Birchmeier C (2000) The role of Lbx1 in migration of muscle precursor cells. Development 127:437–445
Buckingham M (1992) Making muscle in mammals. Trends Genet 8:144–148
Christ B, Ordah, CP (1995) Early stages of chick somite development. Anat Embryol (Berl) 191: 381–396
Cornelison DD, Olwin BB, Rudnicki MA, Wold BJ (2000) MyoD(–/–) satellite cells in single-fiber culture are differentiation defective and MRF4 deficient. Dev Biol 224:122–137
Daston G, Lamar E, Olivier M, Goulding M (1996) Pax-3 is necessary for migration but not differentiation of limb muscle precursors in the mouse. Development 122:1017–1027
Davis RL, Weintraub H, Lassar AB (1987) Expression of a single transfected cDNA converts fibroblasts to myoblasts. Cell 51:987–1000
Deconinck AE, Rafael JA, Skinner JA, Brown SC, Potter AC, Metzinger L, Watt DJ, Dickson JG, Tinsley JM, Davies KE (1997) Utrophin–dystrophin-deficient mice as a model for Duchenne muscular dystrophy. Cell 90:717–727
Dietrich S, Abou-Rebyeh F, Brohmann H, Bladt F, Sonnenberg-Riethmacher E, Yamaai T, Lumsden A, Brand-Saberi B, Birchmeier C (1999) The role of SF/HGF and c-Met in the development of skeletal muscle. Development 126:1621–1629
Edmondson DG, Olson EN (1989) A gene with homology to the myc similarity region of MyoD1 is expressed during myogenesis and is sufficient to activate the muscle differentiation program [published erratum appears in Gene Dev 1990 Aug; 4(8):1450]. Genes Dev 3:628–640
Epstein JA, Shapiro DN, Cheng J, Lam PY, Maas RL (1996) Pax-3 modulates expression of the c-Met receptor during limb muscle development. Proc Natl Acad Sci USA 93:4213–4218
Floss T, Arnold HH, Braun T (1996) Myf-5(m1)/Myf-6(m1) compound heterozygous mouse mutants down-regulate Myf-5 expression and exert rib defects: evidence for long-range *cis* effects on Myf-5 transcription. Dev Biol 174:140–147
Garry DJ, Yang Q, Bassel-Duby R, Williams RS (1997) Persistent expression of MNF identifies myogenic stem cells in postnatal muscles. Dev Biol 188:280–294
Garry DJ, Meeson A, Elterman J, Zhao Y, Yang P, Bassel-Duby R, Williams RS (2000) Myogenic stem cell function is impaired in mice lacking the forkhead/winged helix protein MNF. Proc Natl Acad Sci USA 97:5416–5421
Goulding M, Lumsden A, Paquette AJ (1994) Regulation of Pax-3 expression in the dermomyotome and its role in muscle development. Development 120:957–971
Gross MK, Moran-Rivard L, Velasquez T, Nakatsu MN, Jagla K, Goulding M (2000) Lbx1 is required for muscle precursor migration along a lateral pathway into the limb. Development 127: 413–424
Hasty P, Bradley A, Morris JH, Edmondson DG, Venuti JM, Olson EN, Klein WH (1993) Muscle deficiency and neonatal death in mice with a targeted mutation in the myogenin gene (see comments). Nature 364:501–506

Jagla K, Dolle P, Mattei MG, Jagla T, Schuhbaur B, Dretzen G, Bellard F, Bellard M (1995) Mouse Lbx1 and human LBX1 define a novel mammalian homeobox gene family related to the *Drosophila* lady bird genes. Mech Dev 53:345–356

Jostes B, Walther C, Gruss P (1990) The murine paired box gene, Pax7, is expressed specifically during the development of the nervous and muscular system. Mech Dev 33:27–37

Kablar B, Krastel K, Ying C, Asakura A, Tapscott SJ, Rudnicki MA (1997) MyoD and Myf-5 differentially regulate the development of limb versus trunk skeletal muscle. Development 124: 4729–4738

Kaul A, Koster M, Neuhaus H, Braun T (2000) Myf-5 revisited: loss of early myotome formation does not lead to a rib phenotype in homozygous Myf-5 mutant mice. Cell 102:17–19

Kaushal S, Schneider JW, Nadal-Ginard B, Mahdavi V (1994) Activation of the myogenic lineage by MEF2A, a factor that induces and cooperates with MyoD. Science 266:1236–1240

Lai E, Clark KL, Burley SK, Darnell JE (1993) Hepatocyte nuclear factor 3/fork head or "winged helix" proteins: a family of transcription factors of diverse biologic function. Proc Natl Acad Sci USA 90:10421–10423

Lilly B, Zhao B, Ranganayakulu G, Paterson BM, Schulz RA, Olson EN (1995) Requirement of MADS domain transcription factor D-MEF2 for muscle formation in *Drosophila*. Science 267:688–693

Lin Q, Schwarz J, Bucana C, Olson EN (1997) Control of mouse cardiac morphogenesis and myogenesis by transcription factor MEF2C. Science 276:1404–1407

Ludolph DC, Konieczny SF (1995) Transcription factor families: muscling in on the myogenic program. FASEB J 9:1595–1604

Mak KL, To RQ, Kong Y, Konieczny SF (1992) The MRF4 activation domain is required to induce muscle-specific gene expression. Mol Cell Biol 12:4334–4346

Mankoo BS, Collins NS, Ashby P, Grigorieva E, Pevny LH, Candia A, Wright CV, Rigby PW, Pachnis V (1999) Mox2 is a component of the genetic hierarchy controlling limb muscle development. Nature 400:69–73

Mansouri A, Stoykova A, Torres M, Gruss P (1996) Dysgenesis of cephalic neural crest derivatives in Pax7–/– mutant mice. Development 122:831–838

Mansouri A, Goudreau G, Gruss P (1999) Pax genes and their role in organogenesis. Cancer Res 59:1707s–17010s

Maroto M, Reshef R, Munsterberg AE, Koester S, Goulding M, Lassar AB (1997) Ectopic Pax-3 activates MyoD and Myf-5 expression in embryonic mesoderm and neural tissue. Cell 89: 139–148

Megeney LA, Kablar B, Garrett K, Anderson JE, Rudnicki MA (1996) MyoD is required for myogenic stem cell function in adult skeletal muscle. Genes Dev 10:1173–1183

Mennerich D, Schafer K, Braun T (1998) Pax-3 is necessary but not sufficient for lbx1 expression in myogenic precursor cells of the limb. Mech Dev 73:147–158

Miner JH, Wold B (1990) Herculin, a fourth member of the MyoD family of myogenic regulatory genes. Proc Natl Acad Sci USA 87:1089–1093

Molkentin JD, Black BL, Martin JF, Olson EN (1995) Cooperative activation of muscle gene expression by MEF2 and myogenic bHLH proteins. Cell 83:1125–1136

Nabeshima Y, Hanaoka K, Hayasaka M, Esumi E, Li S, Nonaka I (1993) Myogenin gene disruption results in perinatal lethality because of severe muscle defect (see comments). Nature 364: 532–535

Naya FJ, Olson EN (1999) MEF2: a transcriptional target for signaling pathways controlling skeletal muscle growth and differentiation. Curr Opin Cell Biol 11:683–638

Olson EN (1990) MyoD family: a paradigm for development? Genes Dev 4:1454–1461

Olson EN, Klein WH (1994) bHLH factors in muscle development: dead lines and commitments, what to leave in and what to leave out. Genes Dev 8:1–8

Ordahl CP, Le Douarin NM (1992) Two myogenic lineages within the developing somite. Development 114:339–353

Ott MO, Bober E, Lyons G, Arnold H, Buckingham M (1991) Early expression of the myogenic regulatory gene, myf-5, in precursor cells of skeletal muscle in the mouse embryo. Development 111:1097–1107

Patapoutian A, Yoon JK, Miner JH, Wang S, Stark K, Wold B (1995) Disruption of the mouse MRF4 gene identifies multiple waves of myogenesis in the myotome. Development 121:3347–3358

Rawls A, Morris JH, Rudnicki M, Braun T, Arnold HH, Klein WH, Olson EN (1995) Myogenin's functions do not overlap with those of MyoD or Myf-5 during mouse embryogenesis. (Published erratum appears in Dev Biol 1996 Mar 15; 174(2):453.) Dev Biol 172:37–50

Rhodes SJ, Konieczny SF (1989) Identification of MRF4: a new member of the muscle regulatory factor gene family. Genes Dev 3:2050–2061

Ridgeway AG, Wilton S, Skerjanc IS (2000) Myocyte enhancer factor 2C and myogenin upregulate each other's expression and induce the development of skeletal muscle in P19 cells. J Biol Chem 275:41–46

Rosenblatt JD, Yong D, Parry DJ (1994) Satellite cell activity is required for hypertrophy of overloaded adult rat muscle. Muscle Nerve 17:608–613

Rudnicki MA, Braun T, Hinuma S, Jaenisch R (1992) Inactivation of MyoD in mice leads to up-regulation of the myogenic HLH gene Myf-5 and results in apparently normal muscle development. Cell 71:383–390

Rudnicki MA, Schnegelsberg PN, Stead RH, Braun T, Arnold HH, Jaenisch R (1993) MyoD or Myf-5 is required for the formation of skeletal muscle. Cell 75:1351–1359

Sabourin LA, Girgis-Gabardo A, Seale P, Asakura A, Rudnicki MA (1999) Reduced differentiation potential of primary MyoD-/- myogenic cells derived from adult skeletal muscle. J Cell Biol 144:631–643

Sassoon D, Lyons G, Wright WE, Lin V, Lassar A, Weintraub H, Buckingham M (1989) Expression of two myogenic regulatory factors myogenin and MyoD1 during mouse embryogenesis. Nature 341:303–307

Schafer K, Braun T (1999) Early specification of limb muscle precursor cells by the homeobox gene Lbx1h. Nat Genet 23:213–216

Seale P, Rudnicki MA (2000) A new look at the origin, function, and "stem-cell" status of muscle satellite cells. Dev Biol 218:115–124

Seale P, Sabourin LA, Girgis-Gabardo A, Mansouri A, Gruss P, Rudnicki MA (2000) Pax7 is required for the specification of myogenic satellite cells. Cell 102:777–786

Tajbakhsh S, Bober E, Babinet C, Pournin S, Arnold H, Buckingham M (1996a) Gene targeting the myf-5 locus with nlacZ reveals expression of this myogenic factor in mature skeletal muscle fibres as well as early embryonic muscle. Dev Dyn 206:291–300

Tajbakhsh S, Rocancourt D, Buckingham M (1996b) Muscle progenitor cells failing to respond to positional cues adopt non-myogenic fates in myf-5 null mice. Nature 384:266–270

Tajbakhsh S, Rocancourt D, Cossu G, Buckingham M (1997) Redefining the genetic hierarchies controlling skeletal myogenesis: Pax-3 and Myf-5 act upstream of MyoD. Cell 89:127–138

Tallquist MD, Weismann KE, Hellstrom M, Soriano P (2000) Early myotome specification regulates PDGFA expression and axial skeleton development. Development 127:5059–5070

Thayer MJ, Tapscott SJ, Davis RL, Wright WE, Lassar AB, Weintraub H (1989) Positive autoregulation of the myogenic determination gene MyoD1. Cell 58:241–248

Venuti JM, Morris JH, Vivian JL, Olson EN, Klein WH (1995) Myogenin is required for late but not early aspects of myogenesis during mouse development. J Cell Biol 128:563–576

Wang Y, Jaenisch R (1997) Myogenin can substitute for Myf5 in promoting myogenesis but less efficiently. Development 124:2507–2513

Wang Y, Schnegelsberg PN, Dausman J, Jaenisch R (1996) Functional redundancy of the muscle-specific transcription factors Myf5 and myogenin. Nature 379:823–825

Weintraub H, Davis R, Tapscott S, Thayer M, Krause M, Benezra R, Blackwell TK, Turner D, Rupp R, Hollenberg S et al. (1991) The myoD gene family: nodal point during specification of the muscle cell lineage. Science 251:761–766

Williams BA, Ordahl CP (1994) Pax-3 expression in segmental mesoderm marks early stages in myogenic cell specification. Development 120:785–796

Winter B, Braun T, Arnold HH (1992) Co-operativity of functional domains in the muscle-specific transcription factor Myf-5. EMBO J 11:1843–1855
Wright WE, Sassoon DA, Lin VK (1989) Myogenin, a factor regulating myogenesis, has a domain homologous to MyoD. Cell 56:607–617
Yang Q, Bassel-Duby R, Williams RS (1997) Transient expression of a winged-helix protein, MNF-beta, during myogenesis. Mol Cell Biol 17:5236–5243
Zhang W, Behringer RR, Olson EN (1995) Inactivation of the myogenic bHLH gene MRF4 results in up-regulation of myogenin and rib anomalies. Genes Dev 9:1388–1399

Hypaxial Muscle Development

Gary Parkyn, Roy C. Mootoosamy, Louise Cheng, Colin Thorpe, and Susanne Dietrich[1]

1 Introduction

Chordate larvae show a surprisingly uniform "bauplan", with a front end carrying the sense organs plus the gill and feeding apparatus, and a posterior end used for locomotion (reviewed in Goodrich 1958; Young 1962). Although adult forms frequently give up this organisation when they switch to sessile life styles, motility based on trunk muscles is maintained in acrania, and both in jaw-less and jawed vertebrates (agnathans and gnathostomes). The mesoderm on either side of the neural canal is subdivided into metameric blocks of muscle. As the notochord, and in vertebrates the vertebral column, prevent telescoping of the body, the serial action of the muscles on either side leads to an undulating movement.

During vertebrate evolution, two events brought about substantial changes to the basic organisation of trunk muscles. When vertebrates left the water to live on land, the dorsal and ventral halves of the metameric muscle blocks acquired specialised functions (reviewed in Goodrich 1958; Young 1962): the dorsal halves united to form the longitudinally orientated, deep muscles of the back, which were used to erect the spine. These so-called epaxial muscles are innervated by the dorsal branch of the spinal nerves (ramus dorsalis). The ventral halves of the muscle blocks, however, were recruited to serve multiple purposes. They gave rise to all the ventrally, laterally and superficially located muscles, including the muscles of the limbs and the body wall. These so-called hypaxial muscles are innervated by the ventral branches of the spinal nerves (ramus ventralis). Interestingly, separate innervation of ventral and dorsal halves of the muscle blocks is found already in zebrafishes, but not in lampreys, suggesting that the differences between epaxial and hypaxial muscles are an original feature of bony fishes and higher vertebrates, possibly even of all gnathostomes (reviewed by Eisen 1999; S. Kuratani, pers. comm.).

The second event changing the organisation of trunk muscles relates to the evolution of paired extremities (reviewed in Goodrich 1958; Young 1962). Initially, the paired fins present in gnathostome fishes were used to stabilise the body, propelled forward by the action of the metameric trunk muscles as

[1] King's College London, Department of Craniofacial Development, Floor 27 Guy's Tower, Guy's Hospital, London SE1 9RT, UK

Results and Problems in Cell Differentiation, Vol. 38
B. Brand-Saberi (Ed.): Vertebrate Myogenesis

seen in cartilaginous fishes. The paired fins received their muscles from the neighbouring muscle blocks, whose hypaxial halves elongated into the fin buds. During the evolution of bony fishes, the paired fins became more elaborate, involved in steering and moving the body. Here, the hypaxial parts of the muscle blocks released muscle precursors that migrated into the fin buds, where they reorganised and differentiated (Neyt et al. 2000). This pattern of muscle formation was refined in tetrapods, whose tongue muscles, diaphragm (mammals) and limb plus associated shoulder/hip muscles develop from migratory muscle precursors (reviewed in Dietrich 1999). However, the muscles of the abdominal wall still show the evolutionary older mechanism of development as they form from ventrolaterally elongating muscle anlagen.

In this article, we will review the to-date's knowledge on hypaxial muscle formation. We will concentrate on amniotic tetrapods because, in these animals, hypaxial muscle formation is best understood. We will summarise the regulatory cascades for the specification of hypaxial muscle precursors, the distinction between migratory and non-migratory cells and the initiation of precursor cell migration.

2 Developmental Anatomy of Trunk Skeletal Muscles in Amniotes

The sole source of trunk skeletal muscles in vertebrates is the paraxial mesoderm, which is laid down as a mesenchyme during gastrulation, embedded between neural tube and notochord medially, intermediate and lateral mesoderm laterally, surface ectoderm dorsally and endoderm ventrally (reviewed by Christ and Ordahl 1995; Gossler and Hrabe de Angelis 1998). This paraxial mesoderm forms so-called somites, metameric blocks of epithelially arranged tissue. In lower vertebrates, the somites predominantly give rise to metameric muscles as described before (reviewed in Currie and Ingham 1998). In amniotes, however, the somites first de-epithelialise ventrally to provide the large sclerotomes, destined to give rise to the vertebral column and ribs (reviewed by Christ and Ordahl 1995; Gossler and Hrabe de Angelis 1998). The dorsally located dermomyotomes retain their epithelial organisation and give rise to both muscle and dermis. The first muscle precursors arise in the medial lips of the dermomyotomes (Ordahl and Le Douarin 1992; Denetclaw et al. 1997; Kahane et al. 1998a, 1998b; Venters et al. 1999), induced by dorsalising signals from neural tube and surface ectoderm together with ventralising signals from notochord and floor plate (Münsterberg and Lassar 1995; Münsterberg et al. 1995; Stern et al. 1995; Dietrich et al. 1997; Borycki et al. 1998). During outgrowth of the dermomyotomes, the medial lips continuously deposit myoblasts underneath, which constitute the epaxial aspect of the myotomes and immediately differentiate. Ultimately, these epaxial myotomes give rise to the deep muscles of the back (Denetclaw et al. 1997; Kahane et al. 1998a, 1998b; Venters et al. 1999; reviewed by Christ and Ordahl 1995).

Using chick-quail chimera, it has been demonstrated that the amniote hypaxial muscles are generated by the lateral lips of the dermomyotomes and develop independently from neural tube or notochord (Ordahl and Le Douarin 1992; Rong et al. 1992; reviewed by Christ and Ordahl 1995; Dietrich 1999). As mentioned before, the hypaxial muscles precursors are laid down by two distinct mechanisms. At abdominal levels, muscle precursors are deposited by the lateral lips of the outgrowing dermomyotomes, by a mechanism similar to the epaxial counterpart (Christ et al. 1983; Cinnamon et al. 1999; Denetclaw and Ordahl 2000). These muscle precursors constitute the hypaxial aspect of the myotomes, they immediately differentiate and subsequently develop into intercostal muscles and muscles of the body wall. By contrast, at occipital, cervical and limb levels, muscle precursors do not enter the myotomes. They detach from the dispersing lateral dermomyotomal lips and actively migrate to their target sites (Nishi 1967; Grim 1970; Chevallier et al. 1977; Christ et al. 1977; Jacob et al. 1978, 1979; Noden 1983; Couly et al. 1993; Bladt et al. 1995; Dietrich et al. 1999; Huang et al. 1999; reviewed in Dietrich 1999). Muscle-determining genes remain silent until the target site is reached (Bober et al. 1991; Pownall and Emerson 1992; Smith et al. 1994; Williams and Ordahl 1994). Subsequently, these cells give rise to various muscles including tongue muscles, limb and associated shoulder muscles and, in mammals, the muscular diaphragm (Nishi 1967; Grim 1970; Chevallier et al. 1977; Christ et al. 1977; Jacob et al. 1978, 1979; Noden 1983; Couly et al. 1993; Bladt et al. 1995; Dietrich et al. 1999; Huang et al. 1999).

3 Markers for Hypaxial Muscle Precursors

To unravel the mechanisms underlying hypaxial muscle development, molecular markers that identify the precursor cells within the lateral dermomyotomal lips are important tools. The basic helix-loop-helix gene *Sim1* marks the entire lateral somite half, including the lateral dermomyotomal lip (Fan et al. 1996; Pourquié et al. 1996). The paired and homeobox containing transcription factor *Pax3* is expressed throughout the dermomyotome, with elevated levels of expression confined to the lateral dermomyotomal lip (Goulding et al. 1994; Williams and Ordahl 1994). The tyrosine kinase receptor *cMet* is found in both the medial and the lateral dermomyotomal lips (Bladt et al. 1995; Yang et al. 1996; its ligand scatter factor/hepatocyte growth factor (SF/HGF) is provided by the lateral mesoderm, see below). The T-box containing transcription factor *Tbx3* is expressed in all lateral dermomytomal lips (unpubl. observ.; Huang et al. 1999). Finally, the only known gene exclusively labelling migratory muscle precursors is the homeobox containing transcription factor *Lbx1* (Jagla et al. 1995; Dietrich et al. 1998, 1999; Mennerich et al. 1998; Fig. 1A). Expression of this gene is confined to occipital, cervical and limb levels, where it marks migratory muscle precursors within the lateral dermomyotomal lips and during long-range migration.

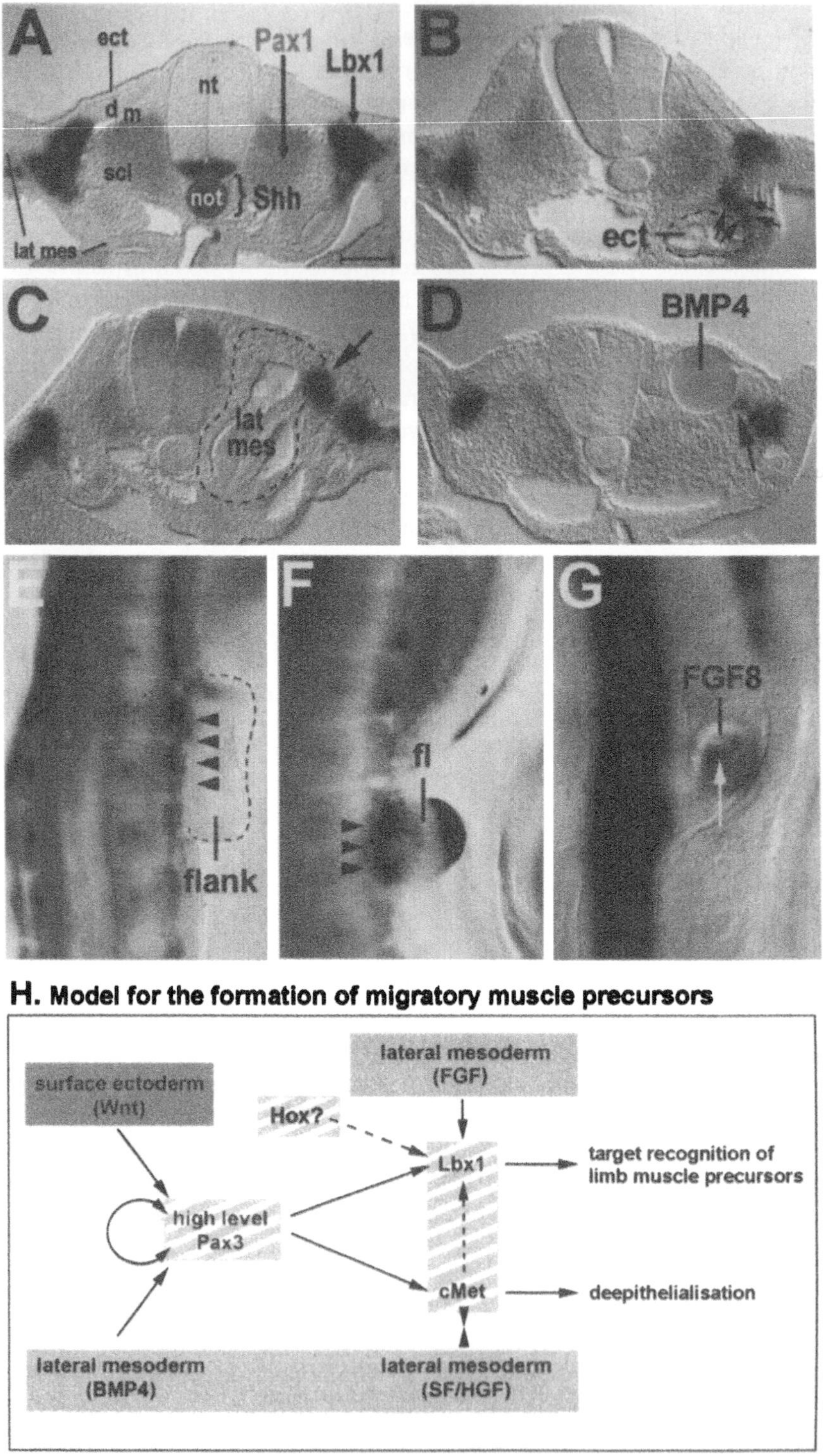
A
ect
Pax1
Lbx1
nt
dm
scl
not
Shh
lat mes
B
ect
C
lat mes
D
BMP4
E
flank
F
fl
G
FGF8
H. Model for the formation of migratory muscle precursors
surface ectoderm (Wnt)
Hox?
lateral mesoderm (FGF)
Lbx1
target recognition of limb muscle precursors
high level Pax3
cMet
deepithelialisation
lateral mesoderm (BMP4)
lateral mesoderm (SF/HGF)

4 Specification of Hypaxial Muscle Precursors

4.1 Cues from the Lateral Mesoderm

Earlier studies had shown that solely epaxial, but not hypaxial muscle development depends on signals from neural tube and notochord (Ordahl and Le Douarin 1992; Rong et al. 1992; reviewed by Christ and Ordahl 1995; Dietrich 1999). However, hypaxial muscle precursors develop in close proximity to the intermediate and lateral mesoderm, suggesting that these tissues may stimulate hypaxial muscle development. Therefore, we separated somites from their lateral environment microsurgically, using the chick embryo as a model (Dietrich et al. 1998). In a second set of experiments, we grafted intermediate and lateral mesoderm medially between neural tube and somites. Whenever somites were deprived of lateral signals, expression of *Sim1*, upregulation of *Pax3* and expression of *Lbx1* was lost, whereas the expression of epaxial markers was expanded, which is in line with earlier reports on medio-lateral somite patterning (Pourquié et al. 1995, 1996). By contrast, when the lateral tissues were transplanted medially, epaxial marker gene expression was lost,

◄

Fig. 1A–H. Cross sections through the fore limb region of chick embryos stage HH17 (Hamburger and Hamilton 1951), dorsal is to the top. The embryos were stained for *Lbx1* (*black*) and the medial sclerotome marker *Pax1* (*grey*). In addition, in **A** notochord and floor plate were stained for *Shh* in (*dark grey*), and in **C** the dorsal neural tube was stained for Wnt4 in (*grey*). **A** Wild-type embryo. *Lbx1* labels the emigrating limb muscle precursors, which are born in the lateral dermomyotomal lip in close proximity to surface ectoderm and intermediate/lateral mesoderm. **B** A strip of ectoderm was inserted ventrally, between the somite and the intermediate/lateral mesoderm. The graft induced an ectopic dermomyotomal epithelium expressing *Lbx1* (*arrows*). **C** Lateral mesoderm was inserted medially, between the somite and neural tube/notochord. The graft induced an additional expression domain for *Lbx1* in the medial dermomyotome (*arrow*). **D** A medially implanted bead loaded with BMP4 caused widening of the *Lbx1* expression domain (*arrow*). **E, F** Dorsal view onto whole embryos, stages as before. Cranial is to the top. The embryos were stained for *Lbx1* (*black*), *Pax3* (**E, F**; *grey*), *Shh* (**G**, black midline signal) and *FGF8* (**E, F**, black signals in the apical ectodermal ridge and the centre of the myotomes). **E** Limb lateral mesoderm was replaced by flank lateral mesoderm as indicated. The somites expressed *Lbx1* regardless (*arrowheads*). **F** Flank lateral mesoderm was replaced by fore limb lateral mesoderm. The graft formed an ectopic limb, which induced *Lbx1* expression in the neighbouring somites (*arrowheads*). **G** A medially implanted bead soaked with FGF8 induced *Lbx1* expression in flank somites (*arrow*). **H** Hierarchy of the known players in the formation of migratory muscle precursors. Signals from surface ectoderm (*grey shading*) and lateral mesoderm (*light grey shading*) are overall regulators of hypaxial muscle development. Within the somite (*hatched*), *Pax3* mediates these signals and controls its own upregulation, which, in turn, is required for *cMet* and *Lbx1* expression. Local signals from the lateral mesoderm stimulate the cMet receptor that acts in the de-epithelialisation of muscle precursors. Further lateral signals, together with a somitic predisposition, account for the production of Lbx1, which is required for target recognition of the limb muscle precursors. *d* Dermomyotome; *ect* ectoderm; *fl* fore limb; *lat mes* lateral mesoderm; *m* myotome; *not* notochord; *nt* neural tube; *scl* sclerotome. The *scale bar* in **A–D, G** represents 50 μm (shown in **A**) and 80 μm in **E, F**

whereas ectopic signals were found for *Sim1*, *Pax3* and *Lbx1* (Fig. 1C). It is important to note that in all experiments performed at cervical or limb levels, upregulation of the *Pax3* gene and expression of *Lbx1* were affected in identical ways. Furthermore, the same cues controlled *Pax3* upregulation at flank levels where *Lbx1* is not expressed. This suggests that at all axial levels, the initial specification of hypaxial muscle progenitors similarly depends on signals from intermediate and lateral mesoderm.

Several studies have demonstrated recently that the overall medio-lateral somite pattern relies on BMP4 produced by the lateral plate mesoderm and its antagonist Noggin, which is expressed in the medial the somite under the control of neural-tube-derived WNT1 (Pourquié et al. 1996; Hirsinger et al. 1997; Marcelle et al. 1997; Capdevila and Johnson 1998; McMahon et al. 1998; Reshef et al. 1998). To test whether BMP4 is sufficient to mimic the role of the lateral mesoderm in the induction of hypaxial muscle precursors, we implanted BMP4 soaked beads medially between somites and neural tube (Dietrich et al. 1998; Fig. 1D). In contrast to medially implanted lateral mesoderm, the beads resulted in widened expression domains for *Lbx1*, but did not induce ectopic signals. Thus, although BMP4 is necessary for somite lateralization, it only in part mimics the function of the lateral mesoderm for the specification of hypaxial muscle progenitors.

4.2 Cues from the Surface Ectoderm

In wild-type as well as in operated embryos, *Lbx1* signals and elevated *Pax3* signals always resided dorsally, close to the overlying surface ectoderm (Dietrich et al. 1998). This suggested that dorsalising signals may be involved in the patterning of the lateral somite half and in the induction of hypaxial muscle progenitors. Therefore, we separated somites from ectoderm or grafted strips of ectoderm ectopically. When the somites were deprived of ectodermal signals, the lateral somite halves looked sclerotome-like and failed to express *Lbx1* and *Pax3*. On the other hand, ectoderm placed ventrally between somites and intermediate/lateral mesoderm led to dermomyotome-like organisation of the somitic tissues that expressed *Lbx1* (Fig. 1B) and, at elevated levels, *Pax3*. However, when the ectoderm was inserted ventromedially between somites and neural tube/notochord, solely *Pax3* was present, at levels typical for the medial dermomyotome (Dietrich et al. 1997). This suggests that the ectoderm acts in the dorsoventral patterning of somites and stimulates dermomyotome development. However, to induce the progenitors for the hypaxial musculature, the dorsalising signals provided by the ectoderm have to act in concert with the intermediate and lateral mesoderm-derived signals.

Candidates for the dorsalising function of the surface ectoderm are members of the wingless/Wnt family of signalling molecules. In vitro, several Wnt genes mimic the function of the ectoderm for dermomyotome induction (Fan et al. 1997). In addition, the WNT receptor Frizzled 1 is expressed in the undifferentiated somite and subsequently in the dermomyotome (Schmidt et al. 2000). WNT7a, which is present in the ectoderm dorsally and overlies the limbs, induces muscle-determining genes in vitro in a temporal pattern that is typical for the hypaxial myotome (Dealy et al. 1993; Smith et al. 1994; Cossu et al. 1996; Kablar et al. 1997; Maroto et al. 1997; Tajbakhsh et al. 1998). Unfortunately, a pan-ectodermal Wnt gene acting in the specification of hypaxial muscle precursors has not been identified yet. However, a candidate may be *Wnt6*, which, in the chick, displays a general ectodermal expression (Schubert et al., in press) and has been shown recently to also induce dermomyotomal markers such as *Pax3* (Fan et al. 1997).

4.3 Master Regulator *Pax3*

In the mouse *Pax3* mutant *Splotch*, the animal model for the human Klein–Waardenburg syndrome (reviewed by Tremblay et al. 1995), the formation of hypaxial muscles is severely affected: the lateral dermomyotomal lips are highly disorganised, and the most ventrally located abdominal muscles are absent, along with the muscular diaphragm and all limb and associated shoulder muscles that stem from migratory muscle precursors (Auerbach 1954; Franz et al. 1993; Bober et al. 1994; Goulding et al. 1994; Daston et al. 1996; Tajbakhsh et al. 1997; Tremblay et al. 1998). Furthermore, in the lateral dermomyotomal lips *Pax3* controls the expression of the tyrosine kinase receptor *cMet* and its own upregulation (Bober et al. 1994; Daston et al. 1996; Epstein et al. 1996; Yang et al. 1996). We found that upregulation of *Pax3* and induction of *Lbx1* depend on the same dorsalising and lateralising signals (Dietrich et al. 1998). Thus, *Pax3* may be the prime target and the mediator of the extrinsic cues for hypaxial muscle development, upstream of *Lbx1*. To determine the hierarchy of both genes, we studied the expression of the mouse *Lbx1* gene in the *Pax3* mutant *Splotch* (Dietrich et al. 1999). At all stages analysed, somites destined to release migratory muscle precursors for the diaphragm and the limb muscles failed to express *Lbx1*, in line with a study by Mennerich et al. (1998). Thus, *Pax3* indeed may act as a master regulator for hypaxial muscle formation. Interestingly, *Lbx1* expression in the tongue muscle precursors was delayed rather than abolished (Dietrich et al. 1999). This suggests that for the occipital somites, which during evolution have been incorporated into the head (reviewed by Goodrich 1958; Young 1962), the loss of *Pax3* function is compensated, and that additional genes may be involved in the mediation of the extrinsic signals.

5 Specification of Migratory Muscle Precursors

5.1 Somitic Competence

Lateralising and dorsalising signals act at all axial levels to specify hypaxial muscle precursors as monitored by the upregulation of *Pax3*. In turn, upregulation of *Pax3* may be a pre-requisite for all hypaxial myogenesis to proceed. However, migratory muscle precursors are born at certain axial levels only, namely at occipital, cervical and limb levels. These migratory cells, but not the non-migratory precursors of the body wall musculature, express the *Lbx1* gene prior to and during migration (Jagla et al. 1995; Dietrich et al. 1998, 1999; Mennerich et al. 1998). Thus, locally restricted control mechanisms for the formation of migratory muscle precursors are in place.

Using *Lbx1* as a marker, we investigated by heterotopic grafting in the chick embryo which lateral mesoderm is capable of inducing migratory muscle precursors (in prep.). We found that lateral mesoderm from all axial levels, including the flank, was able to stimulate *Lbx1* expression in somites that normally release migratory muscle precursors (Fig. 1E). Moreover, when these somites were grafted to the flank, they initiated *Lbx1* expression in a normal fashion, although the muscle precursors did not emigrate. Thus, for the activation of the programme for migratory muscle precursors, somites acted according to their initial location ("herkunftsgemäss", as opposed to "ortsgemäss"). This suggests that somites from occipital, cervical and limb levels may be predisposed to generate migratory muscle precursors whereas somites from flank levels lack this competence. Because the axial identity of somites is controlled by transcription factors of the Hox/HOM gene family (reviewed by Gaunt 1994), it is possible that these genes also control the competence to release migratory muscle precursors. However, when muscle precursors enter the lateral mesoderm, they adopt the Hox expression pattern of their environment (Nowicki and Burke 2000). This suggests that, upon delamination, the migratory muscle precursors closely follow cues from their environment.

5.2 Localised Lateral Signals for the Recruitment of Limb Muscle Precursors

It is established that ectopic limbs can induce the formation of migratory muscle precursors from flank somites that normally contribute non-migratory muscle precursors to the hypaxial myotomes (Hayashi and Ozawa 1995). This suggest that the developing limbs release signals that override the programme for non-migratory hypaxial muscle precursors present in the flank. To test this possibility, we grafted lateral mesoderm from various axial levels next to the flank somites, using conditions that either allow or prevent the formation of ectopic limbs. When the grafts were inserted laterally, mesoderm derived from limb levels generated an ectopic limb, which stimulated the expression of *Lbx1* (Dietrich 1999; Fig. 1F). Likewise, laterally inserted FGF beads triggered ectopic limb development, and *Lbx1* was activated (Mennerich et al. 1998; Dietrich 1999). Grafts from non-limb levels failed to generate ectopic limbs, and *Lbx1* remained suppressed. Finally, when limb-derived mesoderm was grafted medially between neural tube and somites, no ectopic limb developed and *Lbx1* remained silent (Mootoosamy et al., in prep.). Thus, factors provided solely by developing limbs convert non-migratory hypaxial muscle precursors into migratory cells.

Candidates for these factors are members of the fibroblast growth factor (FGF) family. Several FGF molecules are expressed in the limb and can induce ectopic limbs when applied to the flank (reviewed by Cohn and Tickle 1996). Moreover, FGF2 and FGF4 can attract migrating muscle precursors in vitro, and manipulation of the FGF receptor I alters the balance between muscle precursor migration and differentiation (Itoh et al. 1996; Webb et al. 1997). To separate the function of FGF molecules for limb induction and the formation of migratory muscle precursors, we inserted FGF8-loaded beads medially between the somites and neural tube (Fig. 1G; Mootoosamy et al., in prep.). Although limbs did not form, the beads induced ectopic *Lbx1* expression in flank somites. This suggests that FGF molecules act directly in the recruitment of migratory muscle precursors, thereby ensuring the muscularisation of the limbs.

5.3 The Role of Scatter Factor/Hepatocyte Growth Factor and cMet in the Delamination of Migratory Muscle Precursors

Besides FGF molecules, scatter factor/hepatocyte growth factor has been suggested to induce migration of hypaxial muscle precursors. SF/HGF is expressed in occipital, cervical and limb lateral mesoderm along the migratory routes of muscle precursors (Sonnenberg et al. 1993; Bladt et al. 1995; Myokai et al. 1995; Thery et al. 1995; Heymann et al. 1996; Yang et al. 1996; Dietrich et al. 1999). In addition, SF/HGF applied to the flank leads to de-epithelialisation of the lateral dermomyotomal lips, and the molecule is expressed in ectopic limbs, under the control of the limb apical ectodermal ridge (Brand-Saberi et al. 1996a; Heymann et al. 1996, Dietrich 1999; Scaal et al. 1999). SF/HGF acts through the tyrosine kinase receptor cMet, which is expressed in medial and lateral dermomyotomal lips along the antero-posterior axis under the control of *Pax3* (Bladt et al. 1995; Daston et al. 1996; Epstein et al. 1996; Yang et al. 1996). To test whether *Lbx1* is a target of the cMet-SF/HGF pathway, we studied *Lbx1* expression in *SF/HGF* and *cMet* mutant mice that lack all hypaxial muscles derived from migratory precursors (Bladt et al. 1995; Maina et al. 1996; Dietrich et al. 1999). In addition, we and others grafted SF/HGF loaded beads to the flank of chick embryos (Mennerich et al. 1998; Dietrich 1999).

Different from *Splotch* mice, in mutants lacking functional *cMet* and *SF/HGF* genes, *Lbx1* was still expressed (Dietrich et al. 1999). Thus, while *Pax3* regulates both *cMet* and *Lbx1*, the cMet-SF/HGF pathway is not involved in the onset of *Lbx1* expression. However, in *cMet* and *SF/HGF* mutants, the cells labelled with *Lbx1* failed to detach from the lateral dermomyotomal lips and to take up long-range migration. This suggests that the receptor–ligand interaction specifically controls the adhesive properties of hypaxial muscle precursors. In line with this finding, beads loaded with SF/HGF induce de-epithelialisation of the lateral dermomyotomal lips in the flank, but usually not expression of *Lbx1* (Brand-Saberi et al. 1996a; Heymann et al. 1996; Mennerich et al. 1998; Dietrich 1999). However, in rare cases, *Lbx1* signals were found next to the bead, suggesting that SF/HGF may act in a feedback mechanism.

5.4 The Role of *Lbx1* in Target Recognition of Limb Muscle Precursors

The *Lbx1* gene is the only known marker specific for migratory muscle precursors (Jagla et al. 1995; Dietrich et al. 1998, 1999; Mennerich et al. 1998). Knockout experiments in the mouse addressed whether the gene acts in the specification of migratory muscle precursors as suggested earlier (Dietrich et al. 1998). Interestingly, mice deficient for *Lbx1* were deprived of limb muscles only, whereas tongue and diaphragm muscles developed (Schäfer and Braun 1999; Brohmann et al. 2000; Gross et al. 2000). Moreover, migratory muscle precursors from limb levels were diverted into the septum transversum where they contributed to the muscular diaphragm. Thus, migration along the lateral pathway into the limb, but not along the ventral pathways into the septum transversum or towards the mandibular arch was affected. This suggests that *Lbx1*, although expressed in all migratory muscle precursors, specifically acts in the recognition of the limb as a target. Therefore, a *Lbx1*-independent pathfinding mechanism may exist for tongue and diaphragm precursors. Alternatively, because only mammals form a muscular diaphragm, *Lbx1* function may have been restricted to the developing limb muscle precursors during evolution to separate them from the stream of emigrating diaphragm muscle precursors.

6 Building a Regulatory Network for Hypaxial Muscle Development

We reviewed here that in amniotes, hypaxial muscle precursors develop from the lateral lips of the demomyotomes under the control of lateralising signals, in part mimicked by BMP4, and dorsalising signals, possibly represented by Wnt family members (summarised in Fig. 1H). The phenotype of *Splotch* mutant mice suggests that *Pax3* acts a master regulator of hypaxial muscle formation, mediating the lateralising and dorsalising signals. The formation of non-migratory muscle precursors, laid down during somite elongation, may represent the default pathway for vertebrate hypaxial muscle development because migratory muscle precursors evolved at the level of bony fishes. However, because in *Splotch* the flank lateral dermomyotomal lips are severely disorganised, specific regulatory cascades may be in place to maintain their epithelial organisation.

In addition to factors influencing hypaxial muscle development in general, specific extrinsic and intrinsic cues are required for the formation of migratory muscle precursors. The axial identity may contribute to the competence of occipital, cervical and limb somites to readily activate the programme for migratory muscle precursors. FGF molecules may trigger the production of migratory muscle precursors, whereas the interaction of SF/HGF and cMet specifically controls the delamination of migratory muscle precursors, possibly by regulating cell adhesion. Finally, *Lbx1* may act in the homing of limb muscle precursors, although this function may represent a derived condition in mammals. Nevertheless, additional properties of migratory muscle precursors need to be controlled: to migrate, the cells need to have the appropriate cyto-architecture. In addition, during the course of migration, differentiation needs to be suppressed, possibly involving the homeobox gene *Msx1* (Bendall et al. 1999). Cell adhesion has to be kept at bay, but needs to increase at the target site to allow the formation of myotubes. This process is supported by N-cadherin, NCAM and M-Cadherin (Dickson et al. 1990; Knudsen et al. 1990; Donalies et al. 1991; Brand-Saberi et al. 1996b). Moreover, the cell cycle needs to be regulated as the cells proliferate extensively prior to differentiation. Unfortunately, further regulators are not available to date, and the complete regulatory network is yet to be unravelled.

Acknowledgements. We wish to thank A. Lumsden, C. Tickle and P.T. Sharpe for their generous support. We also thank P. Ahlberg, C. Birchmeier, A. Burke, B. Christ, A. Copp, P. Currie, D. Evans, P. Francis-West, S. Gaunt, M. Goulding, A. Graham, C. Healy, R. Hichliffe, S. Hunt, K. Jagla, R. Jeffries, S. Kuratani, A. Mansouri, I. Mason, A.E. Münsterberg, D. Noden, E.N. Olson, C.P. Ordahl, R. Presley, F.R. Schubert, K.K. Smith, M.M. Smith, D. Sosic, C. Stern, I. Thesslef, A.J. Thexton, P. Tremblay, A. Varela-Echavarria, S. Wilson, R. Wingate, A. Yeudall for inspiring discussions and for their collaboration. We are grateful to R. Balling, C.M. Fan, P. Gruss, J.C. Izpisua Belmonte, T. Lints, C. Logan, C. Marcelle, T. Nohno, S. Noji, T. Ogura, B. Paterson, C. Tabin for molecular probes. The work was supported by the EMBO, the HFSPO, the Royal Society, the Special Trustees of Guy's Hospital and the BBSRC.

References

Auerbach R (1954) Analysis of the developmental effects of a lethal mutation in the house mouse. J Exp Zool 127:305–329

Bendall AJ, Ding J, Hu G, Shen MM, Abate-Shen C (1999) *Msx1* anatgonizes the myogenic activity of *Pax3* in migrating limb muscle precursors. Development 126:4965–4976

Bladt F, Riethmacher D, Isenmann S, Aguzzi A, Birchmeier C (1995) Essential role for the c-Met receptor in the migration of myogenic precursor cells into the limb bud. Nature 376:768–771

Bober E, Lyons GE, Braun T, Cossu G, Buckingham M, Arnold HH (1991) The muscle regulatory gene, *Myf-5* has a biphasic pattern of expression during early mouse development. J Cell Biol 113:1255–1265

Bober E, Franz T, Arnold HH, Gruss P, Tremblay P (1994) *Pax-3* is required for the development of limb muscles: a possible role for the migration of dermomyotomal muscle progenitor cells. Development 120:603–612

Borycki A-G, Mendham L, Emerson, CP (1998) Control of somite patterning by Sonic hedgehog and its downstream signal response genes. Development 125:777–790

Brand-Saberi B, Müller TS, Wilting J, Christ B, Birchmeier C (1996a) Scatter factor/ Hepatocyte Growth Factor (SF/HGF) induces emigration of myogenic cells at interlimb in vivo. Dev Biol 179:303–308

Brand-Saberi B, Gamel AJ, Krenn V, Müller TS, Wilting J, Christ B (1996b) N-Cadherin is involved in myoblast migration and muscle differentiation in the avian limb bud. Dev Biol 178:160–173

Brohmann H, Jagla K, Birchmeier C (2000) The role of *Lbx1* in migration of muscle precursor cells. Development 127:437–445

Capdevila J, Johnson RL (1998) Endogenous and ectopic expression of noggin suggests a conserved mechanism for regulation of BMP function during limb and somite patterning. Dev Biol 197:205–217

Chevallier A, Kieny M, Mauger A, Sengel P (1977) Developmental fate of the somitic mesoderm in the chick embryo. In: Ede DA, Hinchliffe JR, Balls M (eds) Vertebrate limb and somite morphogenesis. Cambridge University Press, Cambridge, pp 421–431

Christ B, Ordahl CP (1995) Early stages of chick somite development. Anat Embryol 191:381–396

Christ B, Jacob HJ, Jacob M (1977) Experimental analysis of the origin of the wing musculature in avian embryos. Anat Embryol 150:171–186

Christ B, Jacob M, Jacob HJ (1983) On the origin and development of the ventrolateral abdominal muscles in the avian embryo. An experimental and ultrastructural study. Anat Embryol 166:87–101

Cinnamon Y, Kahane N, Kalcheim C (1999) Characterisation of the early development of specific hypaxial muscles from the ventrolateral myotome. Development 126:4305–4315

Cohn MJ, Tickle C (1996) Limbs: a model for patterning formation within the vertebrate body plan. Trend Genet 12:253–257

Cossu G, Kelly R, Tajbakhsh S, Di Donna S, Vivarelli E, Buckingham M (1996) Activation of different myogenic pathways: *Myf-5* is induced by the neural tube and *MyoD* by the dorsal ectoderm in mouse paraxial mesoderm. Development 122:429–437

Couly GF, Coltey PM, Le Douarin NM (1993) The triple origin of skull in higher vertebrates: a study in quail-chick chimeras. Development 117:409–429

Currie PD, Ingham PW (1998) The generation and interpretation of positional information within the vertebrate myotome. Mech Dev 73:3–21

Daston G, Lamar E, Olivier M, Goulding M (1996) *Pax3* is necessary for migration but not differentiation of limb muscle precursors in the mouse. Development 122:1017–1027

Dealy CN, Roth A, Ferrari D, Brown AMC, Kosher RA (1993) *Wnt-5a* and *Wnt-7a* are expressed in the developing chick limb bud in a manner suggesting roles in the pattern formation along the proximodistal and dorsoventral axis. Mech Dev 43:175–186

Denetclaw WF, Ordahl CP (2000) The growth of the dermomyotome and formation of early myotome lineages in thoracolumbar somites of chicken embryos. Development 127:893–905

Denetclaw WF, Christ B, Ordahl CP (1997) Location and growth of epaxial myotome precursor cells. Development 124:1601–1610

Dickson G, Peck D, Moore SE, Barton CM, Walsh FS (1990). Enhanced myogenesis in NCAM transfected mouse myoblasts. Nature 344:348–351

Dietrich S (1999) Regulation of hypaxial muscle development. Cell Tissue Res 296:175–182

Dietrich S, Schubert FR, Lumsden A (1997) Control of dorsoventral pattern in the chick paraxial mesoderm. Development 124:3895–3908

Dietrich S, Schubert FR, Healy C, Sharpe PT, Lumsden A (1998) Specification of the hypaxial musculature. Development 125:2235–2249

Dietrich S, Abou-Rebyeh F, Brohmann H, Bladt F, Sonnenberg-Riethmacher E, Yamaai T, Lumsden A, Brand-Saberi B, Birchmeier C (1999) The role of SF/HGF and c-Met in the development of skeletal muscle. Development 126:1621–1629

Donalies M, Cramer M, Ringwald M, Starzinski-Powitz A (1991) expression of M-cadherin, a member of the cadherin multigene family, correlates with differentiation of skeletal muscle cells. Proc Natl Acad Sci USA 88:8024–8028

Eisen JS (1999) Patterning motorneurons in the vertebrate nervous system. Trends Neurosci 22: 321–326

Epstein JA, Shapiro DN, Cheng J, Lam PY, Maas RL (1996) *Pax3* modulates expression of the *c-Met* receptor during limb muscle development. Proc Natl Acad Sci USA 93:4213–4218

Fan CM, Kuwana E, Bulfone A, Fletcher CF, Copeland NG, Jenkins NA, Crews S, Martinez S, Puelles L, Rubenstein JL, Tessier-Lavigne M (1996) Expression patterns of two murine homologs of *Drosophila* single-minded suggest possible roles in embryonic patterning and in the pathogenesis of Down syndrome. Mol Cell Neurosci 7:1–16

Fan C-M, Lee CS, Tessier-Lavigne M (1997) A role for WNT proteins in induction of dermomyotome. Dev Biol 191:160–165

Franz T, Kothary R, Surani MA, Halata Z, Grim M (1993) The *Splotch* mutation interferes with muscle development in the limbs. Anat Embryol 187:153–160

Gaunt SJ (1994) Conservation of the Hox code during morphological evolution. Int J Dev Biol 38:549–552

Goodrich ES (1958) Studies on the structure and development of vertebrates. Dover Publications, New York

Gossler A, Hrabe de Angelis M (1998) Somitogenesis. Curr Top Dev Biol 38:225–287

Goulding M, Lumsden A, Paquette AJ (1994) Regulation of *Pax-3* expression in the dermomyotome and its role in muscle development. Development 120:957–971

Grim M (1970) Differentiation of myoblasts and the relationship between somites and wing bud of chick embryos. Z Anat Entwickl Ges 132:260–271

Gross MK, Moran-Rivard L, Velasquez T, Nakatsu MN, Jagla K, Goulding M (2000) *Lbx1* is required for muscle precursor migration along a lateral pathway into the limb. Development 127: 413–424

Hamburger V, Hamilton HL (1951) A series of normal stages in the development of the chick embryo. J Morph 88:49–92

Hayashi K, Ozawa E (1995) Myogenic cell migration from somites is induced by tissue contact with medial region of the presumptive limb mesoderm in chick embryos. Development 121: 661–669

Heymann S, Koudrova M, Arnold HH, Koster M, Braun T (1996) Regulation and function of SF/HGF during migration of limb muscle precursor cells in chicken. Dev Biol 180:566–578

Hirsinger E, Duprez D, Jouve C, Malapert P, Cooke J, Pourquié O (1997) Noggin acts downstream of Wnt and Sonic Hedgehog to antagonize BMP4 in avian somite patterning. Development 124:4605–4614

Huang R, Zhi Q, Izpisua-Belmonte JC, Christ B, Patel K (1999) Origin and development of the avian tongue muscles. Anat Embryol 200:137–52

Itoh N, Mima T, Mikawa T (1996) Loss of fibroblast growth factor receptors is necessary for terminal differentiation of embryonic limb muscle. Development 122:291–300

Jacob M, Christ B, Jacob HJ (1978) On the migration of myogenic stem cells into the prospective wing region of chick embryos. A scanning and transmission electron microscopic study. Anat Embryol 153:179–193

Jacob M, Christ B, Jacob HJ (1979) The migration of myogenic stem cells from the somites into the leg region of avian embryos. Anat Embryol 157:291–309

Jagla K, Dollé P, Mattei M-G, Jagla T, Schuhbaur B, Dretzen G, Bellard F, Bellard M (1995) Mouse *Lbx1* and human *LBX1* define a novel mammalian homeobox gene family related to the *Drosophila* ladybird genes. Mech Dev 53:345–356

Kablar B, Krastel K, Ying C, Asakura A, Tapscott SJ, Rudnicki MA (1997) MyoD and Myf-5 differentially regulate the development of limb versus trunk skeletal muscle. Development 124: 4729–4738

Kahane N, Cinnamon Y, Kalcheim C (1998a) The origin and fate of pioneer myotomal cells in the avian embryo. Mech Dev 74:59–73

Kahane N, Cinnamon Y, Kalcheim C (1998b) The cellular mechanism by which the dermomyotome contributes to the second wave of myotome development. Development 125: 4259–4271

Knudsen KA, McElwee SA, Myers L (1990). A role for the neural cell adhesion molecule NCAM, in myoblast interaction during myogenesis. Dev Biol 138:159–168

Maina F, Casagranda F, Audero E, Simeone A, Comoglio PM, Klein R, Ponzetto C (1996) Uncoupling of Grb2 from the Met receptor in vivo reveals complex roles in muscle development. Cell 87:531–542

Marcelle C, Stark MR, Bronner-Fraser M (1997) Coordinate actions of BMPs, Wnts, Shh and Noggin mediate patterning of the dorsal somite. Development 12:3955–3963

Maroto M, Reshef R, Münsterberg AE, Koester S, Goulding M, Lassar, AB (1997) Ectopic *Pax-3* activates *MyoD* and *Myf-5* expression in embryonic mesoderm and neural tissue. Cell 89: 139–148

McMahon JA, Takada S, Zimmerman LB, Fan CM, Harland RM, McMahon AP (1998) Noggin-mediated antagonism of BMP signaling is required for growth and patterning of the neural tube. Genes Dev 12:1438–1452

Mennerich D, Schäfer K, Braun T (1998) *Pax-3* is necessary but not sufficient for *Lbx1* expression in myogenic precursor cells of the limb. Mech Dev 73:147–158

Münsterberg A, Lassar A (1995) Combinatorial signals from the neural tube, floor plate and notochord induce myogenic bHLH gene expression in the somite. Development 121:651–660

Münsterberg AE, Kitajewski J, Bumcrot DA, McMahon AP, Lassar AB (1995) Combinatorial signaling by Sonic hedgehog and Wnt family members induces myogenic bHLH gene expression in the somite. Genes Dev 9:2911–2922

Myokai F, Washio N, Asahara Y, Yamaai T, Tanda N, Ishikawa T, Aoki S, Kurihara H, Murayama Y, Saito T et al. (1995) Expression of the hepatocyte growth factor gene during chick limb development. Dev Dyn 202:80–90

Neyt C, Jagla, K, Thisse C, Thisse B, Haines L, Currie PD (2000) Evolution of vertebrate appendicular muscle. Nature 408:82–86

Nishi S (1967) Muskeln des Rumpfes. In: Bolk L, Göppert E, Kallius E, Lubosch W (eds) Handbuch der vergleichenden Anatomie der Wirbeltiere, vol 5. A Asher and Co, Amsterdam, pp 341–446

Noden DM (1983) The embryonic origins of avian cephalic and cervical muscles and associated connective tissues. Am J Anat 168:257–276

Nowicki JL, Burke AC (2000) Hox genes and morphological identity: axial versus lateral patterning in the vertebrate mesoderm. Development 127:4265–4275

Ordahl CP, Le Douarin NM (1992) Two myogenic lineages within the developing somite. Development 114:339–353

Pourquié O, Coltey M, Bréant C, Le Douarin NM (1995) Control of somite patterning by signals from the lateral plate. Proc Natl Acad Sci USA 92:3219–3223

Pourquié O, Fan CM, Coltey M, Hirsinger E, Watanabe Y, Bréant C, Francis-West P, Brickell P, Tessier-Lavigne M, Le Douarin NM (1996) Lateral and axial signals involved in avian somite patterning: a role for BMP4. Cell 84:461–471

Pownall ME, Emerson CP (1992) Sequential activation of three myogenic regulatory genes during somite morphogenesis in quail embryos. Dev Biol 151:67–79

Reshef R, Maroto M, Lassar AB (1998) Regulation of dorsal somitic fates: BMPs and Noggin control the timing and pattern of myogenic regulator expression. Genes Dev 12:290–303

Rong PM, Teillet MA, Ziller C, Le Douarin NM (1992) The neural tube/notochord complex is necessary for vertebral but not limb and body wall striated muscle differentiation. Development 115:657–672

Scaal M, Bonafede A, Dathe V, Sachs M, Cann G, Christ B, Brand-Saberi B (1999) SF/HGF is a mediator between limb patterning and muscle development. Development 126:4885–4893

Schäfer K, Braun T (1999) Early specification of limb muscle precursor cells by the homeobox gene Lbx1h. Nat Genet 23:213–216

Schmidt M, Tanaka M, Münsterberg A (2000) Expression of *β-catenin* in the developing chick myotome is regulated by myogenic signals. Development 127:4105–4113

Schubert FR, Mootoosamy RC, Walters EH, Graham A, Tumiotto L, Hünsterberg AE, Lumsden A, Dietrich S (2002) Unit 6 marks sites at epithelial transformations in the chick embryo. Mech Dev, in press

Smith TH, Kachinsky AM, Miller JB (1994) Somite subdomains, muscle cell origins, and the four muscle regulatory factor proteins. J Cell Biol 127:95–105

Sonnenberg E, Meyer D, Weidner KM, Birchmeier C (1993) Scatter factor/hepatocyte growth factor and its receptor, the c-met tyrosine kinase, can mediate a signal exchange between mesenchyme and epithelia during mouse development. J Cell Biol 123:223–235

Stern HM, Brown AM, Hauschka SD (1995) Myogenesis in paraxial mesoderm: preferential induction by dorsal neural tube and by cells expressing Wnt-1. Development 121:3675–3686

Tajbakhsh S, Rocancourt D, Cossu G, Buckingham M (1997) Redefining the genetic hierarchies controlling skeletal myogenesis: Pax3 and Myf5 act upstream of MyoD. Cell 89:127–138

Tajbakhsh S, Borello U, Vivarelli E, Kelly R, Papkoff J, Duprez D, Buckingham M, Cossu G (1998) Differential activation of Myf5 and MyoD by different Wnts in explants of mouse paraxial mesoderm and the later activation of myogenesis in the absence of Myf5. Development 125:4155–4162

Thery C, Sharpe MJ, Batley SJ, Stern CD, Gherardi E (1995) Expression of HGF/SF, HGF1/MSP, and c-met suggests new functions during early chick development. Dev Genet 17:90–101

Tremblay P, Dietrich S, Stoykova A, Stuart ET, Gruss P (1995) Pax genes as pleiotropic regulators of embryonic development. In: Juurlink BHJ, Krone PH, Kulyk WM, Verge VMK, Doucette JR (eds) Neural cell specification: molecular mechanisms and neurotherapeutic implications. Plenum Press, New York

Tremblay P, Dietrich S, Meriskay M, Schubert FR, Li Z, Paulin D (1998) A crucial role for *Pax3* in the development of the hypaxial musculature and the long-range migration of muscle precursors. Dev Biol 203:49–61

Venters SJ, Thorsteinsdottir S, Duxon MJ (1999) Early development of the myotome in the mouse. Dev Dyn 216:219–232

Webb SE, Lee KK, Tang MK, Ede DA (1997) Fibroblast growth factors 2 and 4 stimulate migration of mouse embryonic limb myogenic cells. Dev Dyn 209:206–216

Williams BA, Ordahl CP (1994) *Pax-3* expression in segmental mesoderm marks early stages in myogenic cell specification. Development 12:785–796

Yang XM, Vogan K, Gros P, Park M (1996) Expression of the met receptor tyrosine kinase in muscle progenitor cells in somites and limbs is absent in Splotch mice. Development 122:2163–2171

Young JZ (1962) The life of vertebrates. Oxford University Press, Oxford

Inhibition of Skeletal Muscle Development: Less Differentiation Gives More Muscle

Ernst-Martin Füchtbauer[1]

1 Introduction

Differentiation of cells during organogenesis or regeneration is regulated by the complex interaction of signals conducting the differential activity of transcription factors. Proliferation and differentiation of cells are in many cases mutually exclusive and often seen as two opposing modes. This is particularly obvious in the development of skeletal muscle, in which myoblasts have to leave the cell cycle before they fully differentiate and fuse to become myotubes (Olson 1992). The decision of a cell to either divide or differentiate is, therefore, often the result of a competition between growth promoting factors and tissue-specific differentiation factors. However, whether a factor promotes cell proliferation or differentiation depends on the type and status of the cell that receives the signal. Differentiation of cells, e.g. fusion of myoblasts into myotubes, is often viewed as a cell's final goal, but reaching this goal has its price. To differentiate, i.e. to cease proliferation, is a dramatic step for a cell because no daughter cells will be available to participate in the future life of the organism. For the organism as a whole, namely for an embryo, it is therefore of great importance not only to control proliferation in order to avoid neoplastic growth, but also to fine tune differentiation in order to avoid a lack of stem cells that are needed for further development, growth and regeneration. For several reasons, inhibition of premature or ectopic differentiation seems to be especially important during myogenesis: Firstly, in the vertebrate somite the first muscle differentiation starts before the limbs even develop, yet undifferentiated cells from the somites have to migrate into the limbs and to the body wall in order to form the major muscles of the body. Secondly, muscle development proceeds in waves, where embryonic muscle fibres are followed by foetal muscle fibres, which precede neonatal muscle. Each new wave of myogenesis needs undifferentiated stem cells, which must have escaped differentiation because the pool of myogenic cells is determined early on. Thirdly, muscle has a very high regeneration capacity, which is not only needed to heal injury, but also for repair of the daily wear and tear. To fulfil these different demands for undifferentiated cells, signals are required that protect them from

[1] Institute of Molecular and Structural Biology, Aarhus University, C. F. Møllers Allé, Bygn.130, Århus C, Denmark

Results and Problems in Cell Differentiation, Vol. 38
B. Brand-Saberi (Ed.): Vertebrate Myogenesis

premature differentiation. These signals are produced either in tissues that are adjacent to myogenic cells (as the lateral plate mesoderm is adjacent to the somites) or within the population of differentiating cells. In the latter case, signalling cells might themselves become insensitive to the inhibitory signal, but protect other cells of the pool from differentiating.

In this chapter I will try to outline the importance of inhibitory factors for the proper development of skeletal muscle. Most of the results reviewed here were obtained from tissue culture and in vitro experiments. It has been argued that the various conditions under which these experiments are performed just create various tissue culture artefacts that are irrelevant to normal development and are sometimes in conflict with each other. Although this cannot be ruled out completely, it seems more likely that the results reflect the different aspects of the complex situation in the embryo, where multiple populations of myogenic cells divide, migrate and differentiate in various environments. Whenever possible, I will correlate the data to the situation in the developing embryo. Many of the data originate from "gain of function" experiments, i.e. overexpression of genes or addition of factors to cells or embryos. Although these data are very informative, one would like to see the "loss of function" experiment (e.g. mutation analysis) before a biological function is assumed. However, for many of the genes mentioned, mutations do not lead to a phenotype in muscle development. This is sometimes because of early embryonic lethality, in which case myogenesis cannot be studied. Alternatively, the factor in question might be embedded into a regulatory network in which a single mutation does not result in an observable phenotype.

Differentiation of skeletal muscle is dependent on the expression of one or several members of a group of muscle-specific transcription factors called the MyoD family. MyoD, myogenin, myf-5 and MRF4/herculin/myf-6 are basic helix–loop–helix (bHLH) transcription factors that are all able to induce muscle-specific gene expression in a wide variety of cells from all three germ layers (Davis et al. 1987; Braun et al. 1989, 1990; Edmondson 1989; Rhodes and Konieczny 1989; Wright et al. 1989; Choi et al. 1990; Miner and Wold 1990; Schäfer et al. 1990). Members of the MyoD family act synergistically with MEF2 proteins that belong to the MADS box-containing family of transcription factors (Molkentin et al. 1995). Expression of MEF2 genes is not restricted to skeletal muscle, but, in concert with the muscle-specific bHLH factors, they control muscle-specific gene expression. Members of the MyoD family and MEF2 interact directly by their DNA-binding domains (Black et al. 1998). The knowledge of these factors and the easy accessibility of muscle development in the embryo, in tissue culture, as well as during regeneration in the adult organism, has made myogenesis an intensely studied "prototype" of cellular differentiation. My personal interest in the inhibition of myogenesis originates from our observation that the bHLH transcription factor Twist prevents muscle differentiation in vertebrate cells even though evolutionary arguments might have suggested the opposite function.

2 Secreted Signalling Molecules

The importance of humoral factors to prevent muscle differentiation has been known since the first successful attempts to grow muscle precursor cells in culture. High concentrations of mitogen-rich serum can inhibit the differentiation of myoblasts, whereas removal or reduction of serum induces differentiation (Yaffe and Saxel 1977). This can be explained by the action of basic fibroblast growth factor (FGF2) and transforming growth factor β (TGF-β) because they are the most potent inhibitors of myogenesis known (Olson 1992). FGF and TGF-β can suppress the expression of myogenic bHLH transcription factors (Vaidya et al. 1989), but also silence their transcriptional activity (Davis et al. 1987; Braun et al. 1989).

2.1 Fibroblast Growth Factor Family

The differentiation inhibitory function of FGF2 is not identical to its growth-stimulating activity. Even under conditions where FGF2 is not mitogenic, it can repress muscle-specific gene activation (Schubert et al. 1974; Clegg et al. 1987).

Within the FGF family, a good candidate to negatively control MyoD expression in the embryo is FGF4, which is expressed in the somite a day earlier than MyoD. During a period of co-expression, both genes are found in a salt and pepper pattern, and only in areas without FGF4 is MyoD transcribed at high levels. In the developing limb bud FGF4 is expressed primarily in the ectoderm and not in the mesenchyme (Niswander and Martin 1992), except for some specific muscle anlagen (Drucker and Goldfarb 1993). Another member of the FGF family expressed predominantly in the muscle lineage is FGF6. FGF6 is required for the activation of satellite cells during muscle regeneration (Floss et al. 1997) and seems more important for the regulation of differentiation than of proliferation (Sakuma et al. 2000). Its function might be to keep some cells undifferentiated and proliferating. In the embryo, an additional function of FGF6 could be to signal to neighbouring cells such as the sclerotome cells or connective tissue and prevent them joining into muscle differentiation (Han and Martin 1993). The original hypothesis (Grass et al. 1996) stating that the FGF signal from the early myotome is required to form the distal part of the ribs seems not to hold true because new mutant alleles of myf5 result in a loss of the early myotomes without alteration of rib development (Kaul et al. 2000).

The FGF receptors (FGFR1–4) are transmembrane proteins containing a tyrosine kinase domain on their cytoplasmic side, which, upon binding of the ligand, autophosphorylates the protein. The following signal transduction cascade activates several kinases of the Ras, MAP, ERK pathway (Pena et al. 2000). Expression data suggest that FGFR1 and FGFR4 are involved in inhibition of muscle differentiation in-situ (Stark et al. 1991; Peters et al. 1992). Inactivation of FGFR1 in chicken embryos by expression of a truncated dominant negative protein dramatically reduced the muscle mass (Flanagan-Steet et al. 2000). This indicates that in the embryo FGF signalling is indeed required to protect myogenic cells of from premature differentiation. If this protective signal is blocked, the cells become post-mitotic, differentiate and are finally missing. Because of its persistent expression in satellite cells and its downregulation during differentiation, the so called fibroblast growth factor receptor-like embryonic kinase (FREK) is also a good candidate as an inhibitor of terminal differentiation (Halevy et al. 1994).

The binding of FGFs to FGFRs is strongly regulated by proteoglycans, e.g. syndecans, which are necessary to present the ligands to the receptor (Yayon et al. 1991). Syndecan 3 is downregulated during differentiation (Larrain et al. 1997), and expression of a syndecan 3 antisense construct accelerates differentiation in C_2C_{12} cells (Fuentealba et al. 1999). Accordingly, the forced expression of syndecan 1 inhibits differentiation (Larrain et al. 1998). The dual system of high and low affinity receptors might modulate the cellular response to different levels of FGF in a locally controlled manner (Klagsbrun and Baird 1991).

FGF signalling induces the phosphorylation of a Thr conserved in the DNA-binding domain of all myogenic bHLH proteins. If this Thr is phosphorylated in myogenin, the protein looses its ability to bind DNA, and a myogenin mutant protein lacking this site is resistant to the inactivation by FGF (Li et al. 1992). Phosphorylation of this Thr is probably caused by activation of protein kinase C (PKC; Tsuda et al. 1985). However, there are also other mechanisms by which activated PKC can inhibit muscle specific gene expression. Mutation of the conserved Thr and other phosphorylated amino acids in MRF4 does not make this protein less sensitive to inhibition by PKC and PKA. In this case, modification of co-regulators is a more likely explanation (Hardy et al. 1993).

2.2 Transforming Growth Factor β Superfamily

The other major components of serum-inhibiting myogenesis are members of the TGF-β family. In many reports on the function of TGF-βs the different members were not distinguished, and I will use the generalised form TGF-β. This should not, however, imply that all members of the family have necessarily the same effect. All members of the TGF-β superfamily are synthesised as a pre-pro-proteins, which, upon proteolytic cleavage, will form active homo- or heterodimers.

TGF-β reversibly inhibits expression of muscle specific genes and fusion in myoblasts. It has no effect on DNA binding and heterodimerisation of myogenic bHLH transcription factors with other bHLH proteins, yet the transcriptional activation potential is silenced (Brennan et al. 1991). This indicates that an essential co-factor of myogenic bHLH transcription factors is the target for TGF-β. MEF2 is a good candidate to be that cofactor, because TGF-β specifically prevents it from being imported into the nucleus. This retention to the cytoplasm requires the MEF2 nuclear localisation signal. If this signal is replaced by the one from SV40, MEF2 becomes resistant to TGF-β and is localised in the nucleus even in the presence of TGF-β (De Angelis et al. 1998). If indeed MEF2 is a main target for TGF-β-induced inhibition of differentiation, it explains, why the bHLH region of myogenic factors is required for this silencing (Martin et al. 1992), because that is the domain interacting with MEF2.

Immunohistochemical and mRNA analysis show that TGF-β3, which is differentially expressed in several developing and mature mouse tissues, is the predominantly expressed isoform in developing muscle (Pelton et al. 1991). After fusion of C_2C_{12} cells TGF-β3 expression is increased eight-fold (Lafyatis et al. 1991). The TGF-β3-producing cells are resistant to their own signal because TGF-β receptors are downregulated during normal differentiation. Notably, fusion is required for this down-regulation of TGF-β receptors (Hu and Olson 1990). To keep a pool of undifferentiated stem cells, therefore, is a probable function of the increased TGF-β3.

Embryonic myoblasts are, in contrast to foetal myoblasts, resistant to TGF-β. If mixed primary cultures are treated with TGF-β, only primary (embryonic) muscle fibres appear whereas if the same type of culture is treated with TGF-β-neutralising antibodies, secondary (foetal) myotubes form earlier than in untreated control cultures (Cusella-De Angelis et al. 1994). In the developing embryo this might be one of the mechanisms, by which secondary myoblasts are protected from premature differentiation. Interestingly, the inhibitory action of TGF-β on myoblasts requires cell–cell contact, which is the normal situation in the developing embryo, and does not work on single cells that differentiate even in the presence of TGF-β (De Angelis et al. 1998).

Even though TGF-β is normally a very potent inhibitor of muscle differentiation, its effects can be pleiotropic. C_2C_{12} cells, as well as human rhabdomyosarcoma cells, are inhibited from differentiation by TGF-β at high concentrations, yet they require low levels of TGF-β in order to differentiate and will not do so in TGF-β-depleted medium (Filvaroff et al. 1994; Bouche et al. 2000). In L6E9 rat myoblasts, TGF-β can induce upregulation of myogenin and differentiation, together with a downregulation of c-myc and Id, which both are inhibitors of myogenic differentiation. This effect is only observed in a mitogen-rich medium under low cell density conditions. At least the high serum condition reflects more truly the situation in the developing embryo than does the low mitogen medium normally used to induce differentiation in tissue culture. Under the described conditions, TGF-β apparently prolongs the G1 phase and finally removes the cells from the cell cycle (Zentella and Massagué 1992).

Independent of its pleiotropic effects, TGF-β rapidly upregulates the cellular proto-oncogenes junB and c-jun (Pertovaara et al. 1989; Li et al. 1990). However, the timing and relative upregulation of the two *jun* genes differ and may explain the differences in TGF-β activity. An other explanation why TGF-β has different effects under different conditions could be its action on MEF2, which itself can act differently depending on the co-factors available in a cell. MEF2 not only activates tissue-specific genes, but also stress- and growth-induced genes, which one would expect to prevent rather than to induce differentiation (Naya and Olson 1999). Interference with MEF2 might, therefore, have variable effects on the cells.

Despite the strong effects on muscle differentiation in "gain of function" experiments, none of the TGF-βs mutated so far has a unique function in myogenesis because none of the mutants showed an overt muscle phenotype. Whether any of the mutant mice differ from the wild type with respect to satellite cell numbers or regeneration capacity remains to be seen (Shull et al. 1992; Kaatinen et al. 1995; Proetzel et al. 1995).

The TGF-β superfamily consist of many more members than just TGF-βs. Quite a number of these affect myogenesis. Among them, bone-morphogenic proteins (BMPs) are perhaps the most important. Most BMPs will inhibit muscle differentiation, but some, such as BMP2, are even able to change the differentiation lineage of myoblasts into that of osteoblasts (Katagiri et al. 1994). In the embryo it is, in particular, BMP4 secreted by the lateral plate mesoderm that prevents the premature differentiation of myogenic cells in the somite. If the somite is shielded against this signal, cells, which normally would have migrated into the limb, now differentiate prematurely in the somite (Pourquié et al. 1996). Under normal conditions, only the dorsomedial dermomyotome cells differentiate in the somite to form the myotome. These cells can do so because they express the BMP-binding protein noggin, which locally inactivates the BMP4 signal. Expression of Noggin in somitic cells is induced by Wnt1 (Reshef et al. 1998), which is secreted by the dorsal neural tube (Marcelle et al. 1997, and reviewed in Cossu and Borello 1999).

Probably the only TGF-β-related factor with an exclusive function in skeletal muscle is myostatin (also called GDF-8), which is involved in homeostasis of muscle. Mutation of myostatin in mice results in a significant increase of muscle mass (McPherron et al. 1997). In addition, it has been shown that the double muscle phenotype in cattle is caused by mutations in the myostatin gene (Grobet et al. 1997; Kambadur et al. 1997; McPherron and Lee 1997). The increase in muscle mass seen in myostatin mutant mice is the result of two processes: hyperplasia and hypertrophy. Most probably, myostatin controls hyperplasia during development whereas hypertyrophy is controlled in the maintenance stage of muscle. Expression of myostatin in C_2C_{12} cells indicate that myostatin does not directly inhibit myogenic differentiation, but that it prevents proliferation and, therefore, reduces the number of myogenic precursor cells (Thomas et al. 2000; Rios et al. 2001).

Other members of the TGF-β superfamily, such as activins, can also inhibit muscle differentiation, but their function during embryogenesis is less clear.

TGF-β signals in general are modulated by proteoglycans in the extracellular matrix. For example, removal of decorin from the extracellular matrix of C_2C_{12} cells will reduce the "visibility" of TGF-β to the cells and consequently accelerate differentiation (Riquelme et al. 2000).

3 Extracellular Matrix

With the two most important groups of differentiation inhibiting signals (FGFs and TGF-βs) that are modulated by proteoglycans, it is not surprising that the general composition of the extracellular matrix has a strong influence on myogenic differentiation. Two inhibitors of collagen synthesis (*cis*-4-hydroxyl-L-proline and ethyl-3,4-dihydroxybenzoate) have been shown to inhibit myogenesis in C_2C_{12} cells as well as in myogenin-transfected 10T1/2 cells (Saitoh et al. 1992). It is very likely that the composition of the extracellular matrix, rather than the amount of collagen, is important because type I collagen, which is induced by TGF-β1, inhibits differentiation of L6 myoblasts (Heino and Massagué 1990). In the embryo, the extracellular matrix can present, block, transport or trap the signalling molecules. Therefore, it is part of the complex regulatory system that makes the decision between proliferation, migration and differentiation.

4 Transcription Factors

4.1 Notch

Some signals are not secreted molecules, but are presented on the surface of neighbouring cells. Several differentiation pathways regulated by bHLH transcription factors are inhibited by Notch signalling. Notch is a transmembrane receptor for signals such as Delta or Serrate/Jagged (Bettenhausen et al. 1995; Lindsell et al. 1995), which are expressed on the plasma membrane of neighbouring cells. Upon ligand binding, the intracellular domain of Notch is cleaved off and transported to the nucleus, where it acts as a DNA-binding protein, presumably as a transcription factor (Struhl and Greenwald 2001). Expression of the intracellular domain of Notch thus simulates constitutively activated Notch signalling. The best studied systems of Notch function are the nervous system and the developing pancreas, in which activated Notch keeps cells undifferentiated and where it is also involved in lineage determination mediated by lateral inhibition (Jensen et al. 2000; Kageyama et al. 2000). In these systems, activated Notch induces translation of HES proteins, which are bHLH proteins related to the *Drosophila* Hairy and Enhancer-of-split. HES proteins are direct repressors of tissue-specific bHLH transcription factors such as the neurogenic factors MASH1 and MASH2 (Guillemot 1995). Expression of the constitutively active intracellular domain of Notch in C_2C_{12} cells inhibits differentiation (Kopan et al. 1994; Nye et al. 1994). In contrast to inhibition of neuronal development where Notch activates the inhibitor HES, Notch seems to act directly on muscle-specific transcription factors. It binds to MEF2 and prevents it from DNA binding and co-operation with MyoD (Wilson-Rawls et al. 1999). In *Drosophila*, Notch is also able to inhibit muscle differentiation; however, not all developing muscles are sensitive. The indirect flight muscle will not form under continued expression of activated Notch, whereas differentiation of the direct flight muscle is not impaired (Anant et al. 1998). Inhibition of indirect flight muscle differentiation is mediated by the bHLH transcription factor Twist, which might play a similar role as HES in the vertebrate nervous system. However, the role of *Drosophila* Twist in myogenic differentiation is not as clear and possibly pleiotropic. In addition to its inhibition of indirect flight muscle, *Drosophila* Twist is also able to induce ectopic muscle in ectodermal cells (Baylies and Bate 1996) and it induces the MADS box-containing transcription factor D-Mef2 during normal development (Lilly et al. 1994; Nguyen et al. 1994; Taylor et al. 1995). This activation of D-Mef2 seems remarkable because the mammalian homologue MEF2 is inactivated by Notch in myogenic cells. MEF2, on the other hand, as I have discussed in the context of the pleiotropic effects of TGF-β, might have variable or even opposing activities itself.

In the vertebrate embryo, a possible function of Notch-mediated inhibition is to prevent paraxial mesoderm cells from premature differentiation. These cells express myf5 at a very low level and Notch might be involved in repressing an autoregulatory loop or the activation of MyoD (Kopan et al. 1994; Hirsinger et al. 2001). A likely candidate for a ligand involved in inhibition of myogenesis is the murine delta-like gene 1 (Dll1), which is expressed in the presomitic paraxial mesoderm and in early somites (Bettenhausen et al. 1995).

4.2 Twist

Twist is a potent inhibitor of muscle development in murine cells. However, it is not known whether Notch induces Twist in murine myogenic cells and whether Twist is involved in Notch signalling in a similar way as in *Drosophila* indirect flight muscle. Investigating the expression of the murine Twist gene during embryonic development, we observed widespread expression in mesenchymal mesoderm cells, including cephalic neural crest cells that undergo mesoderm like differentiation. The only epithelial cells expressing Twist are the epithelial cells of the somite, which include the muscle-precursor cells. During development of these cells, we observed a striking mutual exclusion between the expression of Twist and of the early myogenic transcription factors myf-5 and MyoD1 (Fig. 1; Hebrok et al. 1994; Füchtbauer 1995). We also noted the same mutual exclusion in tissue culture cells, where Twist is expressed in a wide variety of mesodermal cell lines, but not found in myogenic cells such as C_2C_{12} and, therefore, reasoned that Twist expression might be incompatible with myogenic differentiation and thus function as an inhibitor. To test this hypothesis we stably transfected tissue culture cells and indeed found that Twist would prevent or delay muscle differentiation in myogenic C_2C_{12} cells as well as in differentiating embryonic stem (ES) cells (Hebrok et al. 1994; Rohwedel et al. 1995). This effect was dosage-dependent and reversible by means of antisense oligonucleotides (Figs. 2 and 3; Hebrok et al.

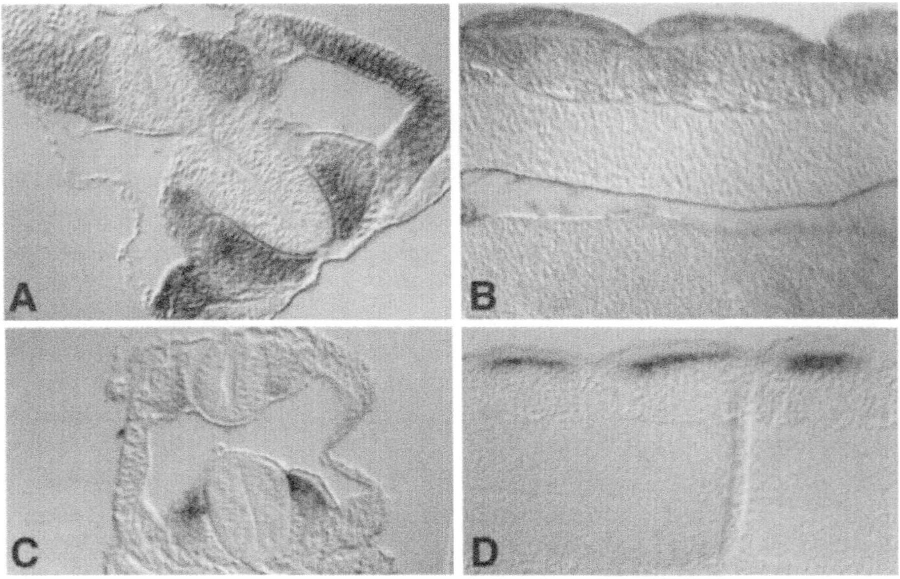

Fig. 1. Comparison of Twist (**A**, **B**) and myf5 (**C**, **D**) expression in early murine somites. **A** and **C** Transverse sections through a 8.5- and 8-day-old embryo respectively. The caudal part of each embryo is towards the top. While Twist is expressed in the entire young (caudal) somites, it becomes restricted to the sclerotomal part of the somites in the maturing (more rostral) somites. Myf5 is expressed only in the dorsomedial part of the somites in which Twist expression has disappeared. **B** and **D** Tangential section through trunk somites of 10.5-day-old embryos. Myf5 expression is confined to the myotome. In contrast, Twist is expressed in both the sclerotome and the dermomyotome, but absent from the myotome. (Modified from Füchtbauer 1995)

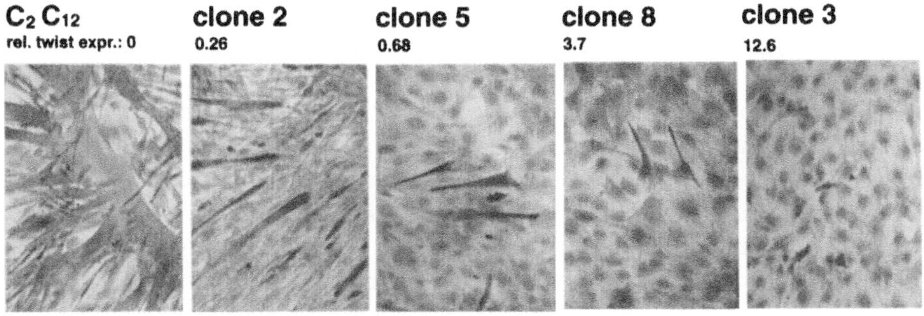

Fig. 2. Myogenic differentiation of C_2C_{12} cells and four clones stably transfected with a Twist expression vector. The relative level of Twist expression, as revealed by RNase protection analysis, is indicated in arbitrary units. C_2C_{12} cells differentiate readily under low serum condition, whereas cells expressing increasing amounts of Twist are increasingly impaired in their differentiation. Note that no increased cell proliferation is observed. Cells were immunostained for skeletal muscle myosin and counterstained with haematoxylin (for experimental details see Hebrok et al. 1994)

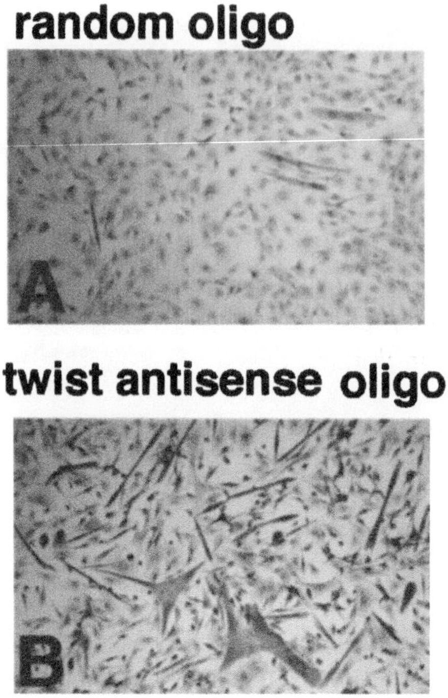

Fig. 3. The inhibitory effect of Twist is reversible. **A** Twist-expressing cells differentiate poorly in the presence of a randomised oligonucleotide or with out any oligonucleotide added (not shown). **B** The presence of Twist antisense oligonucleotides dramatically accelerates the differentiation into myotubes Cells were immunostained for skeletal muscle myosin and counterstained with haematoxylin (for experimental details see Hebrok et al. 1994)

1997). Because the transcription of very early muscle specific genes (i.e. the myogenic bHLH transcription factors) is downregulated in Twist-expressing cells (Fig. 4), we reasoned that Twist directly interferes with muscle-specific gene expression. Myogenic bHLH transcription factors normally require heterodimerisation with more widely expressed activators, e.g. E-proteins. Dominant negative HLH proteins lacking the basic DNA binding domain (e.g. members of the ID family) can sequester the activators and in this way inhibit the transactivation activity of myf-5 or MyoD1 (Benezra et al. 1990). As Twist also belongs to the bHLH family, a simple explanation for its inhibitory activity could be the sequestration of E-proteins. By simultaneously overexpressing E12 and Twist, we could show that this is not the case. In contrast, several domains of the Twist protein participate in different molecular mechanisms by which Twist inhibits muscle-specific gene expression (Spicer et al. 1996; Hebrok et al. 1997). The basic domain, which represents the DNA-binding domain and contains a nuclear localisation signal, plays an important role in the inhibitory function of Twist. This might be partially caused by competition for E-boxes, the consensus DNA sequence to which many bHLH proteins bind. Band shift experiments show, that a fourfold excess of Twist is able to prevent MyoD1 from binding to the E-box (Hebrok et al. 1997). However, more important for the inhibitory function of the basic domain of Twist is its ability to bind directly to the basic domain of MyoD1. This is specific as Twist does not interact with E-proteins in this way, nor does a Twist molecule in which

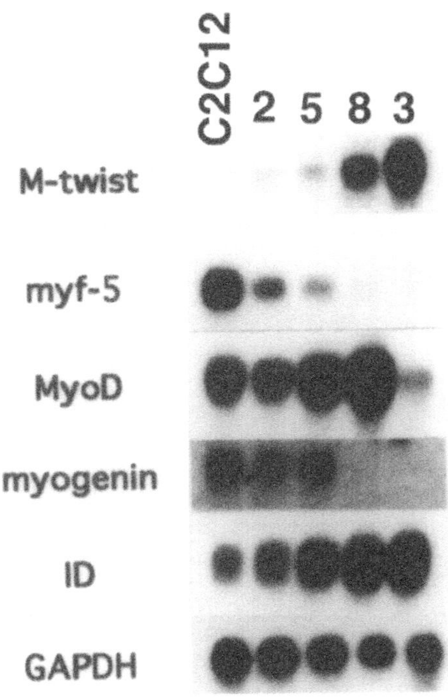

Fig. 4. Northern blot analysis of gene expression in Twist-expressing C_2C_{12} cell clones. The myogenic bHLH transcription factors myf5, MyoD and myogenin are downregulated in Twist-expressing cells. The fact that MyoD expression is upregulated in clone 9-5 and 9-8 is probably explained by the relief from a myf5-mediated repression. The differentiation inhibitor ID is slightly upregulated. Glyceraldehyde-phosphate dehydrogenase (GAPDH) was used as a control for equal loading of the gel (for further discussion of the results see Hebrok et al. 1994)

the basic region is replaced by that of E 12 bind to MyoD1. It appears that Twist, by physical interaction with the DNA-binding domain of MyoD, can inhibit muscle-specific gene expression (Hamamori et al. 1997). The carboxy terminus of Twist is highly conserved among all members of the Twist family, which indicates a functional conserved domain (Scaal et al. 2001). This region can bind to MEF2 and contributes significantly to the inhibitory activity of Twist (Spicer et al. 1996; Hebrok et al. 1997). The inhibitory activity of the carboxy terminus can be neutralised by overexpression of MEF2 indicating that it inhibits by the sequestration of MEF2 factors (unpubl. data). As Twist was shown to inhibit two independent histone acetyl transferases (p300 and p300/CBP-associated factor), it was suggested to be a general inhibitor of transcription in mammalian cells (Hamamori et al. 1999). However, this is unlikely because we have found Twist able to activate a reporter driven by the *Drosophila* tinman enhancer (Fig. 5; Yin et al. 1997), and because the human Gli1 gene has been identified as a direct downstream target gene of Twist (Villavicencio et al.).

4.3 Id

The heterodimerisation of basic helix-loop-helix factors is a prime target for the regulation of myogenesis. Id genes, which code for HLH proteins lacking

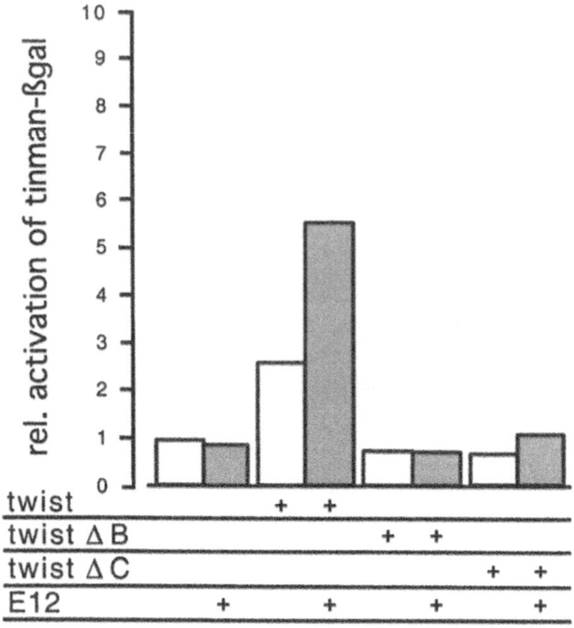

Fig. 5. Twist can act as a transcriptional transactivator. When cotransfected with a β-galactosidase reporter controlled by an enhancer sequence from the *Drosophila* tinman gene (tinman-βgal), Twist activates the reporter about three times above the background level. This activation is synergistically enhanced by the presence of E12. As revealed by the use of deletion mutants, both the basic DNA binding motive (Twist ΔB) and the carboxy-terminus (Twist ΔC) are required for the transactivating activity of Twist. If the E-boxes in the reporter construct are mutated, no transactivation is observed (not shown)

the basic DNA-binding domain, can, by dimerisation with bHLH factors, inhibit the DNA binding of the heterodimers. Id stands originally for "inhibitor of DNA-binding", but is sometimes also referred to as "inhibitor of differentiation". In the mouse, four members of the Id gene family have been described: Id1, Id2, Id3 and Id4 (Benezra et al. 1990; Christy et al. 1991; Sun et al. 1991; Riechmann et al. 1994). In vivo, the most likely binding partners for Id are the E-proteins, which are the gene products of different splice variants of the E2A gene. Stable overexpression of Id1 in C_2C_{12} cells inhibits differentiation. Id and the E-proteins cofractionate and co-immunoprecipitate in whole cell extracts prepared from myoblasts. Accordingly, the loss of Id protein observed during differentiation correlates with the formation of MyoD/E-protein heterocomplexes in whole cell extracts (Jen et al. 1992).

In the mouse embryo, Id1 is expressed upon gastrulation in almost all regions of the embryo, and expression declines as embryogenesis proceeds. In skeletal muscle, Id and the myogenic bHLH factors are expressed in a mutually exclusive manner (Wang et al. 1992). This suggests that myogenic precursor cells do not express both types of factors simultaneously at high levels.

Therefore, it has been questioned whether Id is expressed at high enough levels to consume all available dimerisation partners for MyoD and whether Id alone is sufficient to inhibit differentiation. However, expression of Ids in C_2C_{12} cells, rhabdomyosarcoma cell lines and primary tumour samples suggests a functional role of Id in myogenesis.

4.4 Msx1

If the homeobox containing gene Msx1 is expressed in 10T1/2-derived F3 myoblasts, the cells downregulate MyoD and do not differentiate. In addition, Msx1-expressing cells acquire a transformed phenotype in culture as well as in mice (Song et al. 1992). Msx1 is even able to reactivate cell proliferation in terminally differentiated myotubes derived from C_2C_{12} cells. A transient expression of Msx1 can dedifferentiate cells derived from myotube nuclei into an "embryonic fibroblast-like" stage, from which the cells can differentiate not only into muscle but also into fat, cartilage or bone cells (Odelberg et al. 2000). Msx1 is expressed throughout the limb bud at the time when the outgrowth of the limb begins. Later when the first muscle differentiation starts in the proximal portion of the limb, Msx1 expression becomes restricted to the proliferating distal part. The expression is induced by the apical ectodermal ridge, but can also be evoked by BMP4, FGF2 or FGF4 (Hill et al. 1989; Robert et al. 1989; Wang and Sassoon 1995). Using lacZ integration into the Msx1 locus, expression was also found in the dermomyotome of brachial and thoracic somites as well as in migrating myogenic cells, where it can antagonize the myogenic activity of Pax3 (Bendall et al. 1999; Houzelstein et al. 1999). Therefore, it is very likely that Msx1 has a function in the protection from premature differentiation. Additionally, Msx1 is expressed in early precursor cells of the dorsal dermis. As these cells originate from the dermomyotome, Msx1 might be involved in protecting the future dermis cells from myogenic differentiation.

5 Summary

The fact that stem cells have to be protected from premature differentiation is true for many organs in the developing embryo and the adult organism. However, there are several arguments that this is particularly important for (skeletal) muscle. There are some evolutionary arguments that muscle is a "default" pathway for mesodermal cells, which has to be actively prevented in order to allow cells to differentiate into other tissues. Myogenic cells originate from very small areas of the embryo where only a minor portion of these cells is supposed to differentiate. Differentiated muscle fibres are unconditionally post-mitotic, leaving undifferentiated stem cells as the only source of regeneration. The mechanical usage of muscle and its superficial location in the vertebrate body makes regeneration a frequently used mechanism. Looking at the different inhibitory mechanisms that have been found within the past 10 or so years, it appears as if evolution has taken this issue very serious. At all possible levels we find regulatory mechanisms that help to fine tune the differentiation of myogenic cells. Secreted molecules specifying different populations of somitic cells, diffusing or membrane-bound signals among fellow myoblasts, modulating molecules within the extracellular matrix and last, but not least, a changing set of activating and repressing cofactors. We have come a long way from the simple model of MyoD just to be turned on at the right time in the right cell.

References

Anant S, Roy S, VijayRaghavan K (1998) Twist and Notch negatively regulate adult muscle differentiation in *Drosophila*. Development 125:1361–1369

Baylies MK, Bate M (1996) Twist: a myogenic switch in *Drosophila*. Science 272:1481–1484

Bendall AJ, Ding J, Hu G, Shen MM, Abate-Shen C (1999) Msx1 antagonizes the myogenic activity of Pax3 in migrating limb muscle precursors. Development 126:4965–4976

Benezra R, Davis RL, Lockshon D, Turner DL, Weintraub H (1990) The protein Id: a negative regulator of helix-loop-helix DNA binding proteins. Cell 61:49–59

Bettenhausen B, de Angelis MH, Simon D, Guénet J-L, Gossler A (1995) Transient and restricted expression during mouse embryogenesis of Dll1, a murine gene closely related to *Drosophila* Delta. Development 121:2407–2418

Black BL, Molkentin JD, Olson EN (1998) Multiple roles for the MyoD basic region in transmission of transcriptional activation signals and interaction with MEF2. Mol Cell Biol 18:69–77

Bouche M, Canipari R, Melchionna R, Willems D, Senni MI, Molinaro M (2000) TGF-beta autocrine loop regulates cell growth and myogenic differentiation in human rhabdomyosarcoma cells. FASEB J 14:1147–1158

Braun T, Buschhausen-Denker G, Bober E, Tannich E, Arnold HH (1989) A novel human muscle factor related to but distinct from MyoD1 induces myogenic conversion in 10T1/2 fibroblasts. EMBO J 8:701–709

Braun T, Bober E, Winter E, Rosenthal N, Arnold HH (1990) Myf-6, a new member of the human gene family of myogenic determination factors: evidence for a gene cluster on chromosome 12. EMBO J 9:821–831

Brennan TJ, Edmondson DG, Li L, Olson EN (1991) Transforming growth factor β represses the actions of myogenin through a mechanism independent of DNA binding. Proc Natl Acad Sci USA 88:3822–3826

Choi J, Costa ML, Mermelstein CS, Chagas C, Holtzer S, Holtzer H (1990) MyoD converts primary dermal fibroblasts, chondroblasts, smooth muscle, and retinal pigmented epithelial cells into striated mononucleated myoblasts and multinucleated myotubes. Proc Natl Acad Sci USA 87:7988–7992

Christy BA, Sanders L, Lau LF, Copeland NG, Jenkins NA, Nathans D (1991) An Id-related helix-loop-helix protein encoded by a growth factor-inducible gene. Proc Natl Acad Sci USA 88: 1815–1819

Clegg CH, Linkhart TA, Olwin BB, Hauschka SD (1987) Growth factor control of skeletal muscle differentiation: commitment to terminal differentiation occurs in G1 Phase and is repressed by fibroblast growth factor. J Cell Biol 105:949–956

Cossu G, Borello U (1999) Wnt signaling and the activation of myogenesis in mammals. EMBO J 18:6867–6872

Cusella-De Angelis MG, Molinari S, Le Donne A, Coletta M, Vivarelli E, Bouche M, Molinaro M, Ferrari S, Cossu G (1994) Differential response of embryonic and fetal myoblasts to

TGF-β: a possible regulatory mechanism of skeletal muscle histogenesis. Development 120: 925–933
Davis RL, Weintraub H, Lassar AB (1987) Expression of a single transfected cDNA converts fibroblasts to myoblasts. Cell 51:987–1000
De Angelis L, Borghi S, Melchionna R, Berghella L, Baccarani-Contri M, Parise F, Ferrari S, Cossu G (1998) Inhibition of myogenesis by transforming growth factor beta is density-dependent and related to the translocation of transcription factor MEF2 to the cytoplasm. Proc Natl Acad Sci USA 95:12358–12363
Drucker BJ, Goldfarb M (1993) Murine FGF-4 gene expression is spatially restricted within embryonic skeletal muscle and other tissues. Mech Dev 40:155–163
Edmondson DG (1989) A gene with homology to the *myc* similarity region of *MyoD1* is sufficient to activate the muscle differentiation program. Genes Dev 3:628–640
Filvaroff EH, Ebner R, Derynck R (1994) Inhibition of myogenic differentiation in myoblasts expressing a truncated type II TGF-β receptor. Development 120:1085–1095
Flanagan-Steet H, Hannon K, McAvoy MJ, Hullinger R, Olwin BB (2000) Loss of FGF receptor 1 signaling reduces skeletal muscle mass and disrupts myofiber organization in the developing limb. Dev Biol 218:21–37
Floss T, Arnold HH, Braun T (1997) A role for FGF-6 in skeletal muscle regeneration. Genes Dev 11:2040–2051
Füchtbauer E-M (1995) Expression of M-twist during postimplantation development of the mouse. Dev Dyn 204:316–322
Fuentealba L, Carey DJ, Brandan E (1999) Antisense inhibition of syndecan-3 expression during skeletal muscle differentiation accelerates myogenesis through a basic fibroblast growth factor-dependent mechanism. J Biol Chem 274:37876–37884
Grass S, Arnold H-H, Braun T (1996) Alterations in somite patterning of *Myf-5*-deficient mice: a possible role for FGF-4 and FGF-6. Development 122:141–150
Grobet L, Martin LJ, Poncelet D, Pirottin D, Brouwers B, Riquet J, Schoeberlein A, Dunner S, Menissier F, Massabanda J, Fries R, Hanset R, Georges M (1997) A deletion in the bovine myostatin gene causes the double-muscled phenotype in cattle. Nat Genet 17:71–74
Guillemot F (1995) Analysis of the role of basic-helix-loop-helix transcription factors in the development of neural lineages in the mouse. Biol Cell 84:3–6
Halevy O, Monsonego E, Marcelle C, Hodik V, Mett A, Pines M (1994) A new avian fibroblast growth factor receptor in myogenic and chondrogenic cell differentiation. Exp Cell Res 221: 278–284
Hamamori Y, Wu HY, Sartorelli V, Kedes L (1997) The basic domain of myogenic basic helix-loop-helix (bHLH) proteins is the novel target for direct inhibition by another bHLH protein, Twist. Mol Cell Biol 17:6563–6573
Hamamori Y, Sartorelli V, Ogryzko V, Puri PL, Wu H-Y, Wang JYJ, Nakatani Y, Kedes L (1999) Regulation of histon acetyltransferases p300 and PCFA by the bHLH protein Twist and adenoviral oncoprotein E1A. Cell 96:405–413
Han J-K, Martin GR (1993) Embryonic expression of FGF-6 is restricted to the skeletal muscle lineage. Dev Biol 158:549–554
Hardy S, Kong Y, Konieczny SF (1993) Fibroblast growth factor inhibits MRF4 activity independently of the phosphorylation status of a conserved threonine residue within the DNA-binding domain. Mol Cell Biol 13:5943–5956
Hebrok M, Wertz K, Füchtbauer E-M (1994) M-twist is an inhibitor of muscle differentiation. Dev Biol 165:537–544
Hebrok M, Füchtbauer A, Füchtbauer E-M (1997) Repression of muscle-specific gene activation by the murine twist protein. Exp Cell Res 232:295–303
Heino J, Massagué J (1990) Cell adhesion to collagen and decreased myogenic gene expression implicated in the control of myogenesis by transforming growth factor β. J Biol Chem 265:10181–10184
Hill RE, Jones PF, Rees AR, Sime CM, Justice MJ, Copeland NG, Jenkins NA, Graham E, Davidson DR (1989) A new family of mouse homeo box-containing genes: molecular

structure, chromosomal localisation, and developmental expression of Hox-7.1. Genes Dev 3:26–37

Hirsinger E, Malapert P, Dubrulle J, Delfini MC, Duprez D, Henrique D, Ish-Horowicz D, Pourquie O (2001) Notch signalling acts in postmitotic avian myogenic cells to control MyoD activation. Development 128:107–116

Houzelstein D, Auda-Boucher G, Cheraud Y, Rouaud T, Blanc I, Tajbakhsh S, Buckingham ME, Fontaine-Perus J, Robert B (1999) The homeobox gene Msxl is expressed in a subset of somites, and in muscle progenitor cells migrating into the forelimb. Development 126: 2689–2701

Hu J-S, Olson EN (1990) Functional receptors for transforming growth factor-β are retained by biochemically differentiated C2 myocytes in growth factor deficient medium containing EGTA but down-regulated during terminal differentiation. J Biol Chem 265:7914–7919

Jen Y, Weintraub H, Benezra R (1992) Overexpression of Id protein inhibits the muscle differentiation program: in vivo association of ID with E2A proteins. Genes Dev 6:1466–1479

Jensen J, Heller RS, Funder-Nielsen T, Pedersen EE, Lindsell C, Weinmaster G, Madsen OD, Serup P (2000) Independent development of pancreatic alpha- and beta-cells from neurogenin3-expressing precursors: a role for the notch pathway in repression of premature differentiation. Diabetes 49:163–176

Kaatinen V, Voncken JW, Shuler C, Warburton D, Bu D, Heisterkamp N, Groffen J (1995) Abnormal lung development and cleft palate in mice lacking TGF-β3 indicates defects of epithelial-mesenchymal interaction. Nat Genet 11:415–421

Kageyama R, Ohtsuka T, Tomita K (2000) The bHLH gene Hesl regulates differentiation of multiple cell types. Mol Cells 10:1–7

Kambadur R, Sharma M, Smith TP, Bass JJ (1997) Mutations in myostatin (GDF8) in double-muscled Belgian Blue and Piedmontese cattle. Genome Res 7:910–916.

Katagiri T, Yamaguchi A, Komaki M, Abe E, Takahashi N, Ikeda T, Rosen V, Wozney JM, Fujisawa-Sehara A, Suda T (1994) Bone morphogenetic protein-2 converts the differentiation pathway of C2C12 myoblasts into the osteoblast lineage [published erratum appears in J Cell Biol 1995 Feb;128(4):following 713]. J Cell Biol 127:1755–1766

Kaul A, Koster M, Neuhaus H, Braun T (2000) Myf-5 revisited: loss of early myotome formation does not lead to a rib phenotype in homozygous Myf-5 mutant mice. Cell 102:17–19

Klagsbrun M, Baird A (1991) A dual receptor system is required for basic fibroblast growth factor activity. Cell 67:229–231

Kopan R, Nye JS, Weintraub H (1994) The intracellular domain of mouse Notch: a constitutively activated repressor of myogenesis directed at the basic helix-loop-helix region of MyoD. Development 120:2385–2396

Lafyatis R, Lechleider R, Roberts AB, Sporn MB (1991) Secretion and transcriptional regulation of transforming growth factor-beta 3 during myogenesis. Mol Cell Biol 11:3795–3803

Larrain J, Cizmeci-Smith G, Troncoso V, Stahl RC, Carey DJ, Brandan E (1997) Syndecan-1 expression is down-regulated during myoblast terminal differentiation. Modulation by growth factors and retinoic acid. J Biol Chem 272:18418–18424

Larrain J, Carey DJ, Brandan E (1998) Syndecan-1 expression inhibits myoblast differentiation through a basic fibroblast growth factor-dependent mechanism. J Biol Chem 273:32288–32296

Li L, Hu J-S, Olson EN (1990) Different members of the jun proto-oncogene family exhibit distinct patterns of expression in response to type β transforming growth factor. J Biol Chem 265:1556–1562

Li L, Zhou J, James G, Heller-Harrison R, Czech MP, Olson EN (1992) FGF inactivates myogenic helix-loop-helix proteins through phosphorylation of a conserved protein kinase C site in their DNA binding domains. Cell 71:1181–1194

Lilly B, Galewsky S, Firulli AB, Schulz RA, Olson EN (1994) D-MEF2: a MADS box transcription factor expressed in differentiating mesoderm and muscle cell lineages during *Drosophila* embryogenesis. Proc Natl Acad Sci USA 91:5662–5666

Lindsell CE, Shawber CJ, Boulter J, Weinmaster G (1995) Jagged: a mammalian ligand that activates Notch1. Cell 80:909–917

Marcelle C, Stark MR, Bronner-Fraser M (1997) Coordinate actions of BMPs, Wnts, Shh and noggin mediate patterning of the dorsal somite. Development 124:3955–3963

Martin JF, Li L, Olson EN (1992) Repression of myogenic function by TGF-β1 is targeted at the basic helix-loop-helix motif and is independent of E2A products. J Biol Chem 267: 10956–10960

McPherron AC, Lee SJ (1997) Double muscling in cattle due to mutations in the myostatin gene. Proc Natl Acad Sci USA 94:12457–12461

McPherron AC, Lawler AM, Lee SJ (1997) Regulation of skeletal muscle mass in mice by a new TGF-beta superfamily member. Nature 387:83–90

Miner JH, Wold BJ (1990) Herculin, a fourth member of the MyoD family of myogenic regulatory genes. Proc Natl Acad Sci USA 87:1089–1093

Molkentin JD, Black BL, Martin JF, Olson EN (1995) Cooperative activation of muscle gene expression by MEF2 and myogenic bHLH proteins. Cell 83:1125–1136

Naya FS, Olson E (1999) MEF2: a transcriptional target for signaling pathways controlling skeletal muscle growth and differentiation. Curr Opin Cell Biol 11:683–688

Nguyen HT, Bodmer R, Abmayr SM, McDermott JC, Spoerel NA (1994) D-mef2: a *Drosophila* mesoderm-specific MADS box-containing gene with a biphasic expression profile during embryogenesis. Proc Natl Acad Sci USA 91:7520–7524

Niswander L, Martin GR (1992) Fgf-4 expression during gastrulation, myogenesis, limb and tooth development in the mouse. Development 114:755–768

Nye JS, Kopan R, Axel R (1994) An activated Notch suppresses neurogenesis and myogenesis but not gliogenesis in mammalian cells. Development 120:2421–2430

Odelberg SJ, Kollhoff A, Keating MT (2000) Dedifferentiation of mammalian myotubes induced by msx1. Cell 103:1099–1109

Olson EN (1992) Interplay between proliferation and differentiation within the myogenic lineage. Dev Biol 154:261–272

Pelton RW, Saxena B, Jones M, Moses HL, Gold LI (1991) Immunohistochemical localization of TGF-β1, TGF-β2, and TGF-β3 in the mouse embryo: expression pattern suggest multiple roles during embryonic development. J Cell Biol 115:1091–1105

Pena TL, Chen SH, Konieczny SF, Rane SG (2000) Ras/MEK/ERK up-regulation of the fibroblast KCa channel FIK is a common mechanism for basic fibroblast growth factor and transforming growth factor-beta suppression of myogenesis. J Biol Chem 275:13677–13682

Pertovaara L, Sistonen L, Bos TJ, Vogt PK, Keski-Oja J, Alitalo K (1989) Enhanced *jun* gene expression is an early genomic response to transforming growth factor β stimulation. Mol Cell Biol 9:1255–1262

Peters KG, Werner S, Chen G, Williams LT (1992) Two FGF receptor genes are differentially expressed in epithelial and mesenchymal tissues during limb formation and organogenesis in the mouse. Development 114:233–243

Pourquié O, Fan C-M, Coltey M, Hirsinger E, Watanabe Y, Bréant C, Francis-West P, Brickell P, Tessier-Lavigne M, Le Douarin NM (1996) Lateral and axial signals involved in avian somite patterning: a role for BMP4. Cell 84:461–471

Proetzel G, Pawlowski SA, Wiles MV, Yin M, Boivin GP, Howles PN, Ding J, Ferguson MWJ, Doetschman T (1995) Transforming growth factor-β3 is required for secondary palate fusion. Nat Genet 11:409–414

Reshef R, Maroto M, Lassar AB (1998) Regulation of dorsal somitic cell fates: BMPs and Noggin control the timing and pattern of myogenic regulator expression. Genes Dev 12: 290–303

Rhodes SJ, Konieczny SF (1989) Identification of MRF4: a new member of the muscle regulatory factor gene family. Genes Dev 3:2050–2061

Riechmann V, van Crüchten I, Sablitzky F (1994) The expression pattern of Id4, a novel dominant negative helix-loop-helix protein, is distinct from Id1, Id2 and Id3. Nucleic Acids Res 22: 749–755

Rios R, Carneiro I, Arce VM, Devesa J (2001) Myostatin regulates cell survival during C2C12 myogenesis. Biochem Biophys Res Commun 280:561–566

Riquelme C, Larrain J, Schonherr E, Henriquez JP, Kresse H, Brandan E (2000) Antisense inhibition of decorin expression in myoblasts decreases cell responsiveness to transforming growth factor beta and accelerates skeletal muscle differentiation [published correction appears in J Biol Chem 276:9580–9581]. J Biol Chem 276:3589–3596

Robert B, Sassoon D, Jacq B, Gehring W, Buckingham M (1989) Expression of a novel Hox gene, Hox-7, during mouse embryogenesis, is associated with morphogenetic phenomena. EMBO J 8:91–100

Rohwedel J, Horák V, Hebrok M, Füchtbauer E-M, Wobus AM (1995) M-twist expression inhibits mouse embryonic stem cell-derived myogenic differentiation in vitro. Exp Cell Res 220:92–100

Saitoh O, Periasamy M, Kan M, Matsuda R (1992) *Cis*-4-hydoxy-L-proline and ethyl-3,4-dihydroxybenzoate prevent myogenesis of C2C12 muscle cells and block MyoD1 and myogenin expression. Exp Cell Res 200:70–76

Sakuma K, Watanabe K, Sano M, Uramoto I, Totsuka T (2000) Differential adaptation of growth and differentiation factor 8/myostatin, fibroblast growth factor 6 and leukemia inhibitory factor in overloaded, regenerating and denervated rat muscles. Biochim Biophys Acta 1497: 77–88

Scaal M, Füchtbauer EM, Brand-Saberi B (2001) cDermo-1 expression indicates a role in avian skin development. Anat Embryol (Berl) 203:1–7

Schäfer BW, Blakely BT, Darlington GJ, Blau HM (1990) Effect of cell history on response to helix-loop-helix family of myogenic regulators. Nature 344:454–458

Schubert D, Harris AJ, Devine CE, Heinemann S (1974) Characterization of a unique muscle cell line. J Cell Biol 61:398–413

Shull MM, Ormsby I, Kier AB, Pawlowski S, Diebold RJ, Yin M, Allen R, Sidman C, Proetzel G, Calvin D et al. (1992) Targeted disruption of the mouse transforming growth factor-beta 1 gene results in multifocal inflammatory disease. Nature 359:693–699

Song K, Wang Y, Sassoon D (1992) Expression of Hox-7.1 in myoblast inhibits terminal differentiation and induces cell transformation. Nature 360:477–481

Spicer DB, Rhee J, Cheung WL, Lassar AB (1996) Inhibition of myogenic bHLH and MEF2 transcription factors by the bHLH protein twist. Science 272:1476–1480

Stark KL, McMahon JA, McMahon AP (1991) FGFR-4, a new member of the fibroblast growth factor receptor family, expressed in the definitive endoderm and skeletal muscle lineages of the mouse. Development 113:641–651

Struhl G, Greenwald I (2001) Presenilin-mediated transmembrane cleavage is required for Notch signal transduction in *Drosophila*. Proc Natl Acad Sci USA 98:229–234

Sun X-H, Copeland NG, Jenkins NA, Baltimore D (1991) Id proteins Id1 and Id2 selectively inhibit DNA binding by one class of helix-loop-helix proteins. Mol Cell Biol 11:5603–5611

Taylor MV, Beatty KE, Hunter HK, Baylies MK (1995) *Drosophila* MEF2 is regulated by twist and is expressed in both the primordia and differentiated cells of the embryonic somatic, visceral and heart musculature [published erratum appears in Mech Dev 1995 May; 51(1):139–41]. Mech Dev 50:29–41

Thomas M, Langley B, Berry C, Sharma M, Kirk S, Bass J, Kambadur R (2000) Myostatin, a negative regulator of muscle growth, functions by inhibiting myoblast proliferation. J Biol Chem 275:40235–40243

Tsuda T, Kaibuchi K, Kawahara Y, Fukuzaki H, Taika Y (1985) Induction of proteinkinase C activation and Ca^{2+} mobilization by fibroblast growth factor in Swiss 3T3 cells. FEBS Lett 191:205–210

Vaidya TB, Rhodes SJ, Taparowsky EJ, Konieczny SF (1989) Fibroblast growth factor and transforming growth factor β repress transcription of the myogenic regulatory gene MyoD1. Mol Cell Biol 9:3576–3579

Villavicencio EH, Yoon JW, Frank D, Füchtbauer E-M, Walterhouse DO, Iannaccone PM (2002) Twist is an upstream regulator of human Gli1. Genesis 32:in press

Wang Y, Sassoon D (1995) Ectoderm-mesenchyme and mesenchyme-mesenchyme interactions regulate Msx-1 expression and cellular differentiation in murine limb bud. Dev Biol 168: 374–383

Wang Y, Benezra R, Sassoon DA (1992) Id expression during mouse development: a role in morphogenesis. Dev Dyn 194:222–230

Wilson-Rawls J, Molkentin JD, Black BL, Olson EN (1999) Activated Notch inhibits myogenic activity of the MADS-Box transcription factor myocyte enhancer factor 2C. Mol Cell Biol 19:2853–2862

Wright WE, Sassoon DA, Lin VK (1989) Myogenin, a factor regulating myogenesis, has a domain homologous to MyoD. Cell 56:607–617

Yaffe D, Saxel O (1977) Serial passaging and differentiation of myogenic cells isolated from dystrophic mouse muscles. Nature 270:725–727

Yayon A, Klagsbrun M, Esko JD, Leder P, Ornitz DM (1991) Cell surface, heparin-like molecules are required for binding of basic fibroblast growth factor to its high affinity receptor. Cell 64:841–848

Yin Z, Xu XL, Frasch M (1997) Regulation of the twist target gene tinman by modular *cis*-regulatory elements during early mesoderm development. Development 124:4971–4982

Zentella A, Massagué J (1992) Transforming growth factor β induces myoblast differentiation in the presence of mitogens. Proc Natl Acad Sci USA 89:5176–5180

Control of Muscle Size During Embryonic, Fetal, and Adult Life

Ketan Patel,[1] Bodo Christ,[2] and Frank E. Stockdale[3]

1 Introduction

Size differences of muscles can be found comparing homologous muscles in various species. The biceps brachii muscle of the mouse, for instance, is much smaller than the biceps brachii muscle of a pig or even an elephant. These size differences are mainly caused by differences in the number of muscle cells, meaning that the proliferation rates of the somite-derived muscle cells must be different in these species. In addition to these species-specific differences, muscle sizes vary from one body region to another in a given animal. The musculature in the lower leg is much smaller than it is in the thigh and the back musculature in the thoracic region is much smaller than it is in the lumbar or cervical region. Although the mass of homologous muscles of two differing species may vary by over a 1,000-fold, it is surprising that the sources of the musculature, the somites, do not exhibit such dramatic differences in size. Therefore, muscle mass is determined not by the number of cells in the newly formed somites, but by mechanisms that regulate their proliferation, expansion and differentiation. In this review we have looked at mechanisms that regulate the expansion of the somitic compartment and its derivatives that form the skeletal muscle of the vertebrate body at differing stages in the life of higher vertebrates.

2 Somite Patterning and Specification of Myogenic Cells

All skeletal musculature of the body originates from the segmentally arranged somites (Christ and Ordahl 1995). The somites are epithelial spheres that are formed in pairs in a cranio-caudal direction from the paraxial mesoderm. During maturation they become compartmentalized in a dorsal muscle-forming dermomyotome with proliferating, *Pax3*-expressing, epithelial cells and a ventral skeleton-forming, mesenchymal sclerotome that expresses *Pax1*.

[1] Department of Veterinary Basic Sciences, The Royal Veterinary College, Royal College Street, London NW1 OTU, UK
[2] Anatomisches Institut II, Albert-Ludwigs-Universität, Albertstrasse 17, Freiburg, Germany
[3] Stanford University, School of Medicine, CCSR 1145, Stanford, California 94305-5151, USA

Results and Problems in Cell Differentiation, Vol. 38
B. Brand-Saberi (Ed.): Vertebrate Myogenesis

The medial third of the dermomyotome gives rise to the epaxial myotome, which is the anlage of the intrinsic back muscle (epaxial muscle; Huang and Christ 2000). The myotome is a sheet of differentiating and post-mitotic myoblasts that express myogenic regulatory factor (MRF) genes. At limb levels, the lateral dermomyotomal lip undergoes an epithelial–mesenchymal transition and the detached muscle precursor cells migrate into the limbs where they proliferate and differentiate (Christ et al. 1977). At the interlimb level, the lateral part of the dermomyotomes retains its epithelial structure and forms the hypaxial myotome, which enters the somatopleure where it gives rise to the intercostal and abdominal muscles (Christ et al. 1983). In birds and mice, *myf5* and *MyoD* (Fig. 1) are the first members of the MRF genes to be expressed and *myf5* seems to act upstream of *MyoD* (Hacker and Guthrie 1998; Delfini et al. 2000).

The epaxial and hypaxial muscles are generated by different regions of the somite and the dermomyotome (Ordahl and Le Douarin 1992; Denetclaw et al. 1997; Denetclaw and Ordahl 2000; Huang and Christ 2000). Specification and differentiation of muscle-precursor cells in both the epaxial and hypaxial domains depend on signals emanating from adjacent structures. Signals from the dorsal neural tube and the surface ectoderm (Wnt-proteins, BMP4) have been shown to initiate dermomyotome formation and to maintain its function to yield myoblasts (Fan and Tessier-Lavigne 1994; Marcelle et al. 1997; Schmidt et al. 1998; Amthor et al. 1999). After removal of the dorsal neural tube, the development of the epaxial muscle is deficient. *Pax3* is downregulated and *MyoD* is only expressed in the lateral-most part of the epaxial domain. Wnts from the dorsal neural tube (Wnt1, Wnt3a) are required to start the muscle-differentiation programme in the epaxial domain. It has been shown that removal of the surface ectoderm from early somites results in a downregulation of *Pax3* and a premature expression of *MyoD*. In matured somites, the downregulation of *Pax3* is accompanied by a temporary upregulation of *MyoD*.

In addition to the dorsal signals, Sonic Hedgehog (Shh) produced by the notochord and the floor plate is required for *MyoD* expression in the epaxial domain (Münsterberg et al. 1995). The signal transduction genes in the Shh pathway include *patched (Ptc)*, which encodes the Shh receptor and the *Gli* genes, *Gli1*, *Gli2* and *Gli3*, which encode zinc-finger transcription factors that mediate the nuclear transcriptional activity of Shh signalling. It is suggested that the activation of *Gli2* and *Gli3* is controlled by Wnts whereas the upregulation of *patched* and the activation of *Gli1* is controlled by Shh (Borycki et al. 1998, 2000). These findings might explain the requirement of both Wnts and Shh for the initiation of epaxial myogenesis.

In the hypaxial domain, the activation of myogenesis does not depend on Shh signalling. Ectodermal Wnts and BMP4 from the lateral plate mesoderm have been postulated to be involved in the specification of hypaxial muscle precursor cells (Dietrich et al. 1998; Tajbakhsh et al. 1998). The time and the molecular events of the commitment of hypaxial muscle precursor cells remain to be studied. It has to be acknowledged that a considerable number of lateral

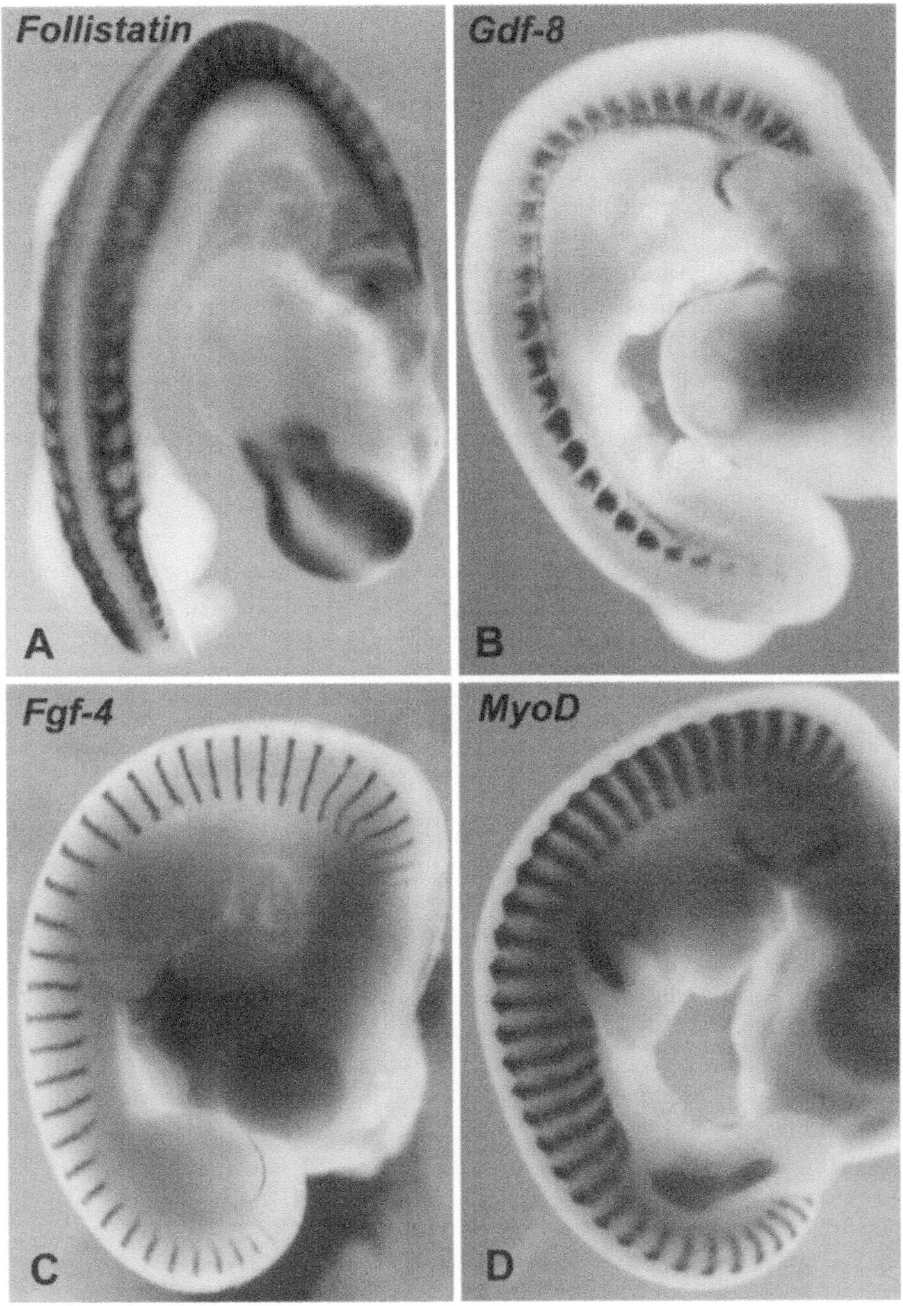

Fig. 1. Whole mount in situ hybridization of 3- to 5-day chick embryos showing the expression of **A** follistatin, **B** myostatin, **C** FGF4 and **D** MyoD

dermomyotome cells do not undergo muscle differentiation during normal development (reviewed by Christ et al. 2001).

3 Allocation of Cells to the Dorsal Somite Compartment

Somites respond to a complex network of ventral and dorsal signals with a compartmentalization leading to a *Pax1* expressing chondrogenic sclerotome, and a dorsal *Pax3* expressing muscle-yielding dermomyotome. Experimental interventions in the normal concert of ventral and dorsal signals alter the allocation of somite cells to the dermomyotomal and sclerotomal lineages (Capdevila et al. 1998; Wagner et al. 2000). Ectopically implanted Wnt1-, Wnt3a- and Wnt4-expressing cells, for instance, considerably enlarge the myotome at the expense of the sclerotome. This experimentally induced scenario in chicks resembles the situation in lower vertebrates and chordates with respect to somite compartmentalization.

4 Migration of Muscle Precursor Cells

Muscle mass development in the limbs depends on the migration of dermomyotome-derived muscle precursor cells. If this migration does not occur, limbs develop with normal skeletal elements and tendons, but without muscle (Christ et al. 1977). Migration requires a de-epithelialization of the lateral dermomyotome which is mediated by a cross talk between c-met, which is expressed in the lateral dermomyotome, and Scatter Factor (SF), which is expressed in the proximal region of the limb (Bladt et al. 1995). Experiments in chick embryos have shown that application of exogenous SF induces ectopic delamination of muscle precursor cells at interlimb level (Brand-Saberi et al. 1996). Another key regulatory gene in the migration of limb muscle precursor cells is *Pax3*. *Splotch (Sp)* mutant mice harbour a mutation at the *Pax3* locus and do not form limb muscles (Franz et al. 1993; Goulding et al. 1994). Muscle precursor cells in *Splotch* mice do not express *c-met*. Recent studies have shown that the homeobox gene *Lbx1*, which is expressed in the lateral lip of the dermomyotomes, is required for the well-ordered migration of muscle precursor cells and the normal patterning and growth of limb muscles (Brohmann et al. 2000; Gross et al. 2000).

5 Balance Between Proliferation and Differentiation

Muscle size is a result of the balance between proliferation and differentiation. During embryonic muscle development, *Pax3*-expressing proliferating muscle precursor cells are located in the dermomyotomes and in the superficial layer of the pre-muscular masses and the muscle blastemata of the limbs. With the beginning of differentiation, the muscle precursor cells express *MyoD* and they become epaxially translocated into the myotomes and into the deep layer of the pre-muscular masses and muscle blastemata of the limbs. Proliferating myoblasts express *Pax3* and *myf5* whereas differentiating myoblasts express *MyoD*. After experimental activation of the Notch pathway, terminal muscle differentiation is inhibited (Delfini et al. 2000). Ectopically applied Shh downregulates *Pax3* and upregulates *MyoD* (Amthor et al. 1999). Under this condition, the muscle precursor cells undergo an accelerated differentiation leading to a rapid exhaustion of the pool of proliferating cells. As a result, the muscle size is reduced. On the other hand, BMPs, FGFs and IGF-I stimulate the proliferation of muscle precursor cells, leading to an enlargement of muscle size (Hannon et al. 1996; Floss et al. 1997; Barton-Davis et al. 1998; Amthor et al. 1999). FGFs-2, 4, 6, 7 and 8 are expressed in the cells of the myotome and might promote the proliferation of the adjacent dermomyotome cells (Joseph-Silverstein et al. 1989; Niswander and Martin 1992, de Lapeyriere et al. 1993; Han and Martin 1993; Dono and Zeller 1994; Mason et al. 1994). The loss of FGF receptor 1 (FGFR1) signalling accelerates muscle differentiation, resulting in a reduction of muscle size (Itoh et al. 1996; Flanagan-Steet 2000). An over-expression of Noggin, an antagonist of BMP signalling, reduces muscle mass. Follistatin, another protein thought to be a BMP antagonist, upregulates *Pax3* expression and stimulates the proliferation of muscle precursor cells. BMP-7 combined with follistatin stimulates muscle growth dramatically. It is suggested that follistatin, which is expressed in muscle, may locally modulate the proliferative effect of BMPs by presenting them to muscle precursor cells (Amthor, Christ, Patel, unpubl. data). BMPs at high concentrations on the other hand, induce apoptosis and muscle loss (Amthor et al. 1998). In the limb, Scatter Factor (SF) increases the motility of muscle precursor cells and keeps them in a proliferating and undifferentiated state (Scaal et al. 1999).

6 Muscle Growth in the Embryonic, Fetal, and Neonatal Periods of Development

It is widely recognized that the size and mass of muscle is controlled by three mechanisms: proliferation of muscle precursor cells, formation of muscle fibres and hypertrophy of existent muscle fibres. All three mechanisms are operative throughout development, with the relative dominance of each differing during the three major phases of development - the embryonic, fetal and neonatal phases. During embryonic development, proliferation and new fibre formation dominate, whereas during fetal development proliferation and new fibre formation continue and hypertrophy begins. In the postnatal period, hypertrophy alone dominates increases in muscle mass. The mechanisms controlling the amplification of muscle cell precursors, myoblasts, and fibre number during embryonic and fetal life are poorly understood, whereas new light has been shed on the process of hypertrophy that predominates in postnatal life.

7 Embryonic and Fetal Muscle Fibres

There is no single period alone in development that defines the number of muscle fibres that will compose the adult muscle. All anatomic muscles of adult vertebrates have their origins in several waves of muscle-fibre formation as development proceeds (Stockdale 1997). The first fibres form during embryonic development and lay down the anlagen of all future muscles. The first muscle fibres are termed primary fibres (Fig. 2) and, in both birds and mammals, are formed from the first population of myoblasts, called embryonic myoblasts, which appear in muscle-forming regions. The amount of muscle mass formed from the embryonic myoblasts is extremely small: their role in muscle formation is to define the type, shape or location of a muscle, rather than to provide mass. On the other hand, much of the muscle mass that ultimately will form in the adult vertebrate is dependent upon the second wave of myogenesis first detected as the transition from embryonic to fetal development occurs. This second wave of myogenesis forms fibres designated as secondary fibres and contributes greatly to the number of fibres that will be found in the mature muscle. Secondary fibres have their origin in fetal myoblasts. The addition of most new fibres to anatomic muscles occurs during fetal development by the addition of secondary fibres, particularly on the periphery of primitive limb muscles with the muscle fibres closest to the limb bones being those that are more likely to be of primary fibre origin.

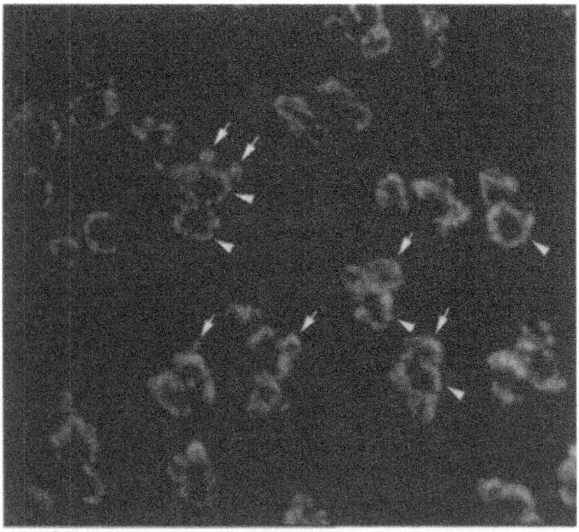

Fig. 2. Primary and secondary skeletal muscle fibres. This is a cryosection of embryonic day-10 pectoral muscle of the quail immunostained with a monoclonal antibody to fast myosin heavy chains. The *arrows* point to secondary fibres of varying diameters in association with primary fibres (*arrowheads*)

Primary muscle fibres appear in all muscle anlagen of the extremities. They traverse the muscles from tendon to tendon and become innervated at multiple endplates prior to formation of secondary fibres in both rats and guinea pigs (Duxson and Usson 1989; Duxson and Sheard 1995). Secondary fibres form (Fig. 2) and insert on the surface of primary fibres beginning near the sites of innervation of primary fibres and, initially, do not traverse the entire length of a muscle. The newly formed secondary fibres are, at first, short fibres on the surface of primary fibres and share a common basal lamina with a primary fibre. Thus, "clusters" are seen in both birds and mammals, the centre of which is occupied by a large diameter fibre, the primary fibre, on whose surface lie one to several much smaller diameter secondary fibres. These secondary fibres increase rapidly in number and nucleation and, subsequently, diameter, and they come to be surrounded by their own basal lamina as they separate from the primary fibre within the cluster. It is this pattern of formation that has lead to the concept that primary fibres form the scaffolding of the future muscle upon which secondary fibres form. One can postulate that primary fibres lay down the location and anatomic shape of a muscle serving to organize all subsequent fibre formation.

8 Embryonic, Fetal, and Adult Myoblasts

All fibres of developing muscles form by fusion of myoblasts with one another (Stockdale and Holtzer 1961) from separable populations of myoblasts that populate growing muscles at specific times during muscle formation. There are three categories of myoblasts: embryonic, fetal and adult or satellite cells (Stockdale 1992, 1997). Embryonic myoblasts differ from fetal and adult myoblasts in several respects. In both the mammals and birds, embryonic myoblasts form fibres containing one or only a very few nuclei and express specific isoforms of myosin heavy chain (MyHC; Miller and Stockdale 1986b; Ontell et al. 1993). In birds, these differences in MyHC expression, both in vivo and in vitro, serve to define the fibre type content or pattern of the fully developed anatomic muscles of the limbs (Crow and Stockdale 1986a). Recent analysis in the quail suggests that the primary fibres of the head muscles may not define the ultimate fibre composition of fully developed head muscles (Marcucio and Noden 1999). It is less clear in mammals that primary fibres define the future fibre-type patterning of muscles (Condon et al. 1990; Dunglison et al. 1999), but fibres formed from mammalian embryonic or fetal myoblasts express different isoforms of MyHC. What is clear is that in both classes of animals the primary fibres are responsible for the formation of what will finally be fibres of the slow type or type I fibres (Crow and Stockdale 1984).

Fetal myoblasts appear near the end of the first week of development of the chick limb (Bonner and Hauschka 1974; Miller and Stockdale 1986b). These myoblasts differ from embryonic myoblasts because they form myotubes with many more nuclei than primary fibres. Thus, in vitro avian embryonic myoblasts form fibres with one to only a few nuclei whereas fetal myoblasts form fibres with many nuclei – sometimes as many as 100 can be found within a single fibre formed from fetal myoblasts. In the bird, fibres formed from fetal myoblasts also differ from those formed from embryonic myoblasts in the isoforms of myosin expressed and the influence of innervation on myosin isoform expression within fibres they form. All fibres formed in vitro from fetal myoblasts express fast isoforms of myosin (Miller and Stockdale 1986b), whereas fibres from embryonic myoblasts can form fibres of differing MyHC expression. Also fetal myoblasts isolated from different muscle anlagen form fibres that respond to innervation differently. Those fetal myoblasts isolated from anlagen that are of slow fibre muscles will express slow as well as fast isoforms of MyHC upon innervated in vitro, whereas those formed from fetal myoblasts from fast muscle anlagen will not (Düsterhoft et al. 1992; Pin and Merrifield 1993; DiMario and Stockdale 1997). Thus, there appear to be different types or lineages of embryonic myoblasts and fetal myoblasts in both mammals and birds limbs based upon their time of appearance, the size of fibres formed, the isoforms of MyHC expressed and the influence of innervation on the fibre-type expressed.

It is the number of fibres that form in a fetal muscle that determines the adult size of every muscle. Therefore, to understand the size of a muscle, the mechanisms that control fetal myoblasts proliferation and fusion into fibres must be understood. As described above, both fetal and the embryonic myoblasts of the limb musculature have their origins in the migratory population of muscle precursors of the lateral dermomyotome (Christ et al. 1977). When they arrive in the limb bud, embryonic myoblasts proliferate, presumably in response to local signals, and begin to fuse into small primary fibres. But the primary myoblast population in the limb musculature is short lived. In the bird it is no longer possible to isolate myoblasts of the embryonic type beyond the first week of development (Miller and Stockdale 1986a, 1986b). Fetal myoblasts can first be isolated from the limb bud about the end of the first week of limb development and dominate the myoblast population until the second week of development, at which time myoblasts with adult or satellite characteristics become the sole population (Miller and Stockdale 1986a,b; Feldman and Stockdale 1992). In both mammals and birds, late fetal muscle fibre formation occurs in the presence of only adult myoblasts (Feldman and Stockdale 1991; Hartley et al. 1992).

9 Number of Embryonic and Fetal Myoblasts and Fibre Formation

There are numbers of unanswered or partially answered questions about myoblasts. It is not known if the precursors of embryonic myoblasts are also the precursors of fetal myoblasts or whether there are separate migratory populations for each within the somite. If retroviral marking of nuclei is performed in the mesenchymal cells of the limb bud, both populations can become labelled suggesting a common origin (Evans et al. 1994). On the other hand, the sizes and the length of muscle of primary myotubes increase contemporaneously with secondary myotube formation, suggesting that there is fusion of fetal myoblasts with primary fibres. If this occurs then one cannot distinguish the myoblast origin of a nucleus within a fibre, making the question of origins unsettled. Although disputed, it does appear that primary fibres grow by the addition of fetal myoblasts. This is borne out by observations on retrovirally-marked clones of dividing cells within the rat limb marked at stages when primary or secondary fibres are forming (Dunglison et al. 1999). They report that embryonic labelled clones appear in primary, but not secondary muscle fibres, whereas fetal labelled clones appear in primary and secondary fibres. On the other hand, work from Harris (Harris et al. 1989a) suggests that primary fibres form synchronously and do not receive continued input from the fetal population of myoblasts. Zhang and McLennan's (1995) recent work supports the observation of Evans and colleagues (1994) using retroviral marking techniques; they showed that primary fibres continue to incorporate nuclei during secondary fibre formation, although the rate of entry of nuclei decreases as fetal development progresses and no further increase in primary fibre marking occurs late in fetal development. The mechanism for restricting primary and secondary fibre formation to particular developmental periods may primarily be based on merely the temporal predominance of embryonic and fetal myoblasts, respectively. Thus, the question becomes what controls the appearance of different categories of myoblasts?

The mechanisms that initiate the expansion of embryonic and fetal myoblasts from as yet unidentified precursors are poorly understood. Several growth factors have been shown to influence the proliferation and differentiation of myoblasts (usually fetal myoblasts). This issue has been most extensively analysed by the Hauschka (Clegg et al. 1987; Seed et al. 1988; Templeton and Hauschka 1992; Pirskanen et al. 2000) and the Olwin group (DeHamer et al. 1994; Olwin et al. 1994a,b; Hannon et al. 1996; Flanagan-Steet 2000). They show that bovine FGF delays the onset of differentiation of myoblasts obtained from day 4–12 chick wing buds while at the same time they show that there is a subset of myoblasts derived from stage 23–27 (days 4–5) embryos that require FGF for myogenic differentiation. Thus the early population of limb myoblasts contains discrete subclasses of cells that are FGF dependent and others that are FGF independent. These subclasses are found in both embryonic and fetal myoblast categories. At later stages, only FGF-independent myoblasts persist. These observations place FGF family members in key roles to control early myoblasts and, thus, fibre number (Fig. 1). Their importance is demonstrated using retroviral infection of early chick limbs using dominant negative constructions for the FGF receptor (FGFR1; Flanagan-Steet 2000). The limbs of such embryos demonstrate an average loss of 30% in skeletal muscle wet weight and a 50% decrease in muscle fibre number. The loss of skeletal muscle mass was accompanied by a loss of myoblasts and of fibroblasts. Flanagan-

Street and colleagues (2000) concluded that signalling mediated via the FGFR1 receptor is necessary to maintain myoblast number and plays a role in myofibre organization.

There are other mechanisms that limit the numbers of embryonic and fetal myoblasts and, thus, the numbers of fibres that can form, as well as the mechanism that regulates the size or diameter and fibre length. It has been suggested that members of the insulin-like growth factors (IGF) and transforming growth factor-beta families of signalling molecules play key roles in the regulation of myoblast cell number. Cusella-De Angelis and colleagues (1994) have shown that TGF-β can inhibit proliferation of fetal, but not embryonic myoblasts. They proposed that regulation of fetal myoblast proliferation involves interplay of TGF-β produced in the surrounding connective tissue and an undefined mitogen that is produced by primary fibres. In mouse C-2 myoblasts, TGF-β inhibited cell fusion and prevented expression of the muscle-specific gene products, creatine kinase and acetylcholine receptor (Olson et al. 1986). However, targeted disruption of the TGF-β-1 gene in mice does not interfere with muscle formation (McLennan et al. 2000). The timing of secondary myotube formation and the total number of myotubes formed were normal in the TGF-β-1-/- mice (McLennan et al. 2000), but other genes such as TGF-β-2 and TGF-β-3 may well be able to substitute for the absence of TGF-β-1 in vivo as they share a receptor with TGF-β-1 (McLennan et al. 2000). Other mediators of growth, IGF-I and II, to be discussed below, act to promote proliferation of myoblasts and appear to be produced systemically and by muscle fibres themselves.

More recently, myostatin, another member of the TGF-β superfamily, has been proposed as a regulator of fetal and later stage myogenesis (Fig. 1). Myostatin acts as an inhibitor of myogenesis. In mice with targeted disruption of the myostatin gene and in cattle with a mutation in this gene (Belgian Blue and Piedmontese cattle), muscle fibres are hypertrophic and perhaps more plentiful (Swatland and Kieffer 1974; McPherron and Lee 1997; McPherron et al. 1997). Thomas has proposed that animals that lack myostatin have more myoblasts and thus can form more fibres (Thomas et al. 2000). However, Zhu et al. (2000) reports that mice with a disruption in the myostatin gene have no more fibres than normal, but exhibited a significant (20–35%) increase in muscle mass, which supports the contention that this signalling molecule has a role in determining fibre size. It can be postulated that the control of fibre formation during fetal life involves the orchestration of myostatin, other TGF-β family members, FGF and IGF family members to modulate the fetal myoblast number and secondary fibre formation and size.

10 Innervation and Muscle Fibre Number and Size

When muscles first form as a grouping of primary fibres, both proliferation and hypertrophy are underway. Innervation plays a key role in these two processes.

Even in *Drosophila* post-pupal development, where new muscle fibres form, the pool of myoblasts is reduced if denervation is performed prior to fibre formation (Fernandes and Keshishian 1998). In a number of reports, Duxson and co-workers (Duxson et al. 1989; Duxson 1992; Duxson and Sheard 1995) have shown that the site of innervation of the primary myotubes serves as a locus around which many new secondary myotubes form, and that each new secondary myotube is centred on a site of innervation of a primary myotube. This body of work indicates that innervation is particularly important for the proliferation of fetal myoblasts and, therefore, the formation of muscle fibres during fetal development. The primary fibres initially are polyneuronally innervated whereas secondary myotubes, even when newly formed, are innervated by a single axon (Duxson 1992). Nerve terminals, particularly on the primary fibres, appear to regulate the number of secondary myotubes by stimulating mitosis of myoblasts in a nerve-dependent population, suggesting that the physical presence of nerve terminals in association with primary myotubes is required for secondary fibre formation (Ross et al. 1987b).

Although less well demonstrated than for secondary fibre formation, there appears to be a role for innervation in primary fibre formation as well. If the mouse, rat or chick limb is denervated or functional innervation is blocked by alpha-bungarotoxin or d-tubocurare during or before primary fibre formation, anatomic muscles form normally, but there is a reduction in the number of primary and fetal muscle fibres (McLennan 1983; Phillips and Bennett 1984; Crow and Stockdale 1986b; Harris et al. 1989b; Hughes and Ontell 1992; Wilson and Harris 1993). But, some workers report that primary and secondary fibres appear at normal times in denervated mouse limbs (Ashby et al. 1993). Although the temporal appearance of secondary fibres may not be altered, they and others (Sheard et al. 1991) conclude that during normal development the generation of additional or normal numbers of secondary myotubes depends on neurally evoked electrical activity in primary myotubes. Primary fibre formation is not nerve dependent, but if they do not become innervated, not all primary fibres persist. Whereas innervation may not be absolutely required to initiate secondary fibre formation, there is general agreement that the number of secondary fibres is reduced (Fredette and Landmesser 1991) following denervation or functional denervation and, thus, there is a need for innervation for normal secondary fibre formation to continue. Some of the interpretative difference reported by different investigators of the relationship of innervation to secondary fibre formation may be because of the timing of denervation. Wilson and Harris (1993) suggest that in the rat there may be a brief critical period during which innervation is an absolute prerequisite for secondary myotube formation. It is fair to conclude that primary fibres will appear in the limb bud of both mammals and birds with or without innervation, that all anatomic muscles will appear in the limb without innervation, that the normal increase in numbers of secondary fibres is diminished if innervation does not occur and that a relationship between innervation and myoblast proliferation is evolutionarily retained.

Increase in secondary fibre size is heavily dependent upon sustained innervation of primary and secondary fibres. Although it is impossible to distinguish fibres of primary or secondary origin later in fetal development, it is also most likely true as well that primary fibre size is not sustained without innervation. Certainly the latter must be the case in developing birds where many of the slow fibres are of primary fibre origin. It has been shown that the slow fibres undergo profound atrophy in the absence of innervation (Crow and Stockdale 1986b). In all the denervation studies cited above there is marked atrophy of the secondary fibres that form in fetal development. Thus, as in adult muscles, maintenance of protein synthesis and fibre diameter is sustained by innervation and activity even at early stages of muscle growth.

11 Muscle Hypertrophy and Regeneration

In vertebrates, the evidence for new muscle fibre formation after fetal development is very limited. However, it is obvious that there is a marked increase in muscle size that begins in neonatal life and continues for months to years depending upon the class of animals. This increase results from an increase in the size of existing fibres, called skeletal muscle hypertrophy. Unless injury occurs, birds and mammals do not form many new fibres in the neonatal period (Goldspink 1972; Ontell and Kozeka 1984; Ontell et al. 1988), although there are limited reports that fibres are added to murine muscles after birth (Ross et al. 1987a). Thus, the fetal period is the principal formative period of muscle fibre number and the neonatal and adult periods are those of hypertrophy and repair. An exception to the generality that normally no new fibres form after fetal life (hatching) may be found in larger fish. Although hypertrophy is a major part of growth, in larger fish there is substantial increase in new fibres as well as the fish increases in size (Zimmerman and Lowery 1999). Thus, in most animals, muscle mass increase beyond the postnatal period is dependent upon protein synthesis and not new fibre formation.

It is difficult to discuss the mechanisms of hypertrophy independently of muscle regeneration (formation of new fibres in response to injury) because both processes are postnatal events and likely to have the activation of adult myoblasts (satellite cells) in common. In some models used to study hypertrophy there is activation of adult myoblasts and increased nucleation of existent fibres. For example, in the hypertrophy that accompanies strength training in humans there is an increase in the number of nuclei in fibres and an increase in adult myoblast cells after training (Kadi and Thornell 2000). In weight overload models of hypertrophy there is damage done to the muscle, thus the molecular analyses of this model include both hypertrophy and regeneration processes (Kennedy et al. 1988).

Recent studies have identified a tremendous potential for postnatal muscle growth through hyperplasia in some vertebrates. In birds, the BUT-T9 strain of turkey is characterized by high body weight (HW) and fast growth rate. At hatching, HW body weight and muscle fibre number were almost identical to those of lightweight (LW) birds (Cherel 1994). However, at 15 weeks, the HW birds weighed twice as much as the LW birds and had enlarged muscle fibres (increased over eight times from hatching stage). More significantly, HW birds contained almost twice the number of muscle fibres present at hatching. However, it cannot be excluded that this increase in the number of muscle fibres was the result of a genetic defect.

In comparison to the doubling of muscle fibre numbers seen in post-natal life of the HW turkeys, large fish show a truly astonishing capacity to generate new fibres after hatching. Atlantic salmon (*Salmo salar*) have approximately 5,000 white muscle fibres (used for high speed swimming, which relies on anaerobic metabolic pathways for energy supplies) in a cross section at the anterior part of the trunk at hatching stage. At first feeding stage, the number of fibres increases to 10,000 rising to 200,000 as the fish enters seawater (smolting). The number of muscle fibres continues to increase until, by 2 years, a cross section can reveal over 1 million white muscle fibres (Zimmerman and Lowery 1999). Hyperplasia is accompanied by fibre hypertrophy, with newly formed myofibres measuring 3–5 μm expanding to over 200 μm in diameter. Interestingly, species of small fish, growing no bigger than a few centimetres do not have the capacity to continually recruit muscle fibres in adult life (Weatherby and Gill 1995).

Therefore, in large fish, satellite cells are deployed for hypertrophy and hyperplasia, both of which occur simultaneously. This raises the important issue of the mechanism that regulates the decision-making process that allows cells to participate in more than one process leading to muscle growth. One possibility is that the satellite pool is not a homogeneous population, but is composed of subpopulations, similar to the *Pax-7+* and *Side population (SP)* cells found in mice, and that these groups have a preferential developmental programme. The use of large fish as an animal model to address this issue could be extremely advantageous as various genetic and environmental parameters (water temperature, oxygen concentration, etc.) have already been identified that influence the balance between hypertrophy and hyperplasia.

Hypertrophy in the postnatal period in most vertebrates is under the control of mechanical processes of muscle stretch, and activities that are transduced by a number of signalling molecules. Prominent among these molecules are the IGF family, FGF, interleukin-6, leukemia inhibitory factor (LIF) and TGF-β family members including myostatin (Barnard et al. 1994; Florini et al. 1996). Because hypertrophy is likely to involve the process of adult myoblast activation in mature fibres, an additional signalling and receptor molecule, hepatocyte growth factor (HGF), and c-met its receptor, are important as well.

Perhaps the growth factors most closely associated with skeletal muscle hypertrophy are members of the IGF family (Adams and McCue 1998; Awede et al. 1999). During mammalian fetal muscle development, vertebrates have circulating blood levels of IGF-I and II, and both IGF-I and II are produced by muscle fibres themselves, indicating that, in vivo, the IGFs are likely to have effects that are endocrine and auto- or paracrine in nature (Gerrard et al. 1998). This conclusion is supported by work on transgenic mice. Transgenic mice that overexpress *IGF-I* demonstrate marked hypertrophy of muscle fibres without elevation of blood levels of IGF-I, supporting the contention that the effect of IGF-I can be an auto- or paracrine rather than just an endocrine effect (Coleman et al. 1995). Furthermore, targeted disruption of *IGF-I* expression in the mouse leads to a reduction in cross-sectional area of all fibre types and muscle hypoplasia (Fournier and Lewis 2000).

Although FGFs have important roles in early muscle development they also have demonstrated effects at later stages of muscle formation, particularly in hypertrophy and regeneration (Floss et al. 1997). In stretch-induced hypertrophy models in the bird, members of the FGF family are elevated as the muscle hypertrophies (Mitchell et al. 1999). Immediately following injury to mammalian muscles there is also an abrupt increase in *FGF* expression (Anderson et al. 1995). However, the exogenous application of bFGF to regenerating muscle has no effect (Mitchell et al. 1996). Differences reported in the role of members of the FGF family in the process of hypertrophy may reflect that different members of the FGF family have different effects. For example, *FGF-4* and *FGF-10* expression increases with stretch of avian adult muscles whereas *FGF-2* and *FGF-4* do not (Mitchell et al. 1999).

Some of the factors that initiate hypertrophy have mechanisms in common. Because calcineurin, a calcium-regulated phosphatase, has also been implicated in signalling related to overload hypertrophy of the heart, calcineurin has been pursued as an intermediate in the transducing pathway in skeletal muscle hypertrophy. Increases in intracellular calcium ion concentration activates calcineurin, which, in turn, has been shown to increase the expression of a number of factors that regulate the expression of myofibrillar proteins (Olson and Williams 2000). That calcineurin can mediate hypertrophy of skeletal muscle was suggested in studies on work overload in mice, where calcineurin inhibitors were shown to prevent a doubling of muscle mass and individual fibre size, while producing a marked increase in the number of slow fibres (Dunn et al. 1999). On the other hand, in transgenic mice that expressed activated calcineurin under control of the muscle creatine kinase enhancer there was confirmation of an increase in the numbers of slow muscle fibres, but no evidence for skeletal muscle hypertrophy (Naya et al. 2000). Another initiator of muscle hypertrophy, motor nerve stimulation, just as forced expression of activated calcineurin, upregulates *MEF2* expression and, in turn, expression of skeletal-muscle-specific genes (Wu et al. 2000).

That IGF-I acts by increasing intracellular ionized calcium levels is demonstrated by studying myogenic cells in culture. *IGF-I*-transfected mammalian muscle cell cultures form myotubes that are larger than normal (Semsarian et al. 1999b). Although there is an increase in the amount of protein in these cultures, DNA synthesis and the number of nuclei within fibres is unchanged. These transfected *IGF-I* myogenic cultures exhibit increased calcineurin transcripts and nuclear localization of calcineurin protein (Musaro et al. 1999).

Likewise, activated calcineurin mimics the effects of IGF-I on cultured myogenic cells, whereas expression of a dominant-negative calcineurin mutant or addition of cyclosporin, a calcineurin inhibitor, represses myotube differentiation and hypertrophy (Musaro et al. 1999). A similar response is found when IGF-I or medium conditioned by *IGF-I*-transfected myogenic cells is added to mammalian myogenic cell cultures. The mechanism responsible for these effects appears to be mobilization of intracellular calcium, which activates calcineurin that initiates nuclear translocation of the transcription factor NF-Atc1 followed by gene activation (Semsarian et al. 1999a).

As with IGF-I, calcium ion signalling appears to play a central role in hypertrophic growth of skeletal muscle in response to mechanical load (Olson and Williams 2000). In response to mechanical load, calcium ions are sensed by calcineurin, which couples the mechanical work to muscle gene expression. This results in changes in contractile and metabolic properties of skeletal muscle by particularly activating the slow muscle fibre-specific gene programme. Both the hypertrophic and the control of contractile and metabolic functions of muscle fibres are mediated by transcription factors of the NFAT and MEF2 families (Olson and Williams 2000).

Finally, the hepatocyte growth factor (HGF) and its receptor c-met must be considered in a discussion of growth of muscle in the postnatal period. Under conditions of work overload, hypertrophy and injury, this growth factor and its receptor are likely to activate to promote adult myoblasts proliferation and ultimately fusion of adult myoblasts with fibres or with one another to form new fibres (Gal-Levi et al. 1998). The c-met receptor is present on quiescent adult myoblasts in normal muscle tissue (Tatsumi et al. 1998). These investigators have demonstrated that HGF is the factor in extracts of damaged muscle that leads to proliferation of adult myoblasts and muscle repair.

12 Programmed Cell Death During Muscle Development

To understand how muscle mass development is regulated, mechanisms have to consider that not only cause an increase, but also a reduction in muscle size. It is only in the past few years that the importance of programmed cell death (PCD), which includes apoptosis, has begun to be realized and, together with cell proliferation and differentiation, it is now thought to be a major determinant in controlling cell number during development.

With the advent of reliable techniques capable of detecting early stages of PCD, recent studies have found that PCD occurs at all stages of muscle development. Amthor et al. (1998) studied the migration of muscle precursors during embryogenesis, which become localized at the centre of the limbs and conspicuously absent at the margins. The latter regions are characterized by the expression of high levels of BMP that can induce PCD (Graham et al. 1994). Amthor et al. (1998) demonstrated that ectopic application of high levels of BMPs resulted in PCD of muscle precursors, but not loose mesenchymal cells.

They proposed a mechanism by which BMP-mediated PCD could be used to position muscle and, therefore, determine muscle patterning. Migration of muscle precursors would only be permissible in regions lacking high levels of BMP expression and any cells that migrated into such regions would never differentiate into muscle because they would be eliminated through PCD.

PCD is a powerful mechanism used by an organism during normal development to sculpt entire muscles (Grim 1977, 1978). This has been best characterized in amphibians during metamorphosis where PCD not only removes the entire tail, but also selectively eliminates dorsal muscles of the body. PCD during metamorphosis is characterized by fragmentation of the muscle fibres into oval apoptotic bodies or sacrolytes (Weber 1969). These are approximately 10–100 µm in length and are found in tail and dorsal muscles. Their location appears to be random and surrounded by normal looking fibres (Nishikawa and Hayashi 1995). Whereas the entire tail is removed through PCD, only selective fibres are eliminated in the body. All tail muscle is of larval muscle type whereas the dorsal body muscle consists of two forms: larval and adult. These can be distinguished on the basis of differential tropomyosin expression (Nishikawa and Hayashi 1994). Therefore, PCD selectively removes larval muscle. PCD in amphibian metamorphosis showed all the characteristics of apoptosis, including DNA fragmentation. The appearance of apoptotic bodies always preceded DNA breakdown. The entire PCD programme during amphibian metamorphosis is regulated by the circulating hormone tri-iodothyronine, and it remains to be determined how a systemic agent can selectively cause the death of some muscle fibres whilst not affecting their immediate neighbours.

Selective PCD in myotubes has also been observed during avian fetal development particularly in the biventer cervicis, a head extensor muscle (McClearn et al. 1995) and the extensor medius brevis muscle (EMB) of the quail (Svagr et al. 1995). In this scenario, degenerating cells had densely packed myofibrils. Longitudinal sections revealed areas of hypercontracted myofibrils alternating with regions devoid of filaments and organelles. In this muscle, myotubes were eliminated through PCD. The sphincter ani muscle in female rats normally undergoes involution. Its degeneration can be prevented by pre- or perinatal administration of testosterone (Cihak et al. 1970).

13 Recruitment of Myogenic Cells from Adult Pluripotent Stem Cells

Recent data suggest that after bone marrow transplantation, donor cells differentiated into many non-haematopoietic tissues including bone (Horwitz et al. 1999), hepatocytes (Alison et al. 2000) and skeletal muscle (Parrish et al. 1996; Bittner et al. 1999). In an elegant study, Ferrari et al. (1998) examined the involvement of non-muscle cells during the re-population of damaged muscle. Previous observations of Grounds et al. (1992) suggested that the number of committed myogenic precursors found soon after injury was far greater than the resident satellite population, implying that these cell were of non-myogenic origin. Injection of bone marrow cells containing the LacZ gene under the control of a skeletal muscle-specific promoter into experimentally damaged tibialis anterior muscle led to β-Gal+ activity in muscle fibres after 2 weeks, demonstrating that haematopoietic cells could populate muscle (Ferrari et al 1998). Interestingly, injected satellite cells were able to participate in muscle regeneration within only 5 days. The time difference could be significant and may reflect bone marrow cells having to undergo a multi-step developmental process in order to become muscle.

In a reciprocal study, Jackson et al. (1999) provided evidence that skeletal muscle contains populations of mononuclear cells that have the capacity to reconstitute the haematopoietic system of lethally irradiated mice. The muscle-derived cells gave rise to all the major lineages found in adult blood. Furthermore the reconstituted bone marrow cells from the irradiated recipients could be used to re-populate secondary recipients. Characterization of mononuclear cells from adult muscle indicated the presence of a heterogeneous population. One of these populations shared characteristics found in bone-marrow-derived haematopoietic stem cells, including the expression of stem cell antigens, Sca-1 and c-kit. The same population, called the Side Population (SP), displayed high efflux of the fluorescent dye Hoechst 33342, a characteristic found in haematopoietic stem cells.

Evidence of at least two populations of mononuclear cells residing in muscle comes from studies of the *Pax7* null mouse. This gene was identified as a gene specifically expressed in satellite cells. Its expression was restricted to cells under the basal lamina of muscle fibres, in positions typical for quiescent satellite cells (Seale et al. 2000). The numbers of *Pax7* cells was elevated in muscle of *Mdx* mice (unable to express dystrophin) and there was an increased number of satellite cells found in this mutant. In new-born mice lacking *Pax7*, there was a reduction in muscle fibre size as opposed to fibre number and, 2 weeks after birth, the null mice were only a third of the weight of heterozygous littermates. Morphological examination of *Pax7* null muscle failed to identify any satellite cells. These studies suggest that *Pax7* is required for the development of satellite cells that will participate in postnatal muscle hypertrophy. Nevertheless, the SP population identified by Jackson et al. (1999) was present at normal levels in the *Pax7* nulls, suggesting that satellite cells and the *SP* cells are distinct entities. The *Pax7* null *SP* had a far greater potential for haematopoietic differentiation compared to wildtype cells, implying that *Pax7* normally serves to restrict the developmental options available to the pluripotent stem cells found in muscle. On the other hand, it cannot be excluded that *Pax7* expression is required for the transition of *SP* cells to satellite cells.

The use of pluripotent stem cells able to move through the circulatory system has been evaluated as a means of repopulating regions of muscle damage produced by genetic disease. Gussoni et al. (1999) have reported that following transplantation of the *SP* from wild-type mice into the bone marrow of *Mdx* mice (unable to express dystrophin) lead to the expression of dystrophin in muscle. In some cases, 9% of all myofibres were dystrophin positive. This approach could overcome the problem of delivering the restitutive muscle cells throughout the entire body in individuals with degenerative muscle diseases.

These recent studies unveil a novel means of muscle repair (Seale and Rudnicki 2000). Adult stem cells appear to be remarkably pluripotent and their differentiation paths are determined not by lineage, but by environmental cues. The contribution of non-muscle stem cells to normal muscle development remains to be assessed (reviewed by Cossu and Mavilio 2000).

Acknowledgements. We thank our colleagues from our laboratories for their contribution to this work. We would also like to thank U. Uhl, Freiburg, for typing the manuscript. Work on the subject of this review is supported by grants (Ch 44/14-1 and SFB 592, A1) from the Deutsche Forschungsgemeinschaft to B.C. This work was also supported by grants USDA 03280, NIH HD-31956 and HL60696.

References

Adams GR, McCue SA (1998) Localized infusion of IGF-I results in skeletal muscle hypertrophy in rats. J Appl Physiol 84:1716–1722

Alison MR, Poulsom R, Jeffery R, Dhillon AP, Quaglia A, Jacob J, Novelli M, Prentice G, Williams J, Wright NA (2000) Hepatocytes from non-hepatic adult stem cells. Nature 406:257

Amthor H, Christ B, Weil M, Patel K (1998) The importance of timing differentiation during limb muscle development. Curr Biol 8:642–652

Amthor H, Christ B, Patel K (1999) A molecular mechanism enabling continuous embryonic muscle growth - a balance between proliferation and differentiation. Development 126: 1041–1053

Anderson JE, Mitchell CM, McGeachie JK, Grounds MD (1995) The time course of basic fibroblast growth factor expression in crush-injured skeletal muscles of SJL/J and BALB/c mice. Exp Cell Res 216:325–334

Ashby PR, Wilson SJ, Harris AJ (1993) Formation of primary and secondary myotubes in aneural muscles in the mouse mutant peroneal muscular atrophy. Dev Biol 156:519–528

Awede B, Thissen J, Gailly P, Lebacq J (1999) Regulation of IGF-I, IGFBP-4 and IGFBP-5 gene expression by loading in mouse skeletal muscle. FEBS Lett 461:263–267

Barnard W, Bower J, Brown MA, Murphy M, Austin L (1994) Leukemia inhibitory factor (LIF) infusion stimulates skeletal muscle regeneration after injury: injured muscle express lif mRNA. J Neurol Sci 123:108–113

Barton-Davis E, Shoturna DI, Musaro A, Rosenthal N, Sweeney HL (1998) Viral mediated expression of insulin-like growth factor blocks the aging-related loss of skeletal muscle function. Proc Natl Acad Sci USA 95:15603–15607

Bittner RE, Schöfer C, Weipoltshammer K, Ivanova S, Streubel B, Hauser E, Freilinger M, Höger H, Elbe-Bürger A, Wachtler F (1999) Recruitment of bone-marrow-derived cells by skeletal and cardiac muscle in adult dystrophic mdx mice. Anat Embryol 199:391–396

Bladt F, Riethmacher D, Isenmann S, Aguzzi A, Birchmeier C (1995) Essential role for the c-met receptor in the migration of myogenic precursor cells into the limb bud. Nature 376:768–771

Bonner PH, Hauschka SD (1974) Clonal analysis of vertebrate myogenesis. I. Early developmental events in the chick limb. Dev Biol 37:317–328

Borycki AG, Mendham L, Emerson CP Jr (1998) Control of somite patterning by Sonic hedgehog and its down stream signal response genes. Development 125:770–790

Borycki AG, Brown AMC, Emerson CP (2000) Shh and Wnt signaling pathways converge to control Gli gene activation in avian somites. Development 127:2075–2087

Brand-Saberi B, Müller TS, Wilting J, Christ B, Birchmeier C (1996) Scatter Factor/Hepatocyte Growth Factor (SF/HGF) induces emigration of myogenic cells at interlimb level in vivo. Dev Biol 179:303–308

Brohmann H, Jagla K, Birchmeier C (2000) The role of Lbx1 in migration of muscle precursor cells. Development 127:437–445

Capdevila J, Tabin C, Johnson RL (1998) Control of dorsoventral somite patterning by Wnt-1 and b-catenin. Dev Biol 193:182–194

Cherel Y, Hurtrel M, Gardahaut MF, Merly F, Magras-Resch C, Fontaine-Perus J, Wyers M (1994) Comparison of postnatal development of anterior latissimus dorsi (ALD) muscle in heavy- and light-weight strains of turkey (*Meleagris gallopavo*). Growth Dev Aging 279:1528–1530

Christ B, Ordahl CP (1995) Early stages of somite development. Anat Embryol 191:381–396

Christ B, Jacob HJ, Jacob M (1977) Experimental analysis of the origin of the wing musculature in avian embryos. Anat Embryol 150:171–186

Christ B, Jacob M, Jacob HJ (1983) On the origin and development of the ventroabdominal muscles in the avian embryo. Anat Embryol 166:87–101

Christ B, Amthor H, Huang R, Wagner J, Patel K (2001) The origin and fate of somites. In: Sanders EJ, Lash JW, Ordahl CP (eds) Life and behavioural sciences, vol 329. NATO Science Series, Series 1. IOS Press, Amsterdam, pp 124–131

Cihák R, Gutmann E, Hanzliková V (1970) Involution and hormone-induced persistence of the M. sphincter (levator) and in female rats. J Anat 106:93–110

Clegg CH, Linkhart TA, Olwin BB, Hauschka SD (1987) Growth factor control of skeletal muscle differentiation: commitment to terminal differentiation occurs in G1 phase and is repressed by fibroblast growth factor. J Cell Biol 105:949–956

Coleman ME, DeMayo F, Yin KC, Lee HM, Geske R, Montgomery C, Schwartz RJ (1995) Myogenic vector expression of insulin-like growth factor I stimulates muscle cell differentiation and myofiber hypertrophy in transgenic mice. J Biol Chem 270:12109–12116

Condon K, Silberstein L, Blau HM, Thompson WJ (1990) Development of muscle fiber types in the prenatal rat hindlimb. Dev Biol 138:256–274

Cossu G, Mavilio F (2000) Myogenic stem cells for the therapy of primary myopathies: wishful thinking or therapeutic perspective? J Clin Invest 105:1669–1974

Crow MT, Stockdale FE (1984) Myosin isoforms and the cellular basis of skeletal muscle development. Expl Biol Med 9:165–174

Crow MT, Stockdale FE (1986a) The developmental program of fast myosin heavy chain expression in avian skeletal muscles. Dev Biol 118:333–342

Crow MT, Stockdale FE (1986b) Myosin expression and specialization among the earliest muscle fibers of the developing avian limb. Dev Biol 113:238–254

Cusella-De Angelis MG, Molinari S, Le Donne A, Voletta M, Vivarelli E, Bouche M, Molinaro M, Ferrari S, Cossu G (1994) Differential response of embryonic and fetal myoblasts to TGF beta: a possible regulatory mechanism of skeletal muscle histogenesis. Development 120:925–933

DeHamer MK, Guevara JL, Hannon K, Olwin BB, Calof AL (1994) Genesis of olfactory receptor neurons in vitro: regulation of progenitor cell divisions by fibroblast growth factors. Neuron 13:1083–1097

de Lapeyriere O, Olledorff V, Planche J, Ott MO, Pizette S, Coulier F, Birnbaum D (1993) Expression of FGF6 gene is restricted to developing skeletal muscle in the mouse embryo. Development 118:601–611

Delfini M-C, Hirsinger E, Pourquié O, Duprez D (2000) Delta 1-activated Notch inhibits muscle differentiation without affecting *Myf-5* and *Pax3* expression in chick limb myogenesis. Development 127:5213–5224

Denetclaw WF, Ordahl CP (2000) The growth of the dermomyotome and formation of early myotome lineages in thoracolumbar somites of chicken embryos. Development 127:893–905

Denetclaw WJ Jr, Christ B, Ordahl CP (1997) Location and growth of epaxial myotome precursor cells. Development 124:1601–1610

Dietrich S, Schubert FR, Healy C, Sharpe PT, Lumsden A (1998) Specification of the hypaxial musculature. Development 125:2235–2249

DiMario JX, Stockdale FE (1997) Both myoblast lineage and innervation determine fiber type and are required for expression of the slow myosin heavy chain 2 gene. Dev Biol 188:167–180
Dono R, Zeller R (1994) Cell-type-specific nuclear translocation of fibroblast growth factor-2 isoforms during chicken kidney and limb morphogenesis. Dev Biol 163:316–330
Dunglison GF, Scotting PJ, Wigmore PM (1999) Rat embryonic myoblasts are restricted to forming primary fibers while later myogenic populations are pluripotent. Mech Dev 87:11–19
Dunn SE, Burns JL, Michel RN (1999) Calcineurin is required for skeletal muscle hypertrophy. J Biol Chem 274: 21908–21912
Düsterhoft S, Dieterle R, Pette D (1992) Myosin isoforms in rat satellite cell cultures of different origin. Eur J Cell Biol 57:16
Duxson MJ (1992) The relationship of nerve to myoblasts and newly-formed secondary myotubes in the fourth lumbrical muscle of the rat fetus. J Neurocytol 21:574–588
Duxson MJ, Sheard PW (1995) Formation of new myotubes occurs exclusively at the multiple innervation zones of an embryonic large muscle. Dev Dyn 204:391–405
Duxson MJ, Usson Y (1989) Cellular insertion of primary and secondary myotubes in embryonic rat muscles. Development 107:243–251
Duxson MJ, Usson Y, Harris AJ (1989) The origin of secondary myotubes in mammalian skeletal muscles: ultrastructural studies. Development 107:743–750
Evans D, Baillie H, Caswell A, Wigmore P (1994) During fetal muscle development, clones of cells contribute to both primary and secondary fibers. Dev Biol 162:348–353
Fan C-M, Tessier-Lavigne M (1994) Patterning of mammalian somites by surface ectoderm and notochord: evidence for sclerotome induction by a hedgehog homolog. Cell 79:1175–1186
Feldman JL, Stockdale FE (1991) Skeletal muscle satellite cell diversity: satellite cells form fibers of different types in cell culture. Dev Biol 143:320–334
Feldman JF, Stockdale FE (1992) Temporal appearance of satellite cells during myogenesis. Dev Biol 153:217–226
Fernandes JJ, Keshishian H (1998) Nerve-muscle interactions during flight muscle development in *Drosophila*. Development 125:1769–1779
Ferrari G, Cusela-De Angelis G, Coletta M, Paolucci E, Stornaiuolo A, Cossu G, Mavilio F (1998) Muscle regeneration by bone marrow-derived myogenic progenitors. Science 279:1529–1530
Flanagan-Steet H (2000) Loss of FGF receptor 1 signaling reduces skeletal muscle mass and disrupts myofiber organization in the developing limb. Dev Biol 218:21–37
Florini JR, Ewton DZ, Coolican SA (1996) Growth hormone and the insulin-like growth factor system in myogenesis. Endocr Rev 17:481–517
Floss T, Arnold HH, Braun T (1997) A role for FGF-6 in skeletal muscle regeneration. Genes Dev 11:2040–2051
Fournier M, Lewis MI (2000) Influences of IGF-I gene disruption on the cellular profile of the diaphragm. Am J Physiol Endocrinol Metab 278:E707–715
Franz T, Kothary R, Surani MAH, Halata Z, Grim M (1993) The splotch mutation interferes with muscle development in the limbs. Anat Embryol 187:153–160
Fredette BJ, Landmesser LT (1991) A reevaluation of the role of innervation in primary and secondary myogenesis in developing chick muscle. Dev Biol 143:19–35
Gal-Levi R, Leshem Y, Aoki S, Nakamura T, Halevy O (1998) Hepatocyte growth factor plays a dual role in regulating skeletal muscle satellite cell proliferation and differentiation. Biochim Biophys Acta 1402:39–51
Gerrard DE, Okamura CS, Ranalletta MA, Grant AL (1998) Developmental expression and location of IGF-I and IGF-II mRNA and protein in skeletal muscle. J Anim Sci 76:1004–1011
Goldspink G (1972) Post embryonic growth and differentiation of striated muscle: In: Bourne GH (ed) The structure and function of muscle, vol 1. Academic Press, New York, pp 179–236
Goulding MD, Lumsden A, Paquett AJ (1994) Regulation of Pax-3 expression in the dermomyotome and its role in muscle development. Development 120:957–971
Graham A, Francis-West P, Brickell P, Lumsden A (1994) The signaling molecule BMP4 mediates apoptosis in the rhombencephalic neural crest. Nature 372:684–686

Grim M (1977) Die Myotubenrückbildung in den Muskelanlagen der menschlichen Hand. Verh Anat Ges 71:1239–1243
Grim M (1978) Acid phosphatase activity in normal and sarcolytic myotubes in muscle anlagen of the human hand. Histochemistry 56:307–316
Gross M, Moran-Rivard L, Velasquez T, Nakatsu MN, Jagla K (2000) Lbxl is required for muscle precursor migration along a lateral pathway into the limb. Development 127:413–424
Grounds MD, Garrett KL, Lai MC, Wright WE, Beilharz MW (1992) Identification of skeletal muscle precursor cells in vivo by use of MyoDl and myogenin probes. Cell Tissue Res 267:99–104
Gussoni E, Soneoka Y, Strickland CD, Buzney EA, Khan MK, Flint AF, Kunkel LM, Mulligan RC (1999) Dystrophin expression in the mdx mouse restored by stem cell transplantation. Nature 401:390–394
Hacker A, Guthrie S (1998) A distinct developmental programme for the cranial paraxial mesoderm in the chick embryo. Development 125:3461–3472
Han JK, Martin GR (1993) Embryonic expression of Fgf-6 is restricted to the skeletal muscle lineage. Dev Biol 158:549–554
Hannon K, Kudla AJ, McAvoy MJ, Clase KL, Olwin BB (1996) Differentially expressed fibroblast growth factors regulate skeletal muscle development through autocrine and paracrine mechanisms. J Cell Biol 132:1161–1159
Harris AJ, Duxson MJ, Fitzsimons RB, Rieger F (1989a) Myonuclear birth dates distinguish the origins of primary and secondary myotubes in embryonic mammalian skeletal muscles. Development 107:771–784
Harris AJ, Fitzsimons RB, McEwan JC (1989b) Neural control of the sequence of expression of myosin heavy chain isoforms in fetal mammalian muscles. Development 107: 751–769
Hartley RS, Bandmann E, Yablonka-Reuveni Z (1992) Skeletal muscle satellite cells appear during late chicken embryogenesis. Dev Biol 153:206–216
Horwitz EM, Prockop DJ, Fitzpatrick LA, Koo WW, Gordon PL, Neel M, Sussman M, Orchard P, Marx JC, Pyeritz RE, Brenner MK (1999) Transplantability and therapeutic effects of bone marrow-derived mesenchymal cells in children with osteogenesis imperfecta. Nat Med 5: 309–313
Huang R, Christ B (2000) Origin of the epaxial and hypaxial myotome in avian embryos. Anat Embryol 202:369–374
Hughes DS, Ontell M (1992) Morphometric analysis of the developing, murine aneural soleus muscle. Dev Dyn 193:175–184
Itoh N, Mima T, Mikawa T (1996) Loss of fibroblast growth factors receptors is necessary for terminal differentiation of embryonic limb muscle. Development 122:291–300
Jackson KA, Mi T, Goodell MA (1999) Hematopoietic potential of stem cells isolated from murine skeletal muscle. Proc Natl Acad Sci USA 961:4482–4486
Joseph-Silverstein J, Consigli SA, Lyser KM, Ver Pault C (1989) Basic fibroblast growth factor in he chick embryo: immunolocalization to striated muscle cells and their precursors. J Cell Biol 108:2459–2466
Kadi F, Thornell LE (2000) Concomitant increases in myonuclear and satellite cell content in female trapezius muscle following strength training. Histochem Cell Biol 113:99–103
Kennedy JM, Eisenberg BR, Reid SK, Sweeney LJ, Zak R (1988) Nascent muscle fiber appearance in overloaded chicken slow-tonic muscle. Am J Anat 181:203–215
Marcelle C, Stark MR, Bronner-Fraser M (1997) Coordinate action of BMPs, Wnts, Shh and Noggin mediate patterning of the dorsal somite. Development 124:3955–3963
Marcucio RS, Noden DM (1999) Myotube heterogeneity in developing chick craniofacial skeletal muscles. Dev Dyn 214:178–194
Mason I, Fuller-Pace F, Smith R, Dickson C (1994) FGF-7 (keratinocyte growth factor) expression during mouse development suggests roles in myogenesis, forebrain regionalization and epithelial-mesenchymal interactions. Mech Dev 45:15–30

McClearn D, Medville R, Noden D (1995) Muscle cell death during he development of head and neck muscles in the chick embryo. Dev Dyn 202:365–377

McLennan IS (1983) The development of the pattern of innervation in chicken hindlimb muscles: evidence for specification of nerve-muscle connections. Dev Biol 97:229–238

McLennan IS, Poussart Y, Koishi K (2000) Development of skeletal muscles in transforming growth factor-beta 1 (TGF-beta1) null-mutant mice. Dev Dyn 217:250–256

McPherron AC, Lee SJ (1997) Double muscling in cattle due to mutations in the myostatin gene. Proc Natl Acad Sci USA 94:12457–12461

McPherron AC, Lawler Am, Lee SJ (1997) Regulation of skeletal muscle mass in mice by a new TGF-beta superfamily member. Nature 387:83–90

Miller JB, Stockdale FE (1986a) Developmental origins of skeletal muscle fibers: clonal analysis of myogenic cell lineages based on fast and slow myosin heavy chain expression. Proc Natl Acad Sci USA 83:3860–3864

Miller JB, Stockdale FE (1986b) Developmental regulation of the multiple myogenic cell lineages of the avian embryo. J Cell Biol 103:2197–2208

Mitchell CA, McGeachie JK, Grounds MD (1996) The exogenous administration of basic fibroblast growth factor to regenerating skeletal muscle in mice does not enhance the process of regeneration. Growth Factors 13:37–55

Mitchell P, Steenstrup T, Hannon K (1999) Expression of fibroblast growth factor family during postnatal skeletal muscle hypertrophy. J Appl Physiol 86:313–319

Münsterberg AE, Kitajewski J, Bumcrot DA, McMahon AP, Lassar AB (1995) Combinatorial signaling by Sonic hedgehog and Wnt family members induces myogenic bHLH gene expression in the somite. Genes Dev 9:2911–2922

Musaro A, McCullagh KJ, Naya, FJ, Olson EN, Rosenthal N (1999) IGF-1 induces skeletal myocyte hypertrophy through calcineurin in association with GATA-2 and NF-ATc1. Nature 400: 581–585

Naya FJ, Mercer B, Shelton J, Richardson JA, Williams RS, Olson EN (2000) Stimulation of slow skeletal muscle fiber gene expression by calcineurin in vivo. J Biol Chem 275:4545–4548

Nishikawa A, Hayashi H (1994) Isoform transition of contractile proteins related to muscle remodeling with an axial gradient metamorphosis in *Xenopus leavis*. Dev Biol 165:86–94

Nishikawa A, Hayashi H (1995) Spatial, temporal and hormonal regulation of programmed muscle cell death during metamorphosis of the frog *Xenopus laevis*. Differentiation 59: 207–214

Niswander L, Martin G (1992) Fgf-4 expression during gastrulation, myogenesis, limb and tooth development in the mouse. Development 114:755–768

Olson EN, Williams RS (2000) Remodeling muscles with calcineurin. Bioessays 22:510–519

Olson EN, Sternberg E, Hu JS, Spizz G, Wilcox C (1986) Regulation of myogenic differentiation by type beta transforming growth factor. J Cell Biol 103:1799–1805

Olwin BB, Arthur K, Hannon K, Hein P, McFall A, Riley B, Szebenyi G, Zhou Z, Zuber ME, Rapraeger AC (1994a) Role of FGFs in skeletal muscle and limb development. Mol Reprod Dev 39: 90–101

Olwin BB, Hannon K, Kudla AJ (1994b) Are fibroblast growth factors regulators of myogenesis in vivo? Prog Growth Factor Res 5:145–158

Ontell M, Kozeka K (1984) Organogenesis of the mouse extensor digitorum longus muscle: a quantitative study. Am J Anat 171:149–161

Ontell M, Hughes D, Bourke D (1988) Morphometric analysis of the developing mouse soleus muscle. Am J Anat 181:279–288

Ontell M, Ontell MP, Sopper MM, Mallonga R, Lyons G, Buckingham M (1993) Contractile protein gene expression in primary myotubes of embryonic mouse hindlimb muscles. Development 117:1435–1444

Ordahl CP, Le Douarin NM (1992) Two myogenic lineages within the developing somite. Development 114:339–353

Parrish EP, Cifuentes-Diaz C, Li ZL, Vicart P, Pulin D, Dreyfus PA, Peschanski M, Harris AJ, Garcia L (1996) Targeting widespread sites of damage in dystrophic muscle: engrafted macrophages as potential shuttles. Gene Ther 3:13–20

Phillips WD, Bennett MR (1984) Differentiation of fiber types in wing muscles during embryonic development: effect of neural tube removal. Dev Biol 106:457–468

Pin CL, Merrifield PA (1993) Embryonic and fetal rat myoblasts express different phenotypes following differentiation in vitro. Dev Genet 14:356–368

Pirskanen A, Kiefer JC, Hauschka SD (2000) IGFs, insulin, Shh, bFGF, and TGF-beta1 interact synergistically to promote somite myogenesis in vitro. Dev Biol 224:189–203

Ross JJ, Duxson MJ, Harris AJ (1987a) Formation of primary and secondary myotubes in rat lumbrical muscles. Development 100:383–394

Ross JJ, Duxson MJ, Harris, AJ (1987b) Neural determination of muscle fiber numbers in embryonic rat lumbrical muscles. Development 100:395–409

Scaal M, Bonafede A, Dathe V, Sachs M, Cann G, Christ B, Brand-Saberi B (1999) SF/HGF is a mediator between limb patterning and muscle development. Development 126:4885–4893

Schmidt C, Christ B, Patel K, Brand-Saberi B (1998) Experimental induction of BMP-4 expression leads to apoptosis in the paraxial and lateral plate mesoderm. Dev Biol 202:253–263

Seale P, Rudnicki MA (2000) A new look at the origin, function and "stem-cell" status of muscle satellite cells. Dev Biol 218:115–124

Seale P, Sabourin LA, Girgis-Gabardo A, Mansouri A, Gruss P, Rudnicki MA (2000) Pax7 is required for the specification of myogenic satellite cells. Cell 102:777–786

Seed J, Olwin BB, Hauschka SD (1988) Fibroblast growth factor levels in the whole embryo and limb bud during chick development. Dev Biol 128:50–57

Semsarian C, Wu MJ, Ju YK, Marciniec T, Yeoh T, Allen DG, Harvey RP, Graham RM (1999a) Skeletal muscle hypertrophy is mediated by a Ca^{2+}-dependent calcineurin signaling pathway. Nature 400:576–581

Semsarian C, Sutrave P, Richmond DR, Graham RM (1999b) Insulin-like growth factor (IGF-I) induces myotube hypertrophy associated with an increase in anaerobic glycolysis in a clonal skeletal-muscle cell model. Biochem J 339:443–451

Sheard PW, Duxson MJ, Harris AJ (1991) Neuromuscular transmission to identified primary and secondary myotubes: a reevaluation of polyneuronal innervation patterns in rat embryos. Dev Biol 148:459–472

Stockdale FE (1992) Myogenic cell lineages. Dev Biol 154:284–298

Stockdale FE (1997) Mechanisms of formation of muscle fiber types. Cell Struct Funct 22:37–43

Stockdale FE, Holtzer H (1961) DNA synthesis and myogenesis. Exp Cell Res 24:508–520

Švagr V, Grim M, Jacob HJ (1995) Zur Kontrolle der Muskelrückbildung während der Entwicklung. Vers Anat Ges Graz 177:120–121

Swatland HJ, Kieffer NM (1974) Fetal development of the double muscled condition in cattle. J Anim Sci 38:752–757

Tajbakhsh S, Borello U, Vivarelli E, Kelly R, Papkoff J, Duprez D, Buckingham M, Cossu G (1998) Differential activation of Myf5 and MyoD by different Wnts in explants of mouse paraxial mesoderm and the later activation of myogenesis in the absence of Myf5. Development 125: 4155–4162

Tatsumi R, Anderson JE, Nevoret CJ, Halevy O, Allen RE (1998) HGF/SF is present in normal adult skeletal muscle and is capable of activating satellite cells. Dev Biol 194:114–128

Templeton TJ, Hauschka SD (1992) FGF-mediated aspects of skeletal muscle growth and differentiation are controlled by a high affinity receptor, FGFR1. Dev Biol 154:169–181

Thomas M, Langley B, Berry C, Sharma M, Kirk S, Bass J, Kambadur R (2000) Myostatin, a negative regulator of muscle growth, functions by inhibiting myoblast proliferation. J Biol Chem 275:40235–40243

Wagner J, Schmidt C, Nikovits W, Christ B (2000) Compartmentalization of the somite and myogenesis in chick embryos are influenced by Wnt expression. Dev Biol 228:86–94

Weatherby AH, Gill HS (1985) Dynamics of increase in muscle fibres in relation to size and growth. Experimentia 41:353–354

Weber R (1969) Tissue involution and lysosomal enzymes during anuran metamorphosis in biology and pathology. In: Dingle JT, Fell HG (eds) Lysosomes in biology and pathology. North Holland, Amsterdam, pp 437–446
Wilson SJ, Harris AJ (1993) Formation of myotubes in aneural rat muscles. Dev Biol 156:509–518
Wu H, Naya FJ, McKinsey TA, Mercer B, Shelton JM, Chin ER, Simard AR, Michel RN, Bassel-Duby R, Olson EN (2000) MEF2 responds to multiple calcium-regulated signals in the control of skeletal muscle fiber type. EMBO J 19:1963–1973
Zhang M, McLennan IS (1995) During secondary myotube formation, primary myotubes preferentially absorb new nuclei at their ends. Dev Dyn 204:168–177
Zhu X, Hadhazy M, Wehling M, Tidball JG, McNally EM (2000) Dominant negative myostatin produces hypertrophy without hyperplasia in muscle. FEBS Lett 474:71–75
Zimmerman AM, Lowery MS (1999) Hyperplastic development and hypertrophic growth of muscle fibers in the white seabass (*Atractoscion nobilis*). J Exp Zool 284:299–308

Cadherins in Skeletal Muscle Development

Zoe Waibler and Anna Starzinski-Powitz[1]

1 Cadherins

One essential step in the complex process of myogenesis is the fusion of mononuclear myoblast cells into polynuclear myotubes. This fusion process is promoted by cadherins, a multigene family of transmembrane cell adhesion glycoproteins that mediate homophilic interactions in a calcium-dependent manner, thus selectively associating cells into specific tissues. In the absence of calcium the cadherin molecular conformation is changed, leading to proteolytic digestion of the proteins (Ringwald et al. 1987; Takeichi 1991; Geiger and Ayalon 1992; Huber et al. 1996). Because the expression of cadherins is tissue-specific, many of these molecules have been classified according to the tissues from which they were isolated, such as P-cadherin from placenta, E-cadherin first isolated from epithelial cells or M-cadherin from muscle.

The cadherins can be divided into two subfamilies, the classical cadherins and the protocadherins (Tepass 1999). Each of these groups can be further subdivided, but only members of the classical cadherins are involved in vertebrate myogenesis.

1.1 Cadherin Structure and Interactions

All classical cadherins are initially expressed as a precursor with an N-terminal signal and propeptide domain that is proteolytically cleaved off during maturation, a necessary step to obtain active adhesion-competent molecules (Ozawa and Kemler 1990). After cleavage, cadherins possess molecular weights ranging from 105 to 140 kDa (Ringwald et al. 1987; Shapiro et al. 1995; Tamura et al. 1998). The mature protein can be divided into three main domains: the N-terminal extracellular domain, a short transmembrane region and the C-terminal cytoplasmic domain (see Fig. 1). The extracellular domain consists of five subdomains, EC1 to EC5, each comprising approximately 110 amino acids (desmogleins, which may be grouped together with the classical cadherins, possess only four extracellular cadherin domains). Calcium-

[1] Johann Wolfgang Goethe Universität, Institut für Anthropologie und Humangenetik für Biologen, Siesmayerstr. 70, 60323 Frankfurt am Main, Germany

Results and Problems in Cell Differentiation, Vol. 38
B. Brand-Saberi (Ed.): Vertebrate Myogenesis

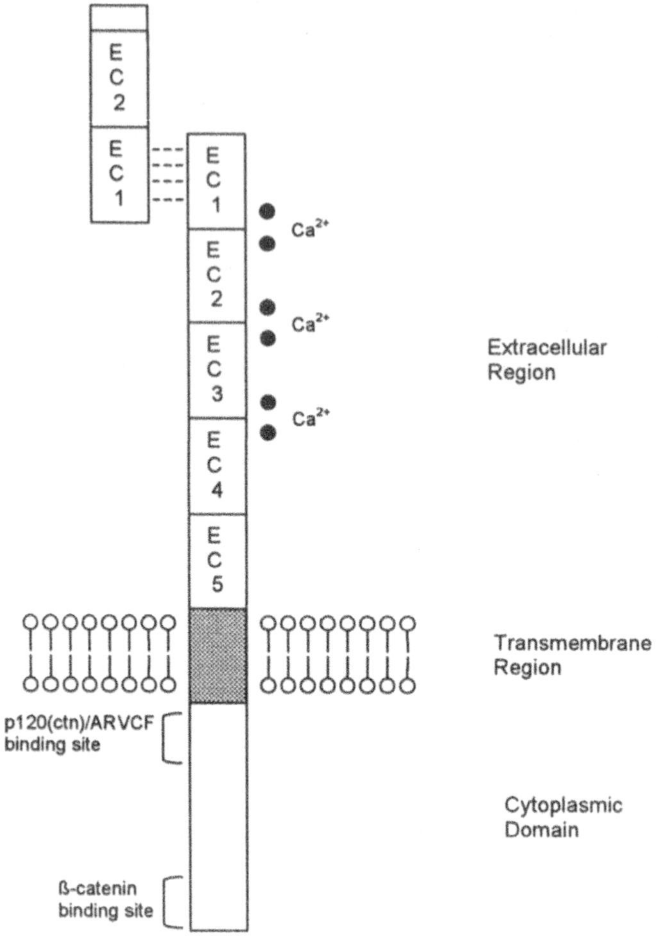

Fig. 1. Cadherin–cadherin interaction and consensus structure of mature classical cadherin. The extracellular domains (EC1–5) and calcium binding sites are indicated (Ca^{2+}). ARVCF or p120(ctn) binding site comprises the cadherin juxtamembrane region, the β-catenin binding site is proximal to the C-terminus

binding sites are located in between these EC domains. The proximal N-terminal EC1 domain contains the HAV (single-letter code) tripeptide required for general interactions with cadherins expressed on neighbouring cells. This HAV motive may be substituted by other amino acids, for instance an FAL motif in the case of M-cadherin (Donalies et al. 1991). The specificity of the homophilic cadherin interaction is regulated by amino acids in EC1 other than the HAV or FAL tripeptide (Nose et al. 1990). In addition to this *trans*-interaction (interaction between cadherins of neighbouring cells), cadherins are also able to form *cis*-dimers (interaction between cadherins in the same cell). *cis*-dimerization is dependent on calcium being bound to the extracellular domain and seems to be necessary for subsequent *trans*-interactions

(Chothia and Jones 1997; Humphries and Newham 1998; Tamura et al. 1998; Steinberg and McNutt 1999; Takeda et al. 1999). Previously, it was thought that *cis*-dimers exist only between cadherins of the same type, but recently it was demonstrated that specific formation of *cis*-heterodimers between N- and R-cadherin is also possible (Shan et al. 2000).

1.2 Cadherins and Catenins

Cell–cell adhesion requires not only cadherin-mediated interaction between two neighbouring cells but also the cadherin cell adhesion molecules linked to the cytoskeleton. The cytoplasmic domain of cadherins, about 150 to 160 amino acids in length, is responsible for connecting the cadherins to the cytoskeleton. However, this interaction is not direct, but mediated by a group of linker molecules called catenins (Latin catena: chain) via a C-terminal binding site on the cadherin protein (Fig. 1). The armadillo repeat protein β-catenin binds to both the proximal C-terminal amino acids of cadherins and to α-catenin, which, in turn, is able to bind to actin. Plakoglobin, also known as γ-catenin, is also able to bind to the cytoplasmic tail of cadherins and can, therefore, exclude β-catenin from the complex (Ozawa et al. 1989; Hirano et al. 1992; Butz and Kemler 1994; Hinck et al. 1994; Knudsen et al. 1995; Hertig et al. 1996; Kuch et al. 1997). Linkage of the cadherin complex to actin via catenins is an essential component for conferring adhesiveness to this complex. Experiments showed that adhesion between cells expressing E-cadherin truncated for the β-catenin binding site is much lower than between cells expressing wild-type E-cadherin (Miller and Moon 1996). Furthermore, in cardiomyocytes the N-cadherin–catenin complex was shown to be a crucial component for forming intact myofibrillar structures in these cells. Disturbance of the adherence complex reduces the integrity of the myofibrils and thus may impair cardiac function (Soler and Knudsen 1994).

In addition to the proximal C-terminal amino acids of the cytoplasmic cadherin tail, the juxtamembrane region of the protein is also important for the interaction with certain members of the catenin family. Amino acids in this region are able to recruit, for example, p120(ctn) or ARVCF (armadillo repeat gene deleted in velo cardio facial syndrome) to the complex (Reynolds et al. 1994; Kaufmann et al. 2000; Mariner et al. 2000). Both molecules are members of the armadillo repeat protein family and at least p120(ctn) is able to modulate cadherin-mediated cell adhesion in a positive and negative manner (Yap et al. 1998; Aono et al. 1999). Depending on the cell context and the phosphorylation status of p120(ctn), the protein is able to promote or reduce adhesiveness mediated by the cadherins. The clustering potential of the juxtamembrane region and p120(ctn) seems to be activated by the homophilic *trans*-interaction occurring between the cadherin extracellular domains. On the other hand, truncation of the p120(ctn) binding site in the respective cadherin leads to weaker adhesion between transfected cells than between cells expressing wildtype cadherin. Truncated cadherin was also no longer able to induce clustering despite the mutated molecule still being able to associate with the cytoskleleton (Anastasiadis and Reynolds 2000). Recent reports also demonstrate that E-cadherin is able to interact indirectly with the immunoglobulin-like Ca^{2+} independent adhesion molecule, nectin, via their cytoplasmic domain-associated proteins (Tachibana et al. 2000).

Clearly, cadherin-mediated intercellular adhesion is regulated by multiple mechanisms. For example, connection of the cadherin–catenin complex to the microfilament system via β-catenin and α-catenin can be controlled by tyrosine kinases (e.g. src, c-met) and phosphatases, which mainly regulate the degree of tyrosine phosphorylation of β-catenin, and probably also of plakoglobin, cadherin and p120(ctn); (reviewed in Provost and Rimm 1999). Thus, the degree of intercellular adhesion is a dynamic process that can be modulated by paracrine mechanisms.

In addition to their function in the adherence complex, catenins and cadherins are themselves important players in signal transduction. One major field of biomedical research in the past few years has concerned investigations where catenins have been shown to act as signalling molecules (Barker and Clevers 2000). Because they are able to interact with transcription factors and other proteins they participate in the regulation of gene expression, thereby influencing cell behaviour such as cell motility and cell cycle (Barth et al. 1997; Behrens 1999). These processes, however, have so far mostly been studied in non-muscle cells.

2 Cadherins in Myogenesis

Cadherins are expressed in a tissue-specific manner. The cadherins important for skeletal muscle development are mainly N-cadherin, R-cadherin and M-cadherin. N-cadherin was initially identified as a neuronal cadherin, but it is also expressed in cardiac and skeletal muscle, lens and endothelial cells (Charlton et al. 1997; George-Weinstein et al. 1997; Linask et al. 1997, 1998; Goichberg and Geiger 1998). R-cadherin (retinal cadherin) is expressed in developing muscle - but not cardiac muscle - and also in developing epithelial in kidney, lung, thymus and in early optic nerve glia (Inuzuka et al. 1991; Hutton et al. 1993; Matsunami et al. 1993; Redies and Takeichi 1993; Rosenberg et al. 1997). The muscular M-cadherin is mainly expressed in developing and regenerating skeletal muscle, satellite cells and in specialized adherence junctions in the granule cell layer of the cerebellar glomerulus (Moore and Walsh 1993; Irintchev et al. 1994; Rose et al. 1995; Zeschnigk et al. 1995).

2.1 M-Cadherin

Using in-situ hybridization on sections of BALB/c mice, M-cadherin mRNA is first detected in somites at day 8.5 of gestation (E8.5). This mRNA seems to

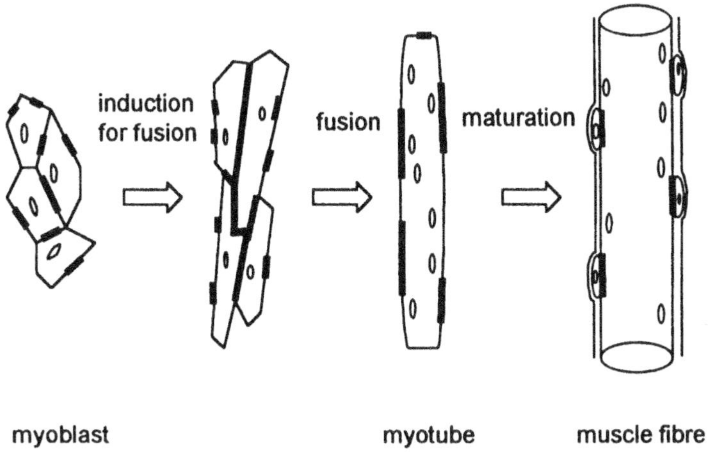

Fig. 2. Fusion of myoblasts into myotubes and subsequent maturation of the muscle fibre. *Black bars* indicate M-cadherin distribution during this process. The mature muscle fibre and the corresponding satellite cells are covered by the basal lamina. (Bornemann and Schmalbruch 1994; Irintchev et al. 1994; Pouliot et al. 1994; Rose et al. 1994; Eng et al. 1997; Kuch et al. 1997)

be translated with a delay of nearly 2 days because protein can first be detected only at day E10 1/3 using specific immunofluorescent antibodies. At these developmental stages M-cadherin is evenly distributed over the surface of the myotomal cells. At day E16, the point when the differentiating muscle precursor consists of large primary myotubes and smaller secondary myotubes (Ontell and Kozeka 1984), M-cadherin appears in clusters at the cell surface of all myotubes (Moore and Walsh 1993; Rose et al. 1994). In the developing limb bud, studied in rat fetuses, M-cadherin shows a developmentally regulated pattern of expression similar to that observed in the axial muscle of mouse. While migrating myoblasts do not express M-cadherin, cells forming small muscle bundles express the transmembrane protein, which is evenly distributed on the cell surface (Kaufmann et al. 1999a). During the fusion of myoblasts into myotubes, M-cadherin forms clusters similar to those seen in axial muscle on day E16 of mouse muscle development (Bornemann and Schmalbruch 1994; Rose et al. 1994). These changes in M-cadherin distribution in vivo are confirmed by investigations carried out in cell culture (Eng et al. 1997). In primary rat myoblasts M-cadherin is again distributed evenly over the cell surface, but is mainly located at cell–cell contacts during induction of fusion. In the fused myotubes M-cadherin is observed at various regions of the plasma membrane, but in the mature muscle fibre it is concentrated at the interface between the fibre and the satellite cells that also express this cadherin (see Fig. 2). M-cadherin expression during myogenesis in vitro was also investigated by Northern and Western blots. During this myogenic development, the transcript is present at low levels in myoblasts, but is upregulated in cells forming myotubes (Donalies et al. 1991; Kuch et al. 1997). In vivo, M-cadherin expression is downregulated shortly before birth, but expression can be transiently reactivated after muscle injury when quiescent satellite cells re-enter the cell cycle to subsequently replace the injured muscle sections (Irintchev et al. 1994). However, expression of M-cadherin in reactivated satellite cells isolated from myoD deficient mice is severely depressed (Sabourin et al. 1999). These myoD(–/–) satellite cells also show unusual aggregate structures, fail to fuse efficiently, and exhibit reduced differentiation efficiency (Cornelison et al. 2000). These observations are in line with the idea that M-cadherin is an important morphoregulatory protein during myogenesis. The important role of M-cadherin during myogenesis, in particular the regulation of myoblast fusion, is further supported by experiments showing that peptide antagonists of the extracellular binding region of M-cadherin, or expression of M-cadherin antisense RNA, reduces the fusion of L6 myoblasts in cell culture. Blocking M-cadherin using peptide antagonists induced a dose-dependent inhibition of myotube formation, but could not prevent biochemical differentiation of the cells (Zeschnigk et al. 1995).

Myoblast fusion is also influenced by the intermediate filaments desmin and vimentin (Tao and Wallace 1991; Li et al. 1994). Other investigations have implied that intact microtubules are important for myogenic differentiation, in particular for the alignment and elongation of the pre-fusion myoblasts (Antin et al. 1981; Toyama et al. 1982; Holtzer et al. 1985; Saitoh et al. 1988). With regard to cadherins, it was shown that the M-cadherin/catenin complex is able to interact specifically with microtubules in myogenic cells, but not in non-muscle cells. This interaction appears to be mediated by a muscle-specific factor (or factors). Thus, the M-cadherin–catenin complex and microtubules are proposed to be joint players in controlling the elongation and alignment of myoblasts during the fusion process (Kaufmann et al. 1999b).

In summary, investigations concerning the expression profile, regulation and function of M-cadherin in vivo (axial muscle, limb bud) and in cell culture implicate a central role for M-cadherin in muscle development, especially in the fusion of myoblasts into myotubes and in attaching satellite cells to the intact muscle fibre.

2.2 N-Cadherin

In addition to its well-established function in heart muscle, N-cadherin also plays a crucial role in skeletal myogenesis. In cell culture N-cadherin is distributed on the surface of the fusing myoblast (Mège et al. 1992). Induction of myogenesis in murine C2C12 cells results in upregulation of N-cadherin, whereas it is downregulated in myotubes (Shin et al. 2000). Previous experiments with myogenic chicken cultures have shown that myoblast fusion can be inhibited with anti-N-cadherin antibodies or by using synthetic peptides containing the HAV binding motive of the N-cadherin extracellular domain EC1 (Knudsen et al. 1990; Mège et al. 1992).

In vivo, N-cadherin was detected in migrating pre-mitotic myoblasts and in stationary mesenchyme cells of the dorsal and ventral myogenic zone as well as in the progress zone. Injection of N-cadherin antibodies into chicken embryos blocked the migration of dermomyotomal cells to the limb bud, which implied that N-cadherin is an important path-finding molecule for myogenic precursor cells (Brand-Saberi et al. 1996). This assumption is in line with recent findings in tumor biology, which indicated that over-expression of N-cadherin might influence invasive and metastatic cell behaviour (Hazan et al. 2000).

N-cadherin(–/–) mice die by day 10 of gestation (E10) because of the heart tube failing to develop normally. In addition, epithelial organization of the somites is disrupted, the mesodermal and endodermal layers of the yolk sac separate and the neural tube is malformed (Radice et al. 1997). Because these mice die before muscle tissue formation, a genetic analysis of N-cadherin in skeletal muscle is not possible with these knockouts. Instead, cultured N-cadherin deficient primary myoblasts were used to genetically test the proposed role of N-cadherin in myoblast fusion. Fusion competence of these cells lacking N-cadherin was found to be equivalent to myoblasts expressing the neuronal adhesion molecule, both in cell culture and in vivo as determined by injecting them into the muscle of adult mice (Charlton et al. 1997). In contrast to this, experiments with cadherin-negative baby hamster kidney cells (BHK) cultured in three-dimensional (3-D) cell aggregates showed that expression of exogenous N-cadherin stimulated differentiation into skeletal muscle, as measured by increased expression of sarcomeric myosin. A non-functional mutant of N-cadherin failed to promote skeletal muscle differentiation, suggesting an adhesion-competent cadherin is required (Redfield et al. 1997). In addition to its proposed role in myoblast fusion, N-cadherin also seems to be involved in the biochemical differentiation of myogenic cells. Introduction of N-cadherin into BHK cells enhances expression of myoD and upregulates the protein level of nuclear myogenin in cells induced to differentiate by 3-D culture (Redfield et al. 1997; Seghatoleslami et al. 2000). Thus, one role of N-cadherin adhesion in myogenesis might be to stimulate the upregulation of myogenic markers.

2.3 R-Cadherin

R-cadherin was first identified as a molecule involved in retinal morphogenesis, but further investigations showed that it also plays a part in myogenic development of the skeletal muscle (Inuzuka et al. 1991; Redies and Takeichi 1993; Rosenberg et al. 1997). In the developing muscle R-cadherin is first detected at E10.5–11.5 in the somitic myotome. Subsequently, the molecule is expressed at the highest levels in the myotome, early skeletal muscle and smooth muscle. During primary myogenesis R-cadherin is expressed ubiquitously in myoblasts, but is absent from primary myotubes. At later stages the retinal adhesion molecule is expressed selectively at the interface between primary myotubes and secondary myoblasts (Hutton et al. 1993; Rosenberg et al. 1997). Experiments with E-cadherin-deficient ES cells (embryonic stem cells) that no longer form specific tissue showed that R-cadherin, but not E- or N-cadherin, is able to rescue formation of striated muscle, although whether the muscle was cardiac or skeletal was not determined (Rosenberg et al. 1997).

Taken together, the results indicate that R-cadherin is involved in the development of skeletal muscle, although its role is less defined than that of M- or N-cadherin during these processes.

3 Summary and Outlook

Cadherins are a group of intercellular adhesion molecules exhibiting morphoregulatory functions. In myogenic morphogenesis the main cadherins involved are M-cadherin, N-cadherin and R-cadherin. They mediate cell–cell finding such as is necessary for myoblast fusion and migration of dermomyotomal cells to the limb bud, tissue maintenance and connection of satellite cells to the myofibre. Although cadherins are obviously needed for myoblast fusion, there is no experimental evidence that they are part of the fusion process itself. Furthermore, little or almost nothing is known about the differential roles of M-, N-, and R-cadherin and how they each perform during skeletal myogenesis. There is, however, evidence that the cadherin/catenin complexes might contribute to establishing and/or maintaining a functional muscle tissue morphology by interacting with cytoskeletal components.

Although the catenins are generally known to be part of signalling processes, little is known about the specific cadherin/catenin-dependent signal transduction pathways during myotube formation and muscle maturation. However, it is likely that the catenins also exhibit signal transduction functions in myogenesis and that growth factor receptors such as c-met may regulate cadherin and catenin function in muscle cells, as has been shown with other cell types, for example, epithelial and neuronal cells. Also, the fact that N-cadherin is able to upregulate myoD when introduced into BHK cells favours the assumption that cadherin/catenin-mediated communication between the plasma membrane and the nucleus is crucial for myogenesis to occur (for references see preceding sections).

In conclusion, it is apparent that cadherins are important players in regulating myogenesis. However, the detailed mechanisms and regulatory loops by which these intercellular adhesion proteins and their associated partners perform their functions in muscle cells remain unclear, providing many intriguing future challenges in this field.

Acknowledgements. We would like to thank Karen Knudsen for critical reading of the manuscript and Carolyne Agnew and Avril Arthur-Goettig for teaching us proper English.

References

Anastasiadis PZ, Reynolds AB (2000) The p120(ctn) catenin family: complex roles in adhesion, signaling and cancer. J Cell Sci 113:1319–1334

Antin P, Forry-Schaudies S, Friedmann TM, Tapscott SJ, Holtzer H (1981) Taxol induces postmitotic myoblasts to assemble interdigitating microtubule-myosin arrays that exclude actin-filaments. J Cell Biol 90:300–308

Aono S, Nakagawa S, Reynolds AB, Takeichi M (1999) p120(ctn) acts as an inhibitory regulator of cadherin function in colon carcinoma cells. J Cell Biol 145(3):551–562

Barker N, Clevers H (2000) Catenins, Wnt signaling and cancer. Bioessays 22(11):961–965

Barth AI, Nathke IS, Nelson WJ (1997) Cadherins, catenins and APC protein: interplay between cytoskeletal complexes and signaling pathways. Curr Opin Cell Biol 9(5):683–690

Behrens J (1999) Cadherins and catenins: role in signal transduction and tumor progression. Cancer Metastasis Rev 18(1):15–30

Bornemann A, Schmalbruch H (1994) Immunocytochemistry of M-cadherin in mature and regenerating rat muscle. Anat Rec 239:119–125

Brand-Saberi B, Gamel AJ, Krenn V, Müller TS, Wilting J, Christ B (1996) N-cadherin is involved in myoblast migration and muscle differentiation in the avian limb bud. Dev Biol 178:160–173

Butz S, Kemler R (1994) Distinct cadherin-catenin complexes in Ca (2+) -dependent cell-cell adhesion. FEBS Lett 355:195–200

Charlton CA, Mohler WA, Radice G, Hynes RO, Blau H (1997) Fusion competence of myoblasts rendered genetically null for N-cadherin in culture. J Cell Biol 138:331–336

Chothia C, Jones EY (1997) The molecular structure of cell adhesion molecules. Annu Rev Biochem 66:823–862

Cornelison DD, Olwin BB, Rudnicki MA, Wold BJ (2000) MyoD(–/–) satellite cells in single-fiber culture are differentiation defective and MRF4 deficient. Dev Biol 15:122–137

Donalies M, Cramer M, Ringwald M, Starzinski-Powitz A (1991) Expression of M-cadherin, a member of the cadherin multigene family, correlates with differentiation of skeletal muscle cells. Proc Natl Acad Sci USA 88:8024–8028

Eng H, Herrenknecht K, Semb H, Starzinski-Powitz A, Ringertz N, Gullberg D (1997) Effects of divalent cations on M-cadherin expression and distribution during primary rat myogenesis in vitro. Differentiation 61(3):169–176

Geiger B, Ayalon O (1992) Cadherins. Annu Rev Cell Biol 8:307–332

George-Weinstein M, Gerhart J, Blitz J, Simak E, Knudsen K (1997) N-cadherin promotes the commitment and differentiation of skeletal muscle precursor cells. Dev Biol 185:14–24

Goichberg P, Geiger B (1998) Direct involvement of N-cadherin-mediated signaling in muscle differentiation. Mol Biol Cell 9(11):3119–3131

Hazan RB, Phillips GR, Qiao RF, Norton L, Aaronson SA (2000) Exogenous expression of N-cadherin in breast cancer cells induces cell migration, invasion, and metastasis. J Cell Biol 148(4):779–790

Hertig CM, Butz S, Koch S, Eppenberger-Ebehardt M, Kemler R, Eppenberger HM (1996) N-cadherin in adult rat cardiomyocytes in culture. II. Spatio-temporal appearance of proteins involved in cell-cell contact communication. Formation of two distinct N-cadherin/catenin complexes . J Cell Sci 109:11–20

Hinck L, Näthke IS, Papkoff J Nelson WJ (1994) Dynamics of cadherin-catenin complex formation: novel protein interactions and pathways of complex assembly. J Cell Biol 125:1327–1340

Hirano S, Kimoto N, Shimoyama Y, Hirohashi S Takeichi M (1992) Identification of a neural alpha-catenin as a key regulator of cadherin function and multicellular organization. Cell 70:293–301

Holtzer H, Forry-Schaudies S, Dlugosz A, Antin P, Dubyak G (1985) Interaction between intermediate filaments, microtubules, and myofibrils in fibrogenic and myogenic cells. Ann NY Acad Sci 455:106–125

Huber O, Bierkamp C, Kemler R (1996) Cadherins and catenins in development. Curr Opin Cell Biol 8:685–691

Humphries MJ, Newham P (1998) The structure of cell-adhesion molecules. Trends Cell Biol 8(2):78–83

Hutton J, Christofori G, Chi W, Edman U, Guest P, Hanahan D, Kelly RB (1993) Molecular cloning of mouse pancreatic islet R-cadherin: differential expression in endocrine and exocrine tissue. Mol Endocrinol 7:1151–1160

Inuzuka H, Miyatani S, Takeichi M (1991) R-cadherin: a novel Ca (2+)-dependent cell-cell adhesion molecule expressed in the retina. Neuron 7:69–79

Irintchev A, Zeschnigk M, Starzinski-Powitz A, Wernig A (1994) Expression pattern of M-Cadherin in normal, denervated and regenerating mouse muscle. Dev Dyn 199:326–337

Kaufmann U, Martin B, Link D, Witt K, Zeitler R, Reinhard S, Starzinski-Powitz A (1999a) M-cadherin and its sisters in development of striated muscle. Cell Tissue Res 296:191–198

Kaufmann U, Kirsch J, Irintchev A, Wernig A, Starzinski-Powitz A (1999b) The M-cadherin catenin complex interacts with microtubules in skeletal muscle cells: implications for the fusion of myoblasts. J Cell Sci 112:55–67

Kaufmann U, Zuppinger C, Waibler Z, Rudiger M, Urbich C, Martin B, Jockusch BM, Eppenberger H, Starzinski-Powitz A (2000) The armadillo repeat region targets ARVCF to cadherin-based cellular junctions. J Cell Sci 113:4121–4135

Knudsen KA, Myers L, McElwere SA (1990) A role for the Ca $^{2+}$-dependent adhesion molecule, N-cadherin, in myoblast interaction during myogenesis. Exp Cell Res 188:175–184

Knudsen KA, Soler AP, Johnson KR, Wheelock MJ (1995) Interaction of alpha-actinin with the cadherin/catenin cell-cell adhesion complex via alpha-catenin. J Cell Biol 130:67–77

Kuch C, Winnekendonk D, Butz S, Unvericht U, Kemler R, Starzinski-Powitz A (1997) M-cadherin-mediated cell adhesion and complex formation with the catenins in myogenic mouse cells. Exp Cell Res 232:331–338

Li H, Choudhary SK, Milner DJ, Munir MI, Kuisk IR, Capetanaki Y (1994) Inhibition of desmin expression blocks myoblast fusion and interferes with myogenic regulator MyoD and myogenin. J Cell Biol 124:827–841

Linask KK, Knudsen KA, Gui YH (1997) N-cadherin-catenin interaction: necessary component of cardiac cell compartmentalization during early vertebrate heart development. Dev Biol 185(2):148–164

Linask KK, Ludwig C, Han MD, Liu X, Radice GL, Knudsen KA (1998) N-cadherin/catenin-mediated morphoregulation of somite formation. Dev Biol 202(1):85–102

Mariner DJ, Wang J, Reynolds AB (2000) ARVCF localizes to the nucleus and adherens junction and is mutually exclusive with p120(ctn)(ctn) in E-cadherin complexes. J Cell Sci 113: 1481–8490

Matsunami H, Miyatani S, Inoue T, Copland NG, Gilbert DJ, Jenkins NA, Takeichi M (1993) Cell binding specificity of mouse R-cadherin and chromosomal mapping of the gene. J Cell Sci 106:401–409

Mège RM, Goudou D, Diaz C, Nicolet M, Garcia L, Geraud G, Rieger F (1992) N-cadherin and N-CAM in myoblast fusion: compared localisation and effect of blockade by peptides and antibodies. J Cell Sci 103:897–906

Miller JR, Moon RT (1996) Signal transduction through beta-catenin and specification of cell fate during embryogenesis. Genes Dev 10(20):2527–2539

Moore R, Walsh F (1993) The cell adhesion molecule M-cadherin is specifically expressed in developing and regenerating, but not denervated skeletal muscle. Development 117:1409–1420

Nose A, Tsuji K, Takeichi M (1990) Localization of specificity determining sites in cadherin cell adhesion molecules. Cell 61(1):147–155

Ontell M, Kozeka K (1984) The organogenesis of murine striated muscle: a cytoarchitectural study. Am J Anat 171:133–148

Ozawa M, Kemler R (1990) Correct proteolytic cleavage is required for the cell adhesive function of uvomorulin. J Cell Biol 111(4):1645–1650

Ozawa M, Baribault H Kemler R (1989) The cytoplasmatic domain of the cell adhesion molecule uvomorulin associates with three independent proteins structurally related in different species. EMBO J 8:1711–1717

Pouliot Y, Gravel M, Holland PC (1994) Developmental regulation of M-cadherin in the terminal differentiation of skeletal myoblasts. Dev Dyn 200:305–312

Provost E, Rimm DL (1999) Controversies at the cytoplasmic face of the cadherin-based adhesion complex. Curr Opin Cell Biol 11(5):567–572

Radice GL, Rayburn H, Matsunami H, Knudsen KA, Takeichi M, Hynes RO (1997) Developmental defects in mouse embryos lacking N-cadherin. Dev Biol 181:64–78

Redfield A, Nieman MT, Knudsen KA (1997) Cadherins promote skeletal muscle differentiation in three-dimensional cultures. J Cell Sci 138:1323–1331

Redies C, Takeichi M (1993) N- and R-cadherin expression in the optic nerve of the chicken embryo. Glia 8(3):161–171

Reynolds AB, Daniel J, McCrea PD, Wheelock MJ, Wu J, Zhang Z (1994) Identification of a new catenin: the tyrosine kinase substrate p120(ctn)cas associates with E-cadherin complexes. Mol Cell Biol 14(12):8333–8342

Ringwald M, Schuh R, Vestweber D, Eistetter H, Lottspeich F, Engel J, Dölz R, Jähnig F, Epplen J, Mayer S, Müller C, Kemler R (1987) The structure of the cell adhesion molecule uvomorulin. Insights into the molecular mechanism of Ca^{2+} dependent cell adhesion. EMBO J 6:3647–3653

Rose O, Rohwedel J, Reinhard S, Bachmann M, Cramer M, Rotter M, Wobus A, Starzinski-Powitz A (1994) Expression of M-cadherin protein in myogenic cells during prenatal mouse development and differentiation of embryonic stem cells in culture. Dev Dyn 201:245–259

Rose O, Grund C, Reinhardt S, Starzinski-Powitz A, Franke WW (1995) Contactus adhaerens, a special type of plaque-bearing adhering junction containing M-cadherin, in the granule layer of the cerebellar glomerulus. Proc Natl Acad Sci USA 92:6022–6026

Rosenberg P, Esni F, Sjodin A, Larue L, Carlsson L, Gullberg D, Takeichi M, Kemler R, Semb H (1997) A potential role of R-cadherin in striated muscle formation. Dev Biol 187:55–70

Sabourin LA, Girgis-Gabardo A, Seale P, Asakura A, Rudnicki MA (1999) Reduced differentiation potential of primary MyoD–/– myogenic cells derived from adult skeletal muscle. J Cell Biol 22:631–643

Saitoh O, Aria T, Obinata T (1988) Distribution of microtubules and other cytoskeletal filaments during myotube elongation as revealed by fluorescence microscopy. Cell Tissue Res 252: 263–273

Seghatoleslami MR, Myers L, Knudsen KA (2000) Upregulation of myogenin by N-cadherin adhesion in three-dimensional cultures of skeletal myogenic BHK cells. J Cell Biochem 77(2): 252–264

Shan WS, Tanaka H, Phillips GR, Arndt K, Yoshida M, Colman DR, Shapiro L (2000) Functional *cis*-heterodimers of N- and R-cadherins. J Cell Biol 148(3):579–590

Shapiro L, Fannon AM, Kwong PD, Thompson A, Lehmann MS, Grübel G, Legrand IP, Als-Nielsen J, Colman DR, Hendrickson WA (1995) Structural basis of cell-cell adhesion by cadherins. Nature 374:327–336

Shin CS, Lecanda F, Sheikh S, Weitzmann L, Cheng SL, Civitelli R (2000) Relative abundance of different cadherins defines differentiation of mesenchymal precursors into osteogenic, myogenic, or adipogenic pathways. J Cell Biochem 78(4):566–577

Soler AP, Knudsen KA (1994) N-cadherin involvement in cardiac myocyte interaction and myofibrillogenesis. Dev Biol 162(1):9–17

Steinberg MS, McNutt PM (1999) Cadherins and their connections: adhesion junctions have broader functions. Curr Opin Cell Biol11(5):554–560

Tachibana K, Nakanishi H, Mandai K, Ozaki K, Ikeda W, Yamamoto Y, Nagafuchi A, Tsukita S, Takai Y (2000) Two cell adhesion molecules, nectin and cadherin, interact through their cytoplasmic domain-associated proteins. J Cell Biol 150(5):1161–1176

Takeda H, Shimoyama Y, Nagafuchi A, Hirohashi S (1999) E-cadherin functions as a cis-dimer at the cell-cell adhesive interface in vivo. Nat Struct Biol 6(4):310–312

Takeichi M (1991) Cadherin cell adhesion receptors as a morphogenic regulator. Science 251: 1451–1455

Tamura K, Shan WS, Hendrickson WA, Colman DR, Shapiro L (1998) Structure-function analysis of cell adhesion by neuronal (N-) cadherin. Neuron 20:1153–1163

Tao J, Wallace I (1991) Site-specific antibodies block kinase A phosphorylation of desmin in vitro and inhibit incorporation of myoblasts into myotubes. Cell Motil Cytoskel 19:109–120

Tepass U (1999) Genetic analysis of cadherin function in animal morphogenesis. Curr Opin Cell Biol 11:540–548

Toyama Y, Forry-Scahudies S, Hoffmann B, Holtzer H (1982) Effects of taxol and colcemid on myofibrillogenesis. Proc Nat Sci USA 79:6556–6560

Yap AS, Niessen CM, Gumbiner BM (1998) The juxtamembrane region of the cadherin cytoplasmic tail supports lateral clustering, adhesive strengthening, and interaction with p120(ctn)ctn. J Cell Biol 141(3):779–789

Zeschnigk M, Kozian D, Kuch C, Schmoll M, Starzinski-Powitz A (1995) Involvement of M-cadherin in fusion and terminal differentiation of myogenic cells. J Cell Sci 108:2973–2981

Slow Myosins in Muscle Development

Frank E. Stockdale, William Nikovits Jr., and Nora R. Espinoza[1]

1 Introduction

The expression of various different isoforms of myosin in skeletal and cardiac muscle serves to define particular types of skeletal muscle fibers and cardiomyocytes, and to describe functional differences among them. This review will focus on those issues related to the developmental expression of different isoforms of myosin heavy chain (MyHC) in the vertebrate embryo with a particular focus on the expression of the MyHCs of the slow type. The slow isoforms of myosin are particularly of interest in myogenesis because they identify the least common fibers in vertebrates, appear principally in specific muscles, are highly dependent on innervation for sustained expression and are commonly expressed both within developing skeletal and cardiac muscle cells. In both cardiac and skeletal muscle development, expressions of slow MyHCs are controlled by a number of molecular and functional cues.

2 Myosin Heavy Chain Genes

The physiological characteristics of a skeletal muscle fiber or cardiomyocyte result largely from the expression of specific isoforms of the many contractile proteins that comprise the sarcomere. In particular, the speed of contraction of single muscle fibers is most highly correlated with the expression of specific isoforms of MyHC (Reiser et al. 1985, 1988). A striated muscle myosin molecule contains two myosin heavy chains associated with two non-identical pairs of myosin light chains. Although in insects alternate splicing of a single MyHC gene accounts for different MyHC isoforms, a separate gene encodes each slow MyHC isoform in vertebrates. The genomes of all vertebrates contain many MyHC genes (Table 1), and in most species at least some of these genes are expressed in both skeletal and cardiac muscle. In the case of the mammals, the MyHC genes expressed primarily in skeletal muscle are located tandemly in a cluster on mouse and human chromosomes 11 and 17, respectively (Leinwand et al. 1983; Weiss et al. 1999), whereas two MyHC genes expressed in the heart (alpha- and beta-cardiac) are located on mouse and human chromosomes 14

[1] Stanford Medical Center, CCSR 1145, Stanford, California 94305-5151, USA

Results and Problems in Cell Differentiation, Vol. 38
B. Brand-Saberi (Ed.): Vertebrate Myogenesis

Table 1. Myosin heavy chains expressed in mammalian and avian striated muscles

Mammals	Birds
Slow/β-cardiac	Slow MyHC1
α-cardiac	Slow MyHC 2
IIa	Slow MyHC 3/AMHC1
IIb	Embryonic MyHC
IId/x	VMHC1
Embryonic	Neonatal
Perinatal	
Extraoccular	Superfast

(Weydert et al. 1985; Saez et al. 1987). The highly fragmentary nature of the avian genome has made localization of the avian MyHC genes more difficult, but a cluster of three MyHCs most closely related to the mammalian cardiac isoforms, has been reported (Chen et al. 1997). Based on their expression in fast and slow contracting muscle fibers, the rate of ATP hydrolysis and sequence homology, two broad classes of MyHC have been defined: slow MyHCs and fast MyHCs. Whereas all striated muscle, skeletal and cardiac, express at least one MyHC gene, an individual skeletal muscle fiber or cardiomyocyte can, and often does, express multiple MyHC genes, including isoforms from both fast and slow class (Wieczorek et al. 1985; Crow and Stockdale 1986b; Taylor and Bandman 1989; Pedrosa-Domellof et al. 2000). The MyHC genes expressed not only vary from muscle to muscle depending upon fiber composition of a muscle, but also are expressed in specific developmentally controlled patterns (Whalen et al. 1981; Bandman 1985; Crow and Stockdale 1986a).

There are many MyHCs expressed in mammalian and avian striated muscles (Table 1). There is extremely high sequence homology in all MyHCs within and between classes of animals. Because MyHC is a critical component of the cellular contractile apparatus, it has been assumed that the diversity of isoforms, despite the homologies, relates to different functional requirements among various muscles. The importance of the evolutionary conservation of multiple MyHC genes has best been explored in MyHCs of the fast class (Stedman et al. 1990; Bandman and Rosser 2000). Leinwand and colleagues have extensively studied the functional importance of different MyHC isoforms in skeletal muscle (see review by Allen et al. 2000). Using strains of mice null for different MyHC isoforms it has become clear that distinct differences in muscle development appear when specific isoforms of MyHC are lacking. In mice null for the IIb MyHC there is significant prenatal lethality, and surviving animals have several cardiac abnormalities, which suggests that noncardiac muscle myosin IIb is required for normal cardiomyocyte development (Tullio et al. 1997). Neonates lacking IIb MyHC have reduced grip strength and muscles that generate reduced force compared with wild-type neonates

(Acakpo-Satchivi et al. 1997). Mice null for IId/x MyHC have a marked increase in IIa MyHC gene expression and a decrease in IIb MyHC expression (Sartorius et al. 1998), but their muscles generate normal amounts of force, although the kinetics of contraction are altered. A sizeable number of IId/x MyHC null mice have severe kyphosis. Both IIb and IId/x null mice show significant decreases in muscle mass and increased interstitial fibrosis between muscle fibers, but there is little evidence that muscle regeneration is occurring. Mice null for α-cardiac MyHC are embryonic lethal, whereas heterozygous mice are viable but have numerous cardiac defects (Jones et al. 1996). Together these observations support the hypothesis that the different isoforms of MyHC are functionally unique and cannot substitute for one another (Allen et al. 2000).

3 Slow Myosin Heavy Chain Genes in Avian Skeletal Muscle

In birds, three slow MyHC isoforms, slow MyHC1, MyHC2 and MyHC3/AMHC1, have been reported (Hoh et al. 1976; Yutzey et al. 1994; Nikovits and Stockdale 1996; Chen et al. 1997; Oana et al. 1998). Each isoform is expressed in a distinct and stereotypical pattern during embryogenesis (Crow and Stockdale 1986b). In birds, slow MyHC1 is the most widely distributed of the slow isoforms both temporally and spatially. An antibody directed against an epitope shared by slow MyHC2 and slow MyHC3 was used to demonstrate that the first slow MyHC-containing hypaxial muscle fibers appear in the dorsal and ventral muscle masses in embryonic chick limbs and subsequently in fibers within specific anatomic muscle fibers in the adult (Crow and Stockdale 1986b; Nikovits and Stockdale 1996). The distribution of these fibers is characteristic of each muscle, and is established by fetal stages of development, forming a "blueprint" for fiber types, which is conserved in muscle development (Crow and Stockdale 1986b). The initial formation of slow muscle fibers in embryonic development is independent of innervation in both birds and mammals (Butler et al. 1982; Crow and Stockdale 1986a; Phillips et al. 1986; Condon et al. 1989).

Because primary fibers serve as anlagen of anatomic muscles that, in the absent of innervation, contain specific arrays of fast and slow fiber types, mechanisms other than innervation were sought to explain subsequent fiber diversification, i.e., differences in expression of MyHC (reviewed by Stockdale 1992). In early avian development, there are differences in the types of fibers that form from cloned myoblasts isolated from the limb bud (Miller and Stockdale 1986) such that muscle fiber clones that express slow MyHC isoforms are derived from different myoblasts than those that express only fast MyHC isoforms. Similar findings apply to satellite cells isolated from muscles of differing functional types in adult birds (Feldman and Stockdale 1991; see Hartley et al. 1992). Thus, diversity of myosins expressed in the embryonic bird is related to different developmental origins of the cells. Recent work by Kardon (pers. comm.), using viral lineage markers, provides evidence that clonally derived myogenic precursors that form fibers in the fetal limb can fuse with fibers of differing types. Thus, unlike the mononucleate or binucleate fibers of the embryo, the multinucleate fibers of the avian fetus are not formed from the progeny of a single myogenic precursor. These fetal fibers are predominately of the fast type. The initial slow embryonic fibers in the limb bud have one or two nuclei, but later become multinucleated, especially as muscles grow during fetal development. It is unclear how MyHC genes that are active within these initially slow MyHC expressing fibers are altered by the later addition of many nuclei contributed by myoblasts that appear in fetal development. However, as discussed below, innervation as well as cellular lineage effects the ultimate MyHC genes that are expressed in fetal and adult development (Stockdale 1997).

4 Slow Myosin Heavy Chain Genes in Mammalian Skeletal Muscle

Mammals are generally accepted to differ from birds in the numbers of genes that encode slow MyHCs. A single MyHC gene, expressed in all adult slow skeletal muscle fibers, encodes type I MyHC, which is identical to the β-cardiac MyHC isoform (Lompre et al. 1984). Two additional MyHC isoforms classified as slow have been identified by immunological methods (Hughes et al. 1993), but the genes encoding these forms have not been isolated. The α-cardiac MyHC gene is expressed in early embryonic slow muscle fibers as well as in adult mammals muscles, and could account for one of the immunologically identified prenatal slow MyHCs (Lefaucheur et al. 1997; Sandri et al. 1999; Stevens et al. 2000). In support of this idea, the well-characterized avian slow MyHC3 gene shows the same degree of nucleotide homology to both α-cardiac and β-cardiac MyHCs (Nikovits and Stockdale 1996). With completion of the human genome sequencing project, additional putative striated muscle MyHCs have been identified (Berg et al. 2001), one of which has been described as being most closely homologous to α-cardiac MyHC. This, and/or the transcript of a second new MyHC locus identified by a search of the human genome, could encode additional slow MyHCs.

5 Slow MyHC Genes in Fish Skeletal Muscle

Apart from birds and mammals the expression of slow myosin in fish has been most extensively investigated at the protein level. Only one MyHC gene has been isolated in the zebrafish and it is expressed in the early heart tube and the ventricle (Yelon et al. 1999). Those MyHC genes expressed in skeletal muscle have not been isolated, but their appearance has been carefully described in studies using immunological probes. These studies, as those in the bird, indicate that fibers that will express slow isoforms of myosin in the embryo have a distinctly different cellular origin than those that will express fast isoforms (Devoto et al. 1996). Early in development of the presomitic mesoderm in the zebrafish, two distinct populations of muscle cell precursors form, which Devoto and colleagues (Devoto et al. 1996) have shown can be distinguished by position, morphology and gene expression. Adjacent to the notochord are adaxial cells that will subsequently migrate radially away from the notochord to become a superficial layer of slow myosin-expressing cells. A subset of the adaxial cells (muscle pioneers) express engrailed, but do not migrate completely to the lateral surface. These cells express a slow as well as a fast MyHC. A second group of muscle fiber precursors, the lateral pre-somitic cells, are those that form only fast myosin fibers of the embryo. Thus, slow and fast muscle fiber types in zebrafish axial muscle arise from distinct populations of cells as in the early avian embryo.

6 Hedgehog Family of Signaling Molecules and Slow Myosin Expression in Skeletal Muscle Development

The family of signaling molecules, the Hedgehog family, has been demonstrated to play important roles in the expression of slow myosins in skeletal muscle of the zebrafish and the bird. The two distinct populations of cells that give rise to the zebrafish musculature are controlled by signals emanating from the notochord. In the no tail zebrafish mutant, embryos lack a notochord, and form abnormal somites that lack muscle pioneer cells (Halpern et al. 1993). The defect in these mutants appears to result from the absence of members of the Hedgehog family of signaling molecules: Echidna Hedgehog and Sonic Hedgehog (Currie and Ingham 1996; Lewis et al. 1999; Norris et al. 2000). This, and other mutants lacking Sonic Hedgehog expression, fail to form slow myosin-expressing fibers, but do form fast myosin expressing fibers, suggesting that the two lineages of muscle cells form in response to different signals (Blagden et al. 1997). A recently described mutant, smu, lacks nearly all slow muscle fibers and has a complete loss of engrailed-expressing muscle pioneers (Barresi et al. 2000). The smu mutant embryos can be rescued by expression of rat smoothened. Smoothened encodes a membrane protein that undergoes an allosteric change upon binding of Sonic Hedgehog to its receptor, patched, thereby permitting signal transduction by Sonic Hedgehog to proceed. This observation suggests that the mutation responsible for this phenotype is within the Shh signaling pathway, probably in patched or smoothened (Barresi et al. 2000). Thus, this mutant substantiates the role that Hedgehog signaling plays in establishing the slow fibers in the zebrafish (Blagden et al. 1997).

Recent reports by Roy and colleagues (Roy et al. 2001) show in the zebrafish that the u-boot gene plays a pivotal role in whether myoblasts of the myotome assume a fast or slow fiber fate. They find that the pathway by which the

Hedgehog family members signal the paraxial mesoderm to form adaxial slow fibers leads to the activation of u-boot in the future muscle pioneer cells that express the slow fiber marker, prox1. Prox1 is down stream of u-boot action and is only expressed in cells that form slow fibers. The mechanism by which u-boot produces a dichotomous switch of fast vs. slow muscle fiber fate is not known, although it appears to control prox1 expression, which is required for proper terminal assembly of slow myofibrils.

Recent work in the chick embryo also demonstrates a role for Shh signaling in the expression of slow myosin within the axial musculature of birds (Cann et al. 1999). It had been shown earlier that the notochord or the ventral neural tube was a source of a signal(s) in vitro and in vivo that initiates myogenesis in isolated chick somites, which is the first site of myogenesis in the bird as well as in the fish (Buffinger and Stockdale 1994, 1995; Münsterberg and Lassar 1995; Stern and Hauschka 1995; Borycki et al. 1999). When analysed for fiber type it was found that there was a marked increase in the expression of slow myosin heavy chain when the somites were exposed to either neural tube or purified Sonic Hedgehog (Cann et al. 2000). In the zebrafish it is generally accepted that Hedgehog signaling is instructive to mesodermal cells (Blagden et al. 1997). Unlike the fish, it is less clear in the bird whether the Hedgehog signaling mechanism is instructive to the muscle cell precursors, or whether it merely acts to amplify existent slow muscle cell precursors at the expense of fast precursors by its proliferation promoting action. In the chicken somite grown in vitro and zebrafish in vivo, excess Shh produces an increase in slow fiber formation at the expense of fast fiber formation (Cann et al. 2000).

Therefore, in bird and fish development it appears that there is early commitment to the formation of slow fibers that form the basis of the pattern of the adult musculature. This commitment is tied to the Hedgehog signaling pathway, with the notochord and perhaps the ventral neural tube acting as the source of the signal. Although the signals that establish the slow lineage may have been identified, the mechanism by which myogenic cells become restricted to the types of myosin that can be expressed is not understood.

7 Innervation and Calcineurin Responsive Pathways and the Control of Slow MyHC Expression in Skeletal Muscle

It has long been known that a switch to a slow fiber type can be elicited in fast fibers by changing the pattern of neural stimulation delivered to neonatal or adult intact muscle (reviewed by Pette 2001). Recently, a mechanism has been proposed that links electrical activity with specific gene expression within muscle fibers (Chin et al. 1998). When predominately fast fiber-type muscle is stimulated for an extended period of time, a switch occurs to a slow muscle-fiber type and to expression of different isoforms of MyHC (Barany 1967; Salmons and Vrbovç 1969; Weeds and Burridge 1975). Tonic motor activity leads to elevated levels of intracellular calcium, which in turn affects the calcineurin–NF-AT signal transduction pathway. A sustained increase in intracellular calcium activates calcineurin, a phosphatase. One target of activated calcineurin is members of the NF-AT family of transcription factors, which, when dephosphorylated, translocate to the nucleus. Once in the nucleus, NF-AT can bind sequence motifs present in slow-muscle-specific promoters and, in collaboration with other transcription factors, activate gene expression. Williams and coworkers (Chin et al. 1998) demonstrated in the mouse myogenic C2 cell line that several slow muscle-specific genes not normally expressed, including myoglobin and slow skeletal troponin I (sTnI), were transcribed following constitutive expression of an activated form of calcineurin. Because the myoglobin and slow sTnI promoters contain NF-AT-binding motifs that respond to calcineurin activation in transfection experiments, it was originally proposed that calcineurin acted to effect fiber-type by altering the subcellular localization of NF-AT (Chin et al. 1998). This was supported by the observation that injection of cyclosporin, an inhibitor of calcineurin activation, into the rat soleus muscle, led to a reduction of slow muscle fiber numbers (Chin et al. 1998). Bigard and colleagues (2000) also reported that rats treated with cyclosporin had fewer slow muscle fibers in the soleus, and reduced levels of detectable type I MyHC. The transition to a slow muscle fiber-type in the mouse plantaris, which is induced by overload hypertrophy, was also prevented by treatment of animals with cyclosporin (Dunn et al. 1999). This observation is supported by studies using a transgenic mouse line in which calcineurin was expressed at high levels from the muscle creatine kinase (MCK) promoter (Naya et al. 2000). They showed an increase in the number of slow myosin-expressing fibers in fast fiber-type muscles using a myosin ATPase staining assay. However, it was noted that this transformation was incomplete, suggesting that additional factors are likely to be involved in slow MyHC gene expression. Not discussed here is an extensive literature demonstrating that this same pathway plays a key role in cardiac hypertrophy where not only are there changes in the expression of β-cardiac MyHC, but a number of other cardiac-specific genes (reviewed by Olson and Williams 2000a, 2000b).

Perhaps the best evidence that a calcineurin signal transduction pathway regulates slow myosin heavy chain gene expression was provided by Delling et al. (Delling et al. 2000). In vitro, both C2 and MyoD-transfected 10T1/2 cells when co-transfected with activated calcineurin, showed accumulation of slow MyHC by Western blotting. They also reported that at 2 weeks following injection of adenovirus expressing activated calcineurin into the primarily fast gastrocnemius muscle of the rat in vivo, slow MyHC was expressed in many fibers, coincident with the expression of the activated calcineurin.

The calcineurin pathways may well interact with those mediated by hormones. An example in skeletal muscle is the hypertrophic response to IGF-I. The effects of IGF-I on skeletal muscle have been observed in vivo and in cell culture, and include hypertrophy and changes in fiber-type expression (Barton-Davis et al. 1998; Musarò et al. 1999; Semsarian et al. 1999). Exposure of C2 cells to IGF-I, or injection of a plasmid expressing IGF-I into leg muscles, resulted in an increase in intracellular calcium, an increase in calcineurin phosphatase activity and a subsequent increase in the translocation of NF-AT into cell nuclei (Semsarian et al. 1999). What followed was hypertrophy of the fibers, and a glycolytic muscle phenotype both in vivo and in the C2 cell line. In these studies the expression of myosin was not examined. Although there is abundant evidence that IGF-I can, and does, initiate events in vivo via the calcineurin signal transduction pathway, the predominant effect of IGF-I is on the hypertrophy of fast glycolytic type IIb muscle fibers (Semsarian et al. 1999), with little alteration in the relative distribution of slow type I, as might have been predicted from the effect of calcineurin on untreated muscle (Chin et al. 1998).

The relationship between IGF-I action and calcineurin pathway effects may be confounded by a direct action of IGF-I on satellite cell proliferation as well as effects of both IGF-I and activated calcineurin on the function of existent and newly formed fibers. Although there is accumulating evidence to support the hypothesis that activated calcineurin and nuclear translocated NF-AT are responsible for the expression of some slow fiber-specific genes, at least in mature muscle fibers trained by neural activity (Chin et al. 1998), there is contradictory evidence regarding the requirement of NF-AT for the activation of other slow genes in skeletal muscle. For example, Calvo and colleagues (1999) have demonstrated that mice lacking the NF-AT binding motif in the sTnI promoter retain slow specific gene expression.

However, others report that the NF-AT-binding motif is required for proper expression and that the response is integrated with calcium ion-dependent pathways mediated by calmodulin-dependent protein kinase IV (Wu et al. 2000). Although activated calcineurin is able to elicit an induction of slow MyHC in vivo and in vitro, expression of NF-ATc3, a primary target of activated calcineurin in differentiating muscle cells, does not appear to produce a change in MyHC expression (Delling et al. 2000), suggesting that a signal transduction pathway(s) involving calcineurin, other than the calcineurin–NF-AT pathway, may be involved in the control of expression of slow MyHC and the slow muscle fiber type. Support that other myogenesis-related pathways may be activated by calcineurin are the observations that activated calcineurin induces expression of the transcription factor GATA-2 (Musarò et al. 1999), and that a calcium-dependent dissociation of histone deacetylase 4 (HDAC) by calcineurin regulates the activity of members of the MEF2 family of muscle transcription factors (Lu et al. 2000).

The role of the calcineurin pathways in skeletal muscle myogenesis in the vertebrate embryo is less defined than in neonatal or adult animals. The studies that have delineated connections between calcineurin activation and NF-AT family member translocations and slow MyHC expression have not been demonstrated in the embryo. Although calcineurin and IGF-I pathways offer mechanistic explanations for the observations that motor neuron-specific impulse patterns, neuromuscular activity and mechanical loading lead to adult skeletal muscle growth and hypertrophy, as well as slow MyHC expression, it is unclear that these mechanisms are important in embryonic slow muscle fiber formation. This is because impulse patterns and mechanical loads are unlikely to be involved in the expression of slow or fast MyHC in primary fibers of the embryo. However, because these mechanisms apply to neonatal and adult slow fiber development it is likely that similar pathways are operative at these early stages of development. Support for this possibility is reported in in vitro studies of myoblasts from satellite cell lines, and at stages of development where satellite cells are important for continued muscle growth (Horsley et al. 2001). That this pathway can have effects on embryonic myoblasts has been most clearly demonstrated in studies in 17.5 pc day mice null for NF-ATC3. In these developing mice, the total number of primary myofibers is reduced, suggesting that alterations in the calcineurin pathway can effect the numbers of fibers formed (Kegley et al. 2001). It remains to be determined if the remaining primary myofibers are of a specific fiber type.

Unlike diversification of skeletal muscle fiber types in fetal or adult muscle, diversification in the embryo occurs in the absence of innervation. This suggests that mechanisms other than those tied to innervation must produce increased intracellular calcium concentrations if calcineurin-mediated events lead to diversification in early development. Even if there is a role for calcineurin pathways in early myogenesis, it is possible that the role relates more to the hypertrophic and proliferative functions of this pathway than to the activation of MyHC genes. Another possible mechanism driving the process of embryonic muscle fiber diversification could rely on the action of hormones. It is known, for example, that many MyHC genes are responsive to thyroid hormone (Izumo et al. 1986). Because more is known about the nuclear hormones and their affects on slow MyHC expression (see below), the nuclear hormones become candidates for signaling MyHC expression. For example, thyroid hormone of maternal origin could effect MyHC gene expression in the early embryonic stages of primary myogenesis. Understanding of hormone action on fiber type at these early stages is limited, but it appears that, in the rat, thyroid hormone effects on slow MyHC expression may not occur until later at fetal stages of myogenesis (Narusawa et al. 1987). The role of nuclear hormones in the initial expression of slow isoforms of MyHC as well as the calcineurin–NF-AT pathways in primary fibers in both mammals and birds need additional study.

8 Slow MyHC Expression in the Developing Heart

In both avian and mammalian species, at least one slow isoform of MyHC is expressed in a subset of cardiomyocytes during heart development. As mentioned, in mammals a single isoform of the slow MyHC class is expressed in slow skeletal muscle fibers where it is named slow or type I MyHC, as well as in cardiomyocytes where it is named β-cardiac MyHC. The α-cardiac MyHC gene is also expressed in the heart and transiently in skeletal muscle development (Lefaucheur et al. 1997). In birds, the embryonic form of slow MyHC, slow MyHC3, is expressed in both the heart and early skeletal muscle (Nikovits and Stockdale 1996; Oana et al. 1998), as is the case of a ventricular form of MyHC (Camoretti-Mercado 1994). Distribution of slow/β-cardiac MyHC in the mammalian heart varies from species to species. In small mammals, such as the mouse, slow/β-cardiac MyHC is initially expressed in the tubular heart, but as development proceeds its expression becomes restricted to the ventricles whereas the α-cardiac MyHC has the reciprocal pattern of expression becoming restricted to the atrium and is downregulated in skeletal muscle. In birds, slow MyHC3/AMHC1 is initially expressed throughout the early heart tube, but becomes restricted to the atria as chamber formation occurs (Yutzey et al. 1994; Wang et al. 1996, 1998). Slow MyHC3 becomes the predominate MyHC isoform expressed in the adult bird atrium. Analyses demonstrate that nuclear hormones play key roles in regulating the developmentally controlled expression of slow MyHCs in the heart chambers of both birds and mammals (Morkin 2000; Wang et al. 2001).

Expression of the slow/β-cardiac MyHC gene is regulated by thyroid hormone (Gambke et al. 1983; Butler-Browne et al. 1984; Everett et al. 1984; Lompre et al. 1984; Gustafson et al. 1985; Izumo et al. 1986), whereas the expression of the slow MyHC3 gene is regulated by other nuclear hormones, Vitamin D and retinoic acid family members (Wang et al. 1996, 1998, 2001). The effects of thyroid hormone on cardiac MyHC expression in the mammalian heart have recently been review by Morkin (2000). As in most vertebrates, there are transitions in the expression of myosin heavy chains in the developing mammalian heart. The two MyHC genes expressed throughout development are α-cardiac and slow/β-cardiac MyHC. The α- and slow/β-MyHC genes are tandemly located on a single chromosome in both the human and mouse genome. The slow/β-cardiac MyHC gene is expressed early in cardiomyocyte formation under the control of several transcription factors, thyroid hormone receptor, NKX-2.5, MEF-2C and GATA-4/5/6. Thyroid hormone acts in a reciprocal fashion on the β- and α-cardiac MyHC gene expression, inhibiting expression of β-cardiac MyHC and promoting expression of α-cardiac MyHC. Thyroid hormone is not only required for expression of α-cardiac-MHC in ventricular muscle, but it also modifies the expression slow/β-cardiac MyHC as well as expression of a number of fast MyHC genes in skeletal muscle (Izumo et al. 1986).

In avian heart development a different set of nuclear hormones and their receptors appear to control the expression of slow MyHC in the embryonic heart. In both the embryonic chicken and quail, the tubular heart expresses slow MyHC3/AMHC1 throughout, but as the heart begins to loop there is a downregulation of this gene in the ventricle and sustained expression in the regions that will give rise to the atrium (Yutzey et al. 1994; Wang et al. 1996; Oana et al. 1998).

The mechanism of repression of slow MyHC3 gene expression in the ventricles has been recently uncovered. The process is mediated by two response elements within a 60-base-pair enhancer within the promoter of this gene. The two elements are a vitamin D response element (VDRE) and a GATA factor-binding element. The VDRE serves to repress expression in the ventricles, whereas the GATA factor-binding site serves as a positive-acting element to drive expression of slow MyHC3 in the atria. Proteins binding to these two elements modulate and control chamber-specific expression in the embryonic heart. Two proteins of the nuclear hormone receptor family, the vitamin D receptor, VDR, and the retinoic acid receptor, RXRα, form a heterodimer that binds at the inhibitory VDRE site to inhibit transcription, whereas GATA 4 (and perhaps other GATA family members) binds at the GATA factor-binding element to foster transcription of the slow MyHC3 gene (Wang et al. 1998). All three of these proteins are expressed in both the primitive chambers of the embryonic heart; thus, their presence alone does not explain differences in expression of slow MyHC3 between the two chambers.

The restricted expression of Irx4, an Iroquois family homeobox gene, limits expression of slow MyHC3 gene to the atrium (Bao et al. 1999). It was demonstrated in mice with targeted disruption of the Irx4 gene and, in the chick embryo with an antisense strategy, that there is inappropriate expression of atrial genes within the ventricle when Irx4 is not present (Bao et al. 1999; Bruneau et al. 2001). When Irx4 expression is prevented, the tubular heart still chamberizes, but there is no distinction in the expression of MyHCs between the atrium and the ventricle. Thus, Irx4 is not involved in the morphological events of heart formation, but controls the chamber-specific expression of genes. Recent analyses reveal the mechanism for this observation (Wang et al. 2001). Repression of slow MyHC3 gene transcription in ventricular cardiomyocytes by Irx4 requires the inhibitory VDRE site, but, unlike the VDR and RXRα proteins, Irx4 does not recognize nor bind directly to the DNA of the VDRE. Rather, Irx4 binds in a protein–protein interaction to the RXRα component of the VDR/RXRα heterodimer complex. Therefore, the mechanism of atrial chamber-specific expression of slow MyHC requires the formation of an inhibitory protein complex composed of at least three proteins - VDR, RXRα and Irx4 - that bind at the VDRE to inhibit slow MyHC3 transcription in the ventricles.

Thus, the regulation of chamber-specific gene expression is based not simply on a transcription activation model, but on a model based on active and ongoing inhibition of gene expression (Wang and Stockdale 1998). Irx4 not only acts as part of the protein complex that inhibits the transcription of slow MyHC3, but it also acts as an activator of VMHC1 (ventricular MyHC1 gene) transcription in the ventricles (Bao et al. 1999). How Irx4 participates in an inhibitory complex as well as in the activation of a gene within the same cell type is not known. The mechanism for the latter could be quite different than its interaction with VDR/RXRα, perhaps binding directly to homeodomain elements within the promoter of VMHC1 or interacting with a different set of cofactors than those involved with slow MyHC 3 regulation (Li et al. 1999). Investigation of how Irx4 regulates expression of cardiac MyHC genes will broaden our understanding of the diversification between atrial and ventricular cardiomyocytes during development, as well as the regulation of MyHC genes in general.

9 Summary

Myogenesis has been a system central to investigations on mechanisms of diversification within groups of differentiating cells. Diversity among cell types has been well described in striated muscle tissue at the protein and enzymatic-function levels for decades, but it is only in recent years that some understanding of the molecular mechanisms responsible for this diversity has begun to emerge. Study of the expression of the slow isoforms of the myosin heavy chain has contributed to our understanding of how cell diversity arises within skeletal and cardiac muscle.

Slow MyHC isoforms are developmentally responsive to a number of cues provided by the nervous system, the endocrine system and, later in development, to functional demands on these developing tissues. Perhaps most informative have been studies on the mechanism for regulation of slow MyHC expression in mammals and birds where studies on the calcineurin–NF-AT pathways and nuclear hormone action have been shown to control MyHC gene expression in skeletal muscle and in the developing heart. The mechanisms involved in cell diversification in myogenesis are undoubtedly more varied and complex than those currently offered to explain cell diversification, but these recent studies have broadened our understanding of the interplay between the nervous system, the endocrine system and cell lineages in controlling cell diversification. Greater focus on the first fibers and cardiomyocytes to form in the embryo are likely to bring additional insights into the mechanisms crucial for establishing the patterns of diversity required for successful formation of embryonic tissues.

Acknowledgements. This work is supported by USDA grant 2000-03280 and NIH grant HL60696.

References

Acakpo-Satchivi LJ, Edelmann W, Sartorius C, Lu BD, Wahr PA, Watkins SC, Metzger JM, Leinwand L, Kucherlapati R (1997) Growth and muscle defects in mice lacking adult myosin heavy chain genes. J Cell Biol 139:1219–1229

Allen DL, Harrison BC, Leinwand LA (2000) Inactivation of myosin heavy chain genes in the mouse: diverse and unexpected phenotypes. Microsci Res Tech 50:492–499

Bandman ER (1985) Myosin isoenzyme transitions in muscle development, maturation, and disease. Int Rev Cytol 97:97–131

Bandman E, Rosser BW (2000) Evolutionary significance of myosin heavy chain heterogeneity in birds. Microsci Res Tech 50:473–491

Bao ZZ, Bruneau BG, Seidman JG, Seidman CE, Cepko CL (1999) Regulation of chamber-specific gene expression in the developing heart by Irx4. Science 283:1161–1164

Barany M (1967) ATPase activity of myosin correlated with speed of muscle shortening. J Genet Physiol 50(Suppl):197–216

Barresi MJ, Stickney HL, Devoto SH (2000) The zebrafish slow-muscle-omitted gene product is required for Hedgehog signal transduction and the development of slow muscle identity. Development 127:2189–2199

Barton-Davis ER, Shoturma DI, Musaro A, Rosenthal N, Sweeney HL (1998) Viral mediated expression of insulin-like growth factor I blocks the aging-related loss of skeletal muscle function. Proc Natl Acad Sci USA 95:15603–15607

Berg JS, Powell BC, Cheney RE (2001) A millennial myosin census. Mol Biol Cell 12:780–794

Bigard X, Sanchez H, Zoll J, Mateo P, Rousseau V, Veksler V, Ventura-Clapier R (2000) Calcineurin Co-regulates contractile and metabolic components of slow muscle phenotype. J Biol Chem 275:19653–19660

Blagden CS, Currie PD, Ingham PW, Hughes SM (1997) Notochord induction of zebrafish slow muscle mediated by Sonic Hedgehog. Genes Dev 11:2163–2175

Borycki AG, Brunk B, Tajbakhsh S, Buckingham M, Chiang C, Emerson CP Jr (1999) Sonic Hedgehog controls epaxial muscle determination through Myf5 activation. Development 126:4053–4063

Bruneau BG, Bao ZZ, Fatkin D, Xavier-Neto J, Georgakopoulos D, Maguire CT, Berul CI, Kass DA, Kuroski-de Bold ML, de Bold AJ, Conner DA, Rosenthal N, Cepko CL, Seidman CE, Seidman JG (2001) Cardiomyopathy in Irx4-deficient mice is preceded by abnormal ventricular gene expression. Mol Cell Biol 21:1730–1736

Buffinger N, Stockdale FE (1994) Myogenic specification in somites: induction by axial structures. Development 120:1443–1452

Buffinger N, Stockdale FE (1995) Myogenic specification of somites is mediated by diffusible factors. Dev Biol 169:96–108

Butler J, Cosmos E, Brierly J (1982) Differentiation of muscle fiber types in aneurogenic brachial muscles of the chick embryo. J Exp Zool 224:65–80

Butler-Browne GS, Herlicoviez D, Whalen RG (1984) Effects of hypothyroidism on myosin isozyme transitions in developing rat muscle. FEBS Lett 166:71–75

Calvo S, Venepally P, Cheng J, Buonanno A (1999) Fiber-type-specific transcription of the troponin I slow gene is regulated by multiple elements. Mol Cell Biol 19:515–525

Camoretti-Mercado B, Jakovcic S, Zak R (1994) Cardiac phenotypic markers expressed in early stages of both cardiac and skeletal muscle development. Cell Mol Biol Res 40:87–91

Cann GM, Lee JW, Stockdale FE (1999) Sonic Hedgehog enhances somite cell viability and formation of primary slow muscle fibers in avian segmented mesoderm. Anat Embryol 200:239–252

Chen Q, Moore LA, Wick M, Bandman E (1997) Identification of a genomic locus containing three slow myosin heavy chain genes in the chicken. Biochim Biophys Acta 7:148–156

Chin ER, Olson EN, Richardson JA, Yang Q, Humphries C, Shelton JM, Wu H, Zhu W, Bassel-Duby R, Williams RS (1998) A calcineurin-dependent transcriptional pathway controls skeletal muscle fiber type. Genes Dev 12:2499–2509

Condon KW, Soileau LC, Silberstein L, Blau HM, Thompson WJ (1989) Development and innervation of muscle fiber types in the rat hindlimb. In: Landmesser L (ed) The assembly of the nervous system. Alan R Liss, New York, pp 51–63

Crow MT, Stockdale FE (1986a) The developmental program of fast myosin heavy chain expression in avian skeletal muscles. Dev Biol 118:333–342

Crow MT, Stockdale FE (1986b) Myosin expression and specialization among the earliest muscle fibers of the developing avian limb. Dev Biol 113:238–254

Currie PD, Ingham PW (1996) Induction of a specific muscle cell type by a Hedgehog-like protein in zebrafish. Nature 382:452–455

Delling U, Tureckova J, Lim HW, De Windt LJ, Rotwein P, Molkentin JD (2000) A calcineurin-NFATc3-dependent pathway regulates skeletal muscle differentiation and slow myosin heavy-chain expression. Mol Cell Biol 20:6600–6611

Devoto SH, Melanìon E, Eisen JS, Westerfield M (1996) Identification of separate slow and fast muscle precursor cells in vivo, prior to somite formation. Development 122:3371–3380

Dunn SE, Burns JL, Michel RN (1999) Caleineurin is required for skeleal muscle hypertrophy. J Biol Chem 274:21908–21912

Everett AW, Sinha AM, Umeda PK, Jakovcic S, Rabinowitz M, Zak R (1984) Regulation of myosin synthesis by thyroid hormone: relative change in the a- and b-myosin heavy chain mRNA levels in rabbit heart. Biochemistry 23:1596–1599

Feldman JL, Stockdale FE (1991) Skeletal muscle satellite cell diversity: satellite cells form fibers of different types in cell culture. Dev Biol 143:320–334

Gambke B, Lyons GE, Haselgrove J, Kelly AM, Rubinstein NA (1983) Thyroidal and neural control of myosin transitions of rat fast and slow muscles. FEBS Lett 156:335–339

Gustafson TA, Markham BE, Morkin E (1985) Analysis of thyroid hormone effects on myosin heavy chain gene expression in cardiac and soleus muscles using a novel dot-blot mRNA assay. Biochem Biophys Res Commun 130:1161–1167

Halpern ME, Ho RK, Walker C, Kimmel CB (1993) Induction of muscle pioneers and floor plate is distinguished by the zebrafish no tail mutation. Cell 75:99–111

Hartley RS, Bandman E, Yablonka-Reuveni Z (1992) Skeletal muscle satellite cells appear during late chicken embryogenesis. Dev Biol 153:206–216

Hoh JFY, McGrath PA, White RI (1976) Electrophoretic analysis of multiple forms of myosin in fast-twitch and slow-twitch muscles of the chick. Biochem J 157:87–95

Horsley V, Friday BB, Matteson S, Kegley KM, Gephart J, Pavlath GK (2001) Regulation of the growth of multinucleated muscle cells by an NFATC2-dependent pathway. J Cell Biol 153: 329–338

Hughes SM, Cho M, Karsch-Mizrachi I, Travis M, Silberstein L, Leinwand LA, Blau HM (1993) Three slow myosin heavy chains sequentially expressed in developing mammalian skeletal muscle. Dev Biol 158:183–199

Izumo S, Nadal-Ginard B, Mahdavi V (1986) All members of the MHC multigene family respond to thyroid hormone in a highly tissue-specific manner. Science 231:597–600

Jones WK, Grupp IL, Doetschman T, Grupp G, Osinska H, Hewett TE, Boivin G, Gulick J, Ng WA, Robbins J (1996) Ablation of the murine alpha myosin heavy chain gene leads to dosage effects and functional deficits in the heart. J Clin Invest 98:1906–1917

Kegley KM, Gephart J, Warren GL, Pavlath GK (2001) Altered primary myogenesis in NFATC3(-/-) mice leads to decreased muscle size in the adult. Dev Biol 232:115–126

Lefaucheur L, Hoffman R, Okamura C, Gerrard D, Läger JJ, Rubinstein N, Kelly A (1997) Transitory expression of alpha cardiac myosin heavy chain in a subpopulation of secondary generation muscle fibers in the pig. Dev Dyn 210:106–116

Leinwand LA, Fournier RE, Nadal-Ginard B, Shows TB (1983) Multigene family for sarcomeric myosin heavy chain in mouse and human DNA: localization on a single chromosome. Science 221:766–769

Lewis KE, Currie PD, Roy S, Schauerte H, Haffter P, Ingham PW (1999) Control of muscle cell-type specification in the zebrafish embryo by Hedgehog signalling. Dev Biol 216:469–480

Li X, Murre C, McGinnis W (1999) Activity regulation of a Hox protein and a role for the homeodomain in inhibiting transcriptional activation. EMBO J 18:198–211

Lompre A-M, Nadal-Ginard B, Mahdavi V (1984) Expression of the cardiac ventricular a- and b-myosin heavy chain genes is developmentally and hormonally regulated. J Biol Chem 259: 6437–6446

Lu J, McKinsey TA, Zhang CL, Olson EN (2000) Regulation of skeletal myogenesis by association of the MEF2 transcription factor with class II histone deacetylases. Mol Cell 6:233–244

Miller JB, Stockdale FE (1986) Developmental origins of skeletal muscle fibers: Clonal analysis of myogenic cell lineages based on fast and slow myosin heavy chain expression. Proc Natl Acad Sci USA 83:3860–3864

Morkin E (2000) Control of cardiac myosin heavy chain gene expression. Microsci Res Tech 50: 522–531

Münsterberg AE, Lassar AB (1995) Combinatorial signals from the neural tube, floor plate and notochord induce myogenic bHLH gene expression in the somite. Development 121:651–660

Musarŷ A, McCullagh KJA, Naya FJ, Olson EN, Rosenthal N (1999) IGF-I induces skeletal myocyte hypertrophy through calcineurin in association with GATA-2 and NF-ATc1. Nature 400: 581–585

Narusawa M, Fitzsimons RB, Izumo S, Nadal-Ginard B, Rubinstein NA, Kelly AM (1987) Slow myosin in developing rat skeletal muscle. J Cell Biol 104:447–459

Naya FJ, Mercer B, Shelton J, Richardson JA, Williams RS, Olson EN (2000) Stimulation of slow skeletal muscle fiber gene expression by calcineurin in vivo. J Biol Chem 275:4545–4548

Nikovits WJ, Stockdale F (1996) Cellular and molecular bases of skeletal muscle fiber diversity. Basic Appl Myol 6:407–415

Norris W, Neyt C, Ingham PW, Currie PD (2000) Slow muscle induction by Hedgehog signalling in vitro. J Cell Sci 113:2695–2703

Oana S, Machida S, Hiratsuka E, Furutani Y, Momma K, Takao A, Matsuoka R (1998) The complete sequence and expression patterns of the atrial myosin heavy chain in the developing chick. Biol Cell 90:605–613

Olson EN, Williams RS (2000a) Remodeling muscles with calcineurin. Bioessays 22:510–519

Olson EN, Williams RS (2000b) Calcineurin signaling and muscle remodeling. Cell 101:689–692

Pedrosa-Domellof F, Holmgren Y, Lucas CA, Hoh JF, Thornell LE (2000) Human extraocular muscles: unique pattern of myosin heavy chain expression during myotube formation. Invest Ophthalmol Vis Sci 41:1608–1616

Pette D (2001) Historical perspectives: plasticity of mammalian skeletal muscle. J Appl Physiol 90:1119–1124

Phillips WD, Everett AW, Bennett MR (1986) The role of innervation in the establishment of the topographical distribution of primary myotube types during development. J Neurocytol 15: 397–405

Reiser PJ, Moss RL, Giulian GG, Greaser ML (1985) Shortening velocity in single fibers from adult rabbit soleus muscles is correlated with myosin heavy chain composition. J Biol Chem 260: 9077–9080

Reiser PJ, Greaser ML, Moss RL (1988) Myosin heavy chain composition of single cells from avian slow skeletal muscle is strongly correlated with velocity of shortening during development. Dev Biol 129:400–407

Roy S, Wolff C, Ingham PW (2001) The u-boot mutation identifies a Hedgehog-regulated myogenic switch for fiber-type diversification in the zebrafish embryo. Genes Dev 15:1563–1576

Saez LJ, Gianola KM, McNally EM, Feghali R, Eddy R, Shows TB, Leinwand LA (1987) Human cardiac myosin heavy chain genes and their linkage in the genome. Nucleic Acids Res 15: 5443–5459

Salmons S, Vrbovç (1969) The influence of activity on some contractile characteristics of mammalian fast and slow muscles. J Physiol 201:535–549

Sandri M, Rizzi C, Rossini K, Catani C, Cantini M, Spina M (1999) Purification of myosin heavy chain isoforms by electroendosmotic preparative gel electrophoresis: characterization of embryonic slow myosin heavy chain. Basic Appl Myol 9:71–78

Sartorius CA, Lu BD, Acakpo-Satchivi L, Jacobsen RP, Byrnes WC, Leinwand LA (1998) Myosin heavy chains IIa and IId are functionally distinct in the mouse. J Cell Biol 141:943–953

Semsarian C, Wu M-J, Ju Y-K, Marciniec T, Yeoh T, Allen DG, Harvey RP, Graham RM (1999) Skeletal muscle hypertrophy is mediated by a Ca^{2+}-dependent calcineurin signalling pathway. Nature 400:576–581

Stedman HH, Eller M, Jullian EH, Fertels SH, Sarkar S, Sylvester JE, Kelly AM, Rubinstein NA (1990) The human embryonic myosin heavy chain. Complete primary structure reveals evolutionary relationships with other developmental isoforms. J Biol Chem 265:3568–3576

Stern HM, Hauschka SD (1995) Neural tube and notochord promote in vitro myogenesis in single somite explants. Dev Biol 167:87–103

Stevens L, Gohlsch B, Mounier Y, Pette D (2000) Upregulation of myosin heavy chain MHClalpha in rat muscles after unweighting and clenbuterol treatment. Biochem Biophys Res Commun 275:418–421

Stockdale FE (1992) Myogenic cell lineages. Dev Biol 154:284–298

Stockdale FE (1997) Mechanisms of formation of muscle fiber types. Cell Struct Funct 22:37–43

Taylor LD, Bandman E (1989) Distribution of fast myosin heavy chain isoforms in thick filaments of developing chicken pectoral muscle. J Cell Biol 108:533–542

Tullio AN, Accili D, Ferrans VJ, Yu ZX, Takeda K, Grinberg A, Westphal H, Preston YA, Adelstein RS (1997) Nonmuscle myosin II-B is required for normal development of the mouse heart. Proc Natl Acad Sci USA 94:12407–12412

Wang GF, Stockdale, FE (1998) Chamber-specific gene expression and regulation during heart development. In: Rosenthal N (ed) Heart development. Academic Press, New York

Wang GF, Nikovits W Jr, Schleinitz M, Stockdale FE (1996) Atrial chamber-specific expression of the slow myosin heavy chain 3 gene in the embryonic heart. J Biol Chem 271:19836–19845

Wang GF, Nikovits W Jr, Schleinitz M, Stockdale FE (1998) A positive GATA element and a negative vitamin D receptor-like element control atrial chamber-specific expression of a slow myosin heavy-chain gene during cardiac morphogenesis. Mol Cell Biol 18:6023–6034

Wang GF, Nikovits WJ, Bao Z-Z, Cann GM, Stockdale FE (2001) Irx4 Forms an inhibitory complex with the vitamin D and retinoic X receptors to regulate cardiac chamber-specific slow MyHC3 expression. J Biol Chem 276:28835–28841

Weeds AG, Burridge K (1975) Myosin from cross-reinnervated cat muscles. Evidence for reciprocal transformation of heavy chains. FEBS Lett 57:203–207

Weiss A, McDonough D, Wertman B, Acakpo-Satchivi L, Montgomery K, Kucherlapati R, Leinwand L, Krauter K (1999) Organization of human and mouse skeletal myosin heavy chain gene clusters is highly conserved. Proc Natl Acad Sci USA 96:2958–2963

Weydert A, Daubas P, Lazaridis I, Barton P, Garner I, Leader DP, Bonhomme F, Catalan J, Simon D, Guenet JL et al. (1985) Genes for skeletal muscle myosin heavy chains are clustered and are not located on the same mouse chromosome as a cardiac myosin heavy chain gene. Proc Natl Acad Sci USA 82:7183–7187

Whalen RG, Sell SM, Butler-Browne GS, Schwartz K, Bouveret P, Pinset-Harstrom I (1981) Three myosin heavy-chain isozymes appear sequentially in rat muscle development. Nature 292: 805–809

Wieczorek DF, Periasamy M, Butler-Browne GS, Whalen RG, Nadal-Ginard B (1985) Co-expression of multiple myosin heavy chain genes, in addition to a tissue-specific one, in extraocular musculature. J Cell Biol 101:618–629

Wu H, Naya FJ, McKinsey TA, Mercer B, Shelton JM, Chin ER, Simard AR, Michel RN, Bassel-Duby R, Olson EN, Williams RS (2000) MEF2 responds to multiple calcium-regulated signals in the control of skeletal muscle fiber type. EMBO J 19:1963–1973

Yelon D, Horne SA, Stainier DY (1999) Restricted expression of cardiac myosin genes reveals regulated aspects of heart tube assembly in zebrafish. Dev Biol 214:23–37

Yutzey KE, Rhee JT, Bader D (1994) Expression of the atrial-specific myosin heavy chain AMHC1 and the establishment of anteroposterior polarity in the developing chicken heart. Development 120:871–883

Molecular Characterization of Early Cardiac Development

Thomas Brand, Birgit Andrée, and Thomas Schlange[1]

1 Introduction

Heart disease is a leading cause of death and disability in the Western world and a major factor in rising health care costs. Despite great efforts and remarkable advances made over the past years, the molecular basis for heart development and the evolution of acquired heart disease are still poorly defined (Chien 2000; Srivastava and Olson 2000). Approximately 10% of spontaneous abortions and stillborns are caused by severe cardiac malformations (Hoffman 1995a, 1995b). In addition, congenital heart defects including aberrant cardiac outflow septation, valve formation, aortic stenosis or hypoplastic left ventricle are prevalent with a frequency of 8 cases in 1000 births. Understanding cardiac development in cellular and molecular terms, therefore, is of major clinical relevance. Key genes in cardiac development are also important components of regulatory circuits in the adult heart (Chien 2000). For example, cardiac-restricted transcription factors such as Nkx2.5 and GATA4 have been implicated, both in cardiac development and under pathophysiological conditions, in the adult heart. GATA4 and its interacting protein partner NFATc have been identified as important components of a signalling pathway involved in generating cardiac hypertrophy (Molkentin et al. 1998). Point mutations in human NKX2.5 (CSX) have been found to cause multiple congenital malformations including atrial and ventricular septal defects, Ebstein's anomaly and other tricuspid valve abnormalities. NKX2.5 mutations are also responsible for familial ventricular arrhythmias in the adult (Schott et al. 1998; Benson et al. 1999). Thus, understanding cardiac development at the molecular level will result in the identification of genetic networks, which are active in various pathophysiological processes in the adult heart as well. For further reading on cardiac development the reader is referred to other recent reviews on these topics (Chen and Fishman 2000; Lough and Sugi 2000; Rosenthal and Xavier-Neto 2000; Srivastava and Olson 2000).

[1] Institute of Biochemistry and Biotechnology, Department of Cell and Molecular Biology, Technical University of Braunschweig, Spielmannstr. 7, 38106 Braunschweig, Germany

Results and Problems in Cell Differentiation, Vol. 38
B. Brand-Saberi (Ed.): Vertebrate Myogenesis

2 Molecular Control of Heart Field and Tubular Heart Formation

2.1 Conserved Regulatory Circuits Control Heart Field Formation in Insects and Vertebrates

The heart is the first organ to be formed in vertebrates. Soon after gastrulation, bilateral heart fields are formed in the anterior lateral plate mesoderm (DeHaan 1965). In the past, it was only possible to define the heart fields operationally based on transplantation and cell-culture experiments; however, in recent years, several signalling molecules and transcription factors have been identified that act within the limits of the heart field and control myocardial specification, differentiation and histogenesis.

2.1.1 Heart Field Formation in Vertebrates

Cell-labelling experiments with DiI revealed that at midgastrula cardiac progenitor cells are present in the upper third of the forming primitive streak. The rostro-caudal position of cells within the streak seems to correspond to their later position in the heart tube (Garcia-Martinez and Schoenwolf 1993). A recently identified transcriptional co-activator, chicken CITED2, demarcates this cell population within the primitive streak at HH stage 3. Most interestingly, labelling of the primitive streak is lost at the time myocardial precursors migrate into the lateral plate mesoderm (Schlange et al. 2000a). Although endocardial and myocardial precursors co-exist within the heart fields of chicken embryos, the cell lineages have already separated at the time the progenitors reside within the primitive streak (Cohen-Gould and Mikawa 1996; Wei and Mikawa 2000). Similar observations have also been made in zebrafish (Lee et al. 1994). Atrial and ventricular progenitor cells are separate cell lineages at the time the heart fields are formed (Mikawa and Fischman 1996). Cell-fate analysis in the chick using a replication-defective retrovirus showed that atrial precursors are present in the caudal part, whereas ventricular progenitor cells are localized to the anterior region of the heart field and no evidence for a bipotential stem cell exists. It is formally possible that stem cells capable of forming both atrial and ventricular myocytes do exist, but are refractile to infection by the retrovirus. However, similar results were obtained in zebrafish with a non-viral lineage tracer (Stainier et al. 1993).

2.1.2 Heart Formation in Insects

Analysis of heart formation in *Drosophila* provided the first opportunity to dissect heart development genetically. In *Drosophila*, a heart-like organ, the dorsal vessel pumps hemolymph through the interstitium. Dorsal vessel progenitors arise from mesodermal cells, which are characterized by the persistent expression of the homeobox gene *tinman* (Bodmer et al. 1990; Bodmer 1993). Mutation of the *tinman* gene results in loss of the dorsal vessel. All cells of the presumptive mesoderm express *tinman* before gastrulation, but its expression becomes limited to the dorsal mesoderm invagination. Maintenance of mesodermal *tinman* expression depends on the expression of the BMP2/4 related gene *decapentaplegic* (*dpp*), and other genes encoding components of the Dpp signal transduction pathway in the adjacent ectoderm (Frasch 1995). Another signalling factor that is required for heart induction in *Drosophila* is *wingless* (*wg*). *Tinman* is induced in mesoderm at the intersection of *dpp* and *wg* ectodermal expression domains (Park et al. 1996). In addition to *tinman*, the GATA zinc finger transcription factor *pannier* (*pnr*) is required for cardiogenesis to occur (Gajewski et al. 1999). Loss of function mutations of *pnr* results in aberrant cardiogenesis. *pnr* functions as a cardiac identity gene because its forced expression results in supernumerary cardial cells at the expense of other derivatives of the dorsal mesoderm. Pnr works synergistically with Tin and ectopic co-expression of both factors results in expanded cardiogenic gene expression domains in ventral and dorsal mesoderm, and in the cephalic region (Gajewski et al. 1999). Pnr is negatively regulated by the zinc-finger protein U-shaped (Ush), which is co-expressed with *pnr* in the cardiogenic mesoderm (Fossett et al. 2000).

2.1.3 Nkx Homeobox Genes

There is strong similarity between heart field formation in vertebrates and dorsal vessel formation in *Drosophila* because transcription factors related to *tinman* and *pnr* are also essential for heart formation in vertebrates. However, no single null mutation in vertebrates has been found that, like *pnr* and *tin*, results in the complete loss of heart formation in vertebrates. Several *tinman*-related NK-homeobox genes are expressed in vertebrate hearts including Nkx2.3, Nkx2.5, Nkx2.6, Nkx2.7, Nkx2.8 and Nkx2.10 (Evans 1999). Although Nkx2.5 is found in all vertebrate hearts, distinct combination of Nkx genes are expressed in different vertebrates. In the chick embryo, Nkx2.5 is present in the heart field shortly after its formation at HH stage 5 (Schultheiss et al. 1995; Buchberger et al. 1996) and Nkx2.8 is transiently co-expressed during tubular heart formation between HH stages 8 and 11 (Boettger et al. 1997; Brand et al. 1997; Reecy et al. 1997). Nkx2.8 is expressed in the dorsal mesocardium for an extended period of time, a region in which parts of the outflow tract myocardium is generated (Markwald et al. 1999). During chamber formation, Nkx2.3 becomes strongly upregulated in the trabecular layer, whereas it is absent from the compact layer myocardium (Buchberger et al. 1996). In the mouse, only two homeobox genes, Nkx2.5 (Lints et al. 1993) and Nkx2.6 (Nikolova et al. 1997; Biben et al. 1998), are expressed in the myocardium. Although Nkx2.5 is expressed in the early heart and remains expressed throughout life, Nkx2.6 is transiently expressed in the inflow and outflow tract myocardium (Nikolova et al. 1997; Biben et al. 1998). Nkx2.3 in mice is expressed in gut mesoderm, but is absent from the myocardium (Pabst et al. 1997). Null-mutations have been created for both Nkx2.5 and Nkx2.6. In the Nkx2.5 null mutation, early cardiac development appears unaltered; however, cardiac development is arrested at the looping stage (Lyons et al. 1995). Thus, early cardiogenesis appears unaffected and the lack of Nkx2.5 seems to become limiting at the time of chamber formation. Based on the spectrum of genes that are lacking in the Nkx2.5 null heart, it has been concluded that Nkx2.5 is specifically required for the formation of the left ventricular cell lineage (Biben and Harvey 1997). The loss-of-function allele of the Nkx2.6 gene does not affect myocardial development in any significant way (Tanaka et al. 2000). Two recent studies in *Xenopus* have provided evidence that NK2 genes are essential factors for the recruitment of cardiac mesoderm in vertebrates. In both studies, dominant-negative Nkx2.5 and Nkx2.3 mutations were generated, either through fusion with an engrailed repressor domain or through mutagenesis of a conserved amino acid in the homeobox domain (Fu et al. 1998; Grow and Krieg 1998). In both studies a strong reduction or complete loss of myocardial gene expression and histogenesis was observed, suggesting that vertebrate NK2 homeobox genes are essential for cardiac mesoderm formation at least in amphibian hearts. At present it is unclear how the Nkx2.3/Nkx2.5 dominant-negative mutants interfere with wild-type protein function. Nkx2.5 genes can homo- and heterodimerize with Nkx2.3 and Nkx2.6 via the carboxy terminus of the homeobox domain (Kasahara et al. 2000). Thus, dominant-interfering mutations either compete for DNA binding or, alternatively, prevent wild-type protein from acting as a transcriptional activator by heterodimerization. In addition, Nkx2.5 was found to interact with various other cardiac-restricted transcription factors, such as GATA4 (Sepulveda et al. 1998), SRF (Chen and Schwartz 1996) and MEF2 (Morin et al. 2000).

2.1.4 GATA Genes

Three GATA genes, GATA4–6, are expressed in vertebrate hearts (Molkentin 2000). Mouse embryos that lack GATA4 have bilateral heart tubes and a reduced number of cardiac myocytes (Kuo et al. 1997; Molkentin et al. 1997). Homozygous GATA4–/– embryonic stem cells can differentiate into contractile myocytes in vitro and in chimeric embryos. Therefore, the observed morphogenetic function of GATA4 is non-cell autonomous. The *cardia bifida* phenotype is probably related to an endoderm deficiency and can be rescued by wild-type endoderm in chimeric embryos (Narita et al. 1997). GATA5 null mutants are viable (Molkentin et al. 2000b), whereas GATA6–/– embryos die soon after implantation (Morrisey et al. 1998). Essential functions of GATA5 for endoderm formation have been described in both zebrafish and *Xenopus* (Weber et al. 2000; Reiter et al. 2001). GATA5 in the zebrafish is encoded by the *faust* locus. Like the GATA4 null mutant in mouse, the *faust* mutant displays a *cardia bifida* phenotype (Reiter et al. 1999). Interestingly, overexpression of GATA5 in the embryo leads to ectopic Nkx2.5 expression and formation of ectopic beating cell clusters in the fish embryo. Likewise, overexpression of GATA4 in embryonic stem cells display enhanced cardiac differentiation (Grepin et al. 1997). Promoter studies reveal a mutually reinforcing regulatory network of Nkx2.5 and GATA transcription factors during cardiogenesis. The GATA6 promoter in both mouse and chick contain functional important Nkx2.5 binding sites (Davis et al. 2000; Molkentin et al. 2000a). Similarly, the murine Nkx2.5 promoter contains GATA sites that are involved in early heart field expression (Searcy et al. 1998; Lien et al. 1999; Reecy et al. 1999).

2.2 Hypoblast and Anterior Endoderm Are Involved in Myocardial Specification and Differentiation

Fate analysis defines the normal developmental pathway a cell will take in a particular spatio-temporal situation; however, it does not indicate its developmental potential. In the chick embryo at HH stage 3, cells within the anterior third of the primitive streak, excluding Hensen's node itself, are fated to become heart muscle (Garcia-Martinez and Schoenwolf 1993). Tissue transplantation of primitive streak cells, however, indicate that these cells have a far greater developmental potential (Inagaki et al. 1993). Explanted epiblast cells from mouse embryos at 6.5 d.p.c. are able to differentiate among other tissues into cardiac muscle (Auda-Boucher et al. 2000). However, if transplanted into chick paraxial mesoderm they do not differentiate into cardiac mesoderm.

Cardiac determination mainly takes place in the anterior lateral plate mesoderm. Both anterior and posterior epiblast cells of the mouse embryo at the early primitive streak stage differentiate into cardiac myocytes, if transplanted into anterior lateral plate mesoderm (Tam et al. 1997). In the avian embryo, the hypoblast can induce heart muscle cells in co-cultured pre-gastrula and mid-gastrula epiblast explants, and has modest inducing ability in the case of HH stage 4 posterior primitive streak (Yatskievych et al. 1997). In contrast, HH stage 5 anterior lateral endoderm does not induce heart development in epiblast cells, but is a potent inducer of cardiac myogenesis in cells from HH stage 4 posterior primitive streak. The hypoblast-derived signal is probably an activin-like molecule (Ladd et al. 1998). At present, it is difficult to distinguish between a specific role for activin or an activin-like signal, such as Nodal for cardiac specification or a general role as a mesoderm and endoderm inducer that secondarily enhances the formation of cardiac mesoderm. In the mouse embryo, cardiac specification occurs between 7.25 and 7.5 d.p.c. (Arai et al. 1997). Explants of cardiogenic mesoderm of 7.25 d.p.c. embryos differentiate into cardiac muscle if both the visceral embryonic endoderm and primitive streak are present. Thus, the hypoblast in chicken and the visceral endoderm in mouse seem to have a similar function in cardiac induction.

According to Slack's definition of specification, a group of cells is called specified if they develop according to their normal fate when isolated and cultured in a neutral environment (Slack 1991). This stage in cardiac induction is reached shortly after presumptive cardiac mesoderm has migrated to the anterior-lateral plate between HH stages 4 and 5 (Antin et al. 1994; Montgomery et al. 1994). Specification, however, is reversible and the plastic state of specification can be tested by transplantation of a specified group of cells into an ectopic location. Mesoderm at HH stage 6 is specified, but not determined because Nkx2.5 and MEF-2C are downregulated when cardiac mesoderm is transplanted into head mesoderm (Lough and Sugi 2000). Thus, cardiac determination is probably not completed before HH stage 8. Two different stages of terminal differentiation can be distinguished: "partial" differentiation can be defined as the stage of development in which structural mRNAs, such as myofibrillar genes, are transcribed, but not yet translated. Transcripts encoding cardiac structural proteins, such as tropomyosin or smooth muscle α-actin (Sma), are first detected at HH stage 5 in the anterior heart fields (Colas et al. 2000; B. Andrée et al. unpubl. observ.), suggesting that "partial" differentiation starts at the time cardiac specification has occurred. "Full" differentiation is the state in which myofibrillar proteins and cell adhesion proteins have been assembled to enable rhythmic contractions. Although Sma mRNA accumulates at HH stage 5, Sma protein is not detectable before HH stage 9 (Colas et al. 2000). There seems to be a tight translational control of myofibrillar protein synthesis preventing full myocyte differentiation occurring before the tubular heart has formed. Several reports consistently describe that myocardial terminal differentiation, like cardiac specification, is dependent on the presence of anterior endoderm (Antin et al. 1994; Sugi and Lough 1994; Gannon and Bader 1995).

2.3 Identification of Signalling Molecules Involved in Cardiac Specification and Differentiation

2.3.1 The Role of BMP2 in Heart Induction

In the chick embryo, anterior lateral endoderm secretes a variety of signalling molecules including various FGFs, activin and insulin-like growth factor II, which all appear to promote cell survival and proliferation of cardiogenic cells and differentiating myocytes (Kokan-Moore et al. 1991; Parlow et al. 1991; Sugi and Lough 1995; Antin et al. 1996; Zhu et al. 1999b). We and others have demonstrated that BMP2 is part of the heart-inducing activity that is derived from pharyngeal endoderm (Lough et al. 1996; Schultheiss et al. 1997; Andrée et al. 1998; Ladd et al. 1998). BMP2 is expressed at HH stages 4 and 5 of chick embryogenesis in lateral mesendoderm adjacent to the heart-forming region and during HH stages 6–9 in pharyngeal endoderm underlying cardiogenic mesoderm. Subsequently, during HH stages 10 and 11, BMP2 is present in the endoderm of the sino-atrial region, but is absent from the tubular heart (Andrée et al. 1998). BMP2 is able to induce the expression of myocardial lineage markers in ectopic locations (Lough et al. 1996; Schultheiss et al. 1997; Andrée et al. 1998; Ladd et al. 1998). Moreover, Noggin which binds BMPs with high affinity prevents myocardial differentiation of lateral mesendoderm cultured in vitro (Schultheiss et al. 1997; Ladd et al. 1998; Schlange et al. 2000b). Multiple cardiac-restricted transcription factors are induced by BMP2 with distinct kinetics (Schlange et al. 2000b). In explanted heart fields Noggin inhibits myocardial marker gene expression. However, full inhibition is only observed at HH stage 4, whereas at stages 5–8 individual genes show a differential sensitivity for Noggin. These data would indicate that cardiac specification might be complete by HH stage 5. Implantation of Noggin-expressing cells into the heart field in vivo also interferes with myocardial determination (Schlange et al. 2000b). Complete loss of NKX2.5 and eHAND expression is observed at HH stages 4–6, whereas, at HH stages 7 and 8, myocardial gene expression and histogenesis are suppressed, suggesting that specification in situ has not been completed before HH stage 6. Myocardial differentiation remains dependent on the presence of BMP2 until HH stage 8, which would be consistent with the view that determination and myocardial differentiation are both controlled by endoderm-derived BMP2. Recently, the concentration dependence of myocardial induction by BMP2 has been analysed (Barron et al. 2000). Myocardial induction in explant cultures is maximal at 50 ng/ml whereas, at 250 ng/ml, a cell type with high alkaline phosphatase activity was induced. This result illustrates the importance of controlling BMP signalling tightly in order to induce myocardial differentiation. Smad6, which interferes with phosphorylation of receptor-regulated Smads, is expressed in the heart field and becomes ectopically induced after implantation of BMP2-expressing cells (Yamada et al. 1999). Thus, Smad6 is part of a negative autoregulatory loop that probably limits the amount of cells that will actually enter the cardiogenic pathway and it probably prevents the induction of mesodermal cell types other than heart.

The dependence of myocardial specification on BMP2 signalling appears to be conserved in vertebrates. In zebrafish, the *BMP2* mutant *swirl* lacks cardiogenic mesoderm (Kishimoto et al. 1997). *Xenopus* embryos with ectopic expression of dominant-negative BMPRII and ALK3 mutations do not form cardiac mesoderm and display an inability to maintain Nkx2.5 expression, whereas initial Nkx2.5 expression is not affected by the mutant receptors (Shi et al. 2000). Murine *BMP2* null mutants display abnormal heart development, but cardiac mesoderm is formed (Zhang and Bradley 1996). Several members of the BMP family with overlapping functions are expressed in murine cardiogenic mesoderm and the essential function in heart formation might only become apparent in double or triple null embryos (Dudley and Robertson 1997; Solloway and Robertson 1999). It has been recently demonstrated that myocardial differentiation in murine P19 embryonic teratocarcinoma cells is dependent on a functional BMP signal transduction pathway

(Monzen et al. 1999). Thus, the observed function of *dpp* in *Drosophila,* which maintains *tinman* expression, seems evolutionary conserved and also utilized in vertebrate heart formation.

2.4 Other Cardiogenic Signals

2.4.1 Wnt Signals Interfere with Heart Formation in Vertebrates

In *Drosophila*, *wg* and *dpp* are both required for *tinman* expression (Wu et al. 1995). A subtractive screen, which was designed to identify mesendodermal genes present in the anterior half of the chick embryo, resulted in the isolation of the secreted wnt inhibitor Crescent (Marvin et al. 2001). Crescent mRNA is abundant in the anterior hypoblast and anterior definitive endoderm from HH stages 2–6. At HH stages 5–6, crescent is expressed in pre-chordal mesendoderm as well, but at HH stages 6–7 its expression begins to decline in the endoderm underlying the presumptive heart and head mesoderm (Pfeffer et al. 1997; Marvin et al. 2001). Crescent binds to several wnt factors, such as Wnt-1, Wnt-3a and Wnt-8c, and prevents binding to the wnt signalling receptors of the Frizzled class. Ectopic expression of Wnt-1 in the heart field blocks myocardial differentiation and induces blood formation (Marvin et al. 2001). Conversely, explants of posterior lateral mesendoderm are induced to form beating cardiac mesoderm in the presence of Crescent at the expense of ρ-globin expression. However, Crescent is unable to induce ectopic cardiogenesis in posterior mesoderm in vivo. Crescent's activity can also be mimicked by other secreted wnt inhibitors, such as Dkk-1 and Frzb, in the chick. Implantation of pellets of Frzb- and BMP4-expressing cells into paraxial mesoderm results into cardiac mesoderm formation and subsequent migration of these cells into the forming heart tube (Tzahor and Lassar 2001). These results suggests that BMP2 and inhibition of Wnt signals are necessary and sufficient components to induce specification of cardiac mesoderm in explant cultures (Marvin et al. 2001; Tzahor and Lassar 2001). The ability of Dkk1 and Crescent to promote cardiogenesis has also been demonstrated in ventral marginal zone explants from *Xenopus* embryos (Schneider and Mercola 2001). Interestingly, various other wnt antagonists lack the cardiogenic potential of Dkk1 and Crescent. This result might either indicate that these factors have a particular weak affinity for a Wnt factor present in the ventral marginal zone or that Dkk1 and Crescent have an additional activity unrelated to their ability to interfere with Wnt signalling. However, the latter interpretation is unlikely because expression of GSK3β is also able to promote cardiac differentiation (Schneider and Mercola 2001). This result indicates that inhibition of cardiogenesis is probably mediated by the β-catenin/Lef-1 signalling pathway. Wnt-11, which is expressed around the posterior edge of the heart field, can induce cardiogenesis in posterior non-precardiac mesoderm (Eisenberg and Eisenberg 1999). Interestingly, Wnt-11 belongs to a different class of Wnt factors that differ in signal transduction and activate PKC and CamKII in a G-protein-dependent manner (Kuhl et al. 2000). As in other systems, Wnt-11 might promote cardiogenesis by antagonizing Wnt-3 and Wnt-8 signal transduction (Torres et al. 1996). However, loss-of function experiments are required to unequivocally demonstrate any role for Wnt-11 in promoting cardiogenesis.

2.4.2 FGF Cooperates with BMP2

Additional cardiogenic signals have been recently identified including FGF8, which is mutated in the zebrafish mutant *acerebellar* (*ace*; Reifers et al. 2000). FGF8 is expressed in the cardiogenic mesoderm itself. Both Nkx2.5 and GATA4 require the presence of FGF8, whereas GATA6 does not depend on FGF8. A similar phenotype is obtained after FGF receptor inhibition using the pharmacological inhibitor SU5402 (Reifers et al. 2000). FGF8 is also present in cardiac mesoderm of mouse and chicken embryos (Crossley and Martin 1995; Crossley et al. 1996). FGF8 deletion mutants display aberrant left–right asymmetry and various cardiac malformations have been observed. However, early heart formation seems not to be affected (Meyers et al. 1998; Meyers and Martin 1999). In posterior mesoderm explants, BMP2, in combination with FGF4, can stimulate heart formation whereas neither factor alone can do so (Lough et al. 1996; Barron et al. 2000). Thus, FGF and BMP signalling seem to synergize to drive mesodermal cells into myocardial differentiation.

2.4.3 Cerberus

Overexpression of Cerberus in *Xenopus* results in the formation of ectopic heads and often duplicated hearts (Bouwmeester et al. 1996). Cerberus is a member of the Dan family of secreted antagonists that interfere with Wnt, Nodal and BMP signalling (Glinka et al. 1997). Recently, microsurgical ablation studies in *Xenopus* demonstrated that the Cerberus expression domain in the anterior endoderm demarcates cells that have heart-inducing activity (Schneider and Mercola 1999). A Cerberus homologue, Cerr 1, has been isolated in the mouse, but has no essential function in development (Belo et al. 2000; Shawlot et al. 2000; Stanley et al. 2000). In the chick, the Cerberus homologue Caronte (Car) is involved in left–right axis formation acting as a BMP antagonist in the left lateral plate mesoderm (Rodriguez Esteban et al. 1999; Yokouchi et al. 1999; Zhu et al. 1999a). In contrast to the Wnt antagonist, Crescent, Caronte fails to promote cardiogenesis in cultured posterior lateral plate mesoderm (Marvin et al. 2001). Thus, at present, it is unclear whether Cerberus-like factors are important for heart development in higher vertebrates.

2.4.4 Cripto

The founding member of the EGF-CFC family of EGF-like signalling molecules, Cripto, has an essential function in mediating Nodal signalling (Gritsman et al. 1999; Salomon et al. 1999; Shen and Schier 2000). Cripto null mutants display a deficit in the conversion of the proximo-distal axis into the embryonic anterio-posterior axis (Ding et al. 1998). A related gene, Cryptic, is also involved in Nodal signal transduction and Cryptic null embryos display aberrant left–right asymmetric morphogenesis (Gaio et al. 1999; Yan et al. 1999). Interestingly, Cripto homozygous embryonic stem cells are deficient in myocardial differentiation, although expression of cardiac-restricted transcription factors such as Nkx2.5, Mef2 or GATA4 appears unaltered, suggesting that Cripto is involved in the transition from myocardial specification to differentiation (Xu et al. 1998, 1999). In the mouse, Cripto is present in the nascent mesoderm, in the heart fields, the tubular heart and between 9.5 and 10.5 d.p.c. in the truncus arteriosus (Dono et al. 1993). In the chick, a related factor, chick CFC, is also expressed in the heart fields and in the tubular heart (Colas and Schoenwolf 2000; Schlange et al. 2001). Although chick CFC appears to be an important factor in generating the left–right axis in the chicken embryo, no functional data are currently available for a role in heart formation.

Recently, it has been shown that treatment of dorsal blastopore lip explants (organizer) with Suramin generates a beating heart, which can even functionally substitute for the endogenous heart in *Xenopus* embryos (Grunz 1999). The molecular mechanism of Suramin action is unknown; however, with several known cardiogenic signalling molecules in the organizer (Dkk1), endoderm (BMP2) and mesoderm (FGF8, Cripto), it might be possible to identify the molecular mechanism of Suramin action.

2.5 Heart Tube Formation

In the chick embryo, myocardial tube formation is dependent on the closure of the anterior intestinal portal (AIP). Incision of the AIP results in formation of two tube-like structures instead one heart tube in the midline. Both tubes acquire contractile activity and undergo some steps of cardiac morphogenesis including heart looping. This structure is called *cardia bifida* (Rosenquist 1970). Several mouse mutants display a *cardia bifida* phenotype including GATA4, the furin proprotein convertase (Roebroek et al. 1998), the helix–loop–helix transcription factor MesP1 (Saga et al. 1999) and the endosome protein Hrs (Komada and Soriano 1999). Similarly, in zebrafish, several mutants were isolated that display *cardia bifida*, including the GATA5 mutant *faust* (Reiter et al. 1999; Reiter et al. 2001) and two mutants that lack endodermal tissue, *casanova* and *one-eyed-pinhead* (Alexander et al. 1999). A common feature of most of these mutants is a deficiency in endoderm formation or function. However, other tissues are also involved in heart-tube formation including cardiac (MesP1; Saga et al. 1999) and paraxial mesoderm in the case of *miles apart* (*mil*). *mil* encodes a sphingosine-1-phosphate receptor (Kupperman et al. 2000), suggesting that sphingolipid signalling, which in other systems has been shown to control cell proliferation, differentiation and cell survival, is involved in cardiac tubulogenesis.

3 Molecular Control of Cardiac Chamber Formation

3.1 Transcriptional Regulators of Chamber Formation

After tubular heart formation is complete and looping has started, transformation of the expression pattern of cardiac genes is observed. Many genes that are initially expressed throughout the tubular myocardium become restricted to the presumptive atrial and ventricular primordia, respectively. Only recently, several transcription factors have been identified that are probably involved in establishing anterior–posterior patterning of the myocardium. Two hairy-related transcription factors, HRT1 and HRT2 (also known as Hey or CHF genes; Leimeister et al. 1999; Nakagawa et al. 1999; Chin et al. 2000), are expressed in atrial and ventricular progenitor cells, respectively. Chamber-specificity of HRT gene expression is maintained throughout development. Like other Hairy-related transcription factors, the HRT family is controlled by a Notch signalling pathway and these proteins act as transcriptional repressors (Nakagawa et al. 2000). The iroquois-related homeobox protein Irx4 labels ventricular progenitor cells in the cardiac crescent and, subsequent to chamber formation, Irx4 is exclusively expressed in the ventricle (Bruneau et al. 2000). In the chick embryo, Irx4 seems to be involved in establishing chamber-restricted gene expression, whereas morphogenesis is apparently not affected by this homeobox gene (Bao et al. 1999). Irx4-deficient mice develop normally in utero; however, null mutants develop a cardiomyopathy in later life (Bruneau et al. 2001). Irx4 may be functionally substituted by other Irx homeobox genes expressed in the myocardium (Bruneau et al. 2001; Christoffels et al. 2000).

The atrial compartment of the myocardium is marked by GATA-4 and the T-box gene Tbx5. Both genes are initially expressed throughout the cardiac crescent, but become localized to the posterior inflow compartment after tubulogenesis is completed. Misexpression of Tbx5 in the ventricle results in loss of ventricular-specific gene expression and retardation of ventricular chamber morphogenesis (Liberatore et al. 2000). The orphan nuclear receptor COUP-TFII is also restricted to the inflow tract myocardium and the null mutation displays grossly retarded development of sinus venosus and atrium (Pereira et al. 1999). GATA4 and Nkx2.5 expression domains overlap during early heart formation and both genes are probably mutually inducing each others expression within the heart field, shortly after GATA4 is down-regulated in the anterior presumptive heart field and is only maintained in the posterior presumptive atrial compartment (Jiang et al. 1998). It is believed that Nkx2.5 is involved in setting up a ventricle-specific gene expression program (Bruneau et al. 2000), whereas GATA4 is probably involved in establishing atrial identity. In retinoic-acid (RA)-deficient quail embryos an oversized ventricle with deficient formation of atrium and omphalomesenteric vein has been observed (Dersch and Zile 1993). In RA-deficient embryos the expression of GATA-4 is severely decreased (Kostetskii et al. 1999), whereas excess administration of RA leads to an anterior expansion of GATA4 expression in frogs (Jiang et al. 1998) and of Tbx 5 in the chick (Liberatore et al. 2000).

The related basic helix-loop-helix transcription factors dHAND and eHAND are involved in ventricular chamber formation. Although both genes are co-expressed in the primary heart tube, they differentially accumulate in the right (dHAND) and left (eHAND) ventricle. Deletion of dHAND in mice results in hypoplasia of the right ventricular chamber. eHAND has been implicated in left ventricular development; however, because of early lethality the function of eHAND can only be inferred from chimeric analysis, which suggests that eHAND is specifically required for the formation of the outer curvature of the left ventricular myocardium (Riley et al. 2000). Mice homozygous for Nkx2.5 fail to express eHAND in the heart, suggesting that eHAND acts downstream of Nkx2.5 in controlling left ventricular chamber formation. In zebrafish, only one HAND gene, dHAND, has been identified. Mutation of dHAND results in loss of ventricular chamber formation (Yelon et al. 2000). Two HAND genes have also been identified in *Xenopus*; however, only dHAND is utilized in the myocardium, whereas eHAND is expressed in vascular tissues (Smith et al. 2000). This suggests that in lower vertebrates with a single ventricular chamber only one HAND gene is expressed in the ventricle. In higher vertebrates with two separate ventricular chambers both HAND genes are expressed in the heart to specify different chamber fates. Right ventricular hypoplasia is also apparent in the MEF2C mutant (Lin et al. 1997) and in the versican mutant, *hdf* (Mjaatvedt et al. 1998).

3.2 Cell–Cell Interaction in Chamber Formation

After chamber formation has progressed, the heart is invaded by several migratory cell populations. The epicardial serosa (derived from the septum transversum in mouse and the early liver anlage in chick) envelopes the myocardium thus forming the outer layer of the heart, the epicardium (Ho and Shimada 1978). Some cells delaminate and migrate into the ventricular wall and differentiate to endothelial and smooth muscle cells of the coronary arteries, and fibroblasts of the cardiac interstitium (Dettman et al. 1998). Another cell population, the cardiac neural crest cells, migrate into the outflow tract to induce septation (Kirby and Waldo 1995). Further maturation of the ventricular myocardium involves the formation of the trabecular layer, which initially, between 9 and 12 d.p.c. in mouse, is more differentiated than the subepicardial compact layer (Kastner et al. 1997). Cardiac trabeculation is dependent on endocardial-myocardial interactions involving neuregulin signalling, which is mediated by a myocardial receptor complex of erbB2 and erbB4 (Gassmann et al. 1995; Lee et al. 1995; Meyer and Birchmeier 1995). The subepicardial compact layer shows several fold higher proliferative capacity than the trabecular layer (Tokuyasu 1990). Only recently, the importance of the epicardium for growth control of the ventricular wall became apparent. Several mouse mutants display a hypoplastic ventricular wall and die between embryonic days 13 and 16. This phenotype has been observed for mutants of the transcription factors N-myc (Moens et al. 1993), RXRα (Kastner et al. 1994; Sucov et al. 1994), WT1 (Kreidberg et al. 1993), Tefl (Chen et al. 1994), FOG2 (Tevosian et al. 2000), the pocket protein p130 (LeCouter et al. 1998), the receptor gp130 (Yoshida et al. 1996), the receptor kinase βARK1 (Jaber et al. 1996) and erythropoietin and its receptor (Wu et al. 1999). In some cases, like the N-myc mutant, the reduced proliferative capacity of cardiac myocytes is probably the cause for the hypoplastic phenotype. For most mutations however, the gene functions in a non-cardiac cell background and the mutation is non-cell-autonomous. In the case of WT1 and the erythropoietin receptor, the genes are expressed in the epicardium whereas the defect is apparent in the subepicardial compact layer (Moore et al. 1999; Wu et al. 1999). Similarly, in the RXRα mutant, the mutation seems not to be cell autonomous and cannot be rescued by cardiac myocyte-specific overexpression of RXRα (Chen et al. 1998; Tran and Sucov 1998). The mutual interaction of the epicardium and the compact layer myocytes is best seen in the case of the FOG2 mutant, which has both a hypoplastic ventricular wall and a deficiency of coronary artery formation (Tevosian et al. 2000). In this case, both phenotypes can be rescued by myocyte-specific overexpression of FOG2. Thus, FOG2 is involved cell-autonomously in the hypoplastic growth of the compact layer, and non-cell autonomously in the differentiation of the coronary arteries. In two mutations in the mouse, VCAM and α-4 integrin, defective growth of the epicardial layer is present and in both mutations a hypoplastic ventricle has been observed (Moens et al. 1993; Kwee et al. 1995). This dependence of the ventricular compact layer on a mitogenic signal from the epicardium is also seen in the avian embryo. Ablation of the epicardium by surgical or pharmacological interventions in the chick results in a hypoplastic ventricle (Männer 1993). Interestingly, one of the retinoic acid synthesizing enzymes, RALDH2, is expressed in the forming epicardium of mice and chicken and might be an indicator of localized RA production in the epicardium (Moss et al. 1998; Xavier-Neto et al. 1999, 2000). In the RXRα mutation a hypoplastic ventricular wall is histologically first detectable in 12 d.p.c. fetuses (Sucov et al. 1994). Normally, myocytes of the compact layer mature around day 16.5, however, in the absence of RXRα or, in vitamin A-deficient embryos, precocious differentiation becomes already apparent at 8.5 d.p.c. (Kastner et al. 1997). A current hypothesis suggests that the epicardium requires an autocrine or paracrine RA signal, which maintains its ability to send out yet unknown mitogenic signal(s) to the myocytes of the compact layer (Tran and Sucov 1998).

As an alternative to the hypothesis of premature myocardial differentiation as a reason for the hypoplastic expansion of the compact layer, evidence for an endocrine role of the placenta for myocardial growth has been recently obtained. The PPARγ receptor is required for placental development, but also displays a thin ventricular phenotype that is more pronounced than the one observed in the RXRα mutant (Barak et al. 1999). Rescue of the placenta deficiency by chimera formation with tetraploid wild-type cells that participate in extra-embryonic tissue development with no contribution to the embryo itself, rescued the myocardial phenotype. A similar combined phenotype in placenta and myocardium is observed in case of the null mutant for TRAP220, a co-factor of PPARγ, and also in the RXRα mutant (Kodera et al. 2000). Thus, an endocrine or nutritional function of the placenta might also be important for myocardial compact layer expansion. Therefore, several different mechanisms are controlling myocardial histogenesis and chamber formation.

3.3 *Popeye* Genes – a Novel Family of Muscle-Restricted Genes

By means of a subtractive hybridization approach, a novel heart-restricted gene family was isolated from day 7 embryonic chick hearts (Andrée et al. 2000). Because of their restricted expression pattern in heart and skeletal muscle and in a few smooth-muscle-containing organs such as bladder and uterus, the family was named *Popeye* (Pop) genes. A total of three cDNAs were identified in mouse and humans whereas, in the chick embryo, a total of six cDNAs were identified (Andrée et al. 2000). Related cDNAs are also found in *Xenopus*, zebrafish and *Drosophila* (T. Brand, unpubl. observ.). Chromosomal mapping indicates that Pop1 and Pop3 genes are clustered on mouse chromosome 10, whereas Pop2 maps to mouse chromosome 16. In the chick, all six cDNAs are generated from a single gene by means of alternative splicing (Andrée et al. 2000). The POP genes encode proteins with three potential transmembrane domains that are conserved in all family members. Pop1 in mouse and POP3 in chicken embryos have similar expression patterns (Fig. 1). Both genes are first expressed only in atrium. Coincident with epicardial covering of the ventricle, Pop1/POP3 are upregulated in a subset of cells belonging to the non-trabeculated part of the ventricle, the subepicardial compact layer (Andrée et al. 2000). Pop2 was found to be expressed in the entire myocardium with the exception of the outflow tract. By whole-mount ISH, expression was first detected at embryonic day 7.5 and was found in the cardiac crescent. Pop2 is the predominant Pop mRNA in the postnatal myocardium. Transfection studies in both Cos cells and cardiac myocytes revealed localization of Pop protein in the endoplasmic reticulum and plasma membrane. Recently, a Pop1 LacZ knock-in mutant was engineered by homologous recombina-

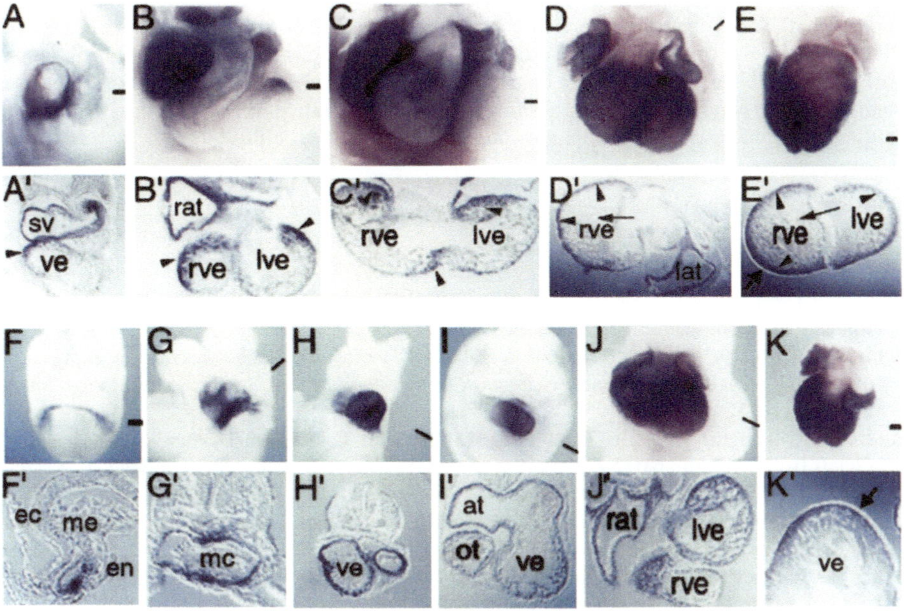

Fig. 1A–K. Cardiac expression of POP1 and POP2 during embryonic development of the mouse. A Right lateral view; **B–D, F–H, J, K** ventral views; I left lateral view. **A–E** Embryos hybridized with Pop1 probe; **F–K** embryos hybridized with Pop2 probe. **A–C** Between E 9.5 and E 11.5 Pop1 was expressed in the atria and dorsal half of the ventricle. **D, E** At E 12.5 and E 14.5, both atria and ventricle display prominent expression of Pop1. **A′–E′** Transversal sections through the heart of the embryos shown in **A–E**. **A′** Pop1 is expressed in the sinus venosus (*sv*) and in the dorsal aspect of the right ventricle (*ve*; *arrowhead*). **B′** Pop1 is strongly expressed in the right atrium (*rat*) and in the dorsal aspects of both right (*rve*) and left (*lve*) ventricle (*arrowheads*), and in **C′** also in the forming muscular septum (*arrowhead*). **D′, E′** At E 12.5 and E 14.5, both atria and ventricles show Pop1 expression. In the ventricle, expression is confined to the subepicardial compact layer (*arrowheads*) and absent from the trabeculated area (*arrow*). **F** At E 7.5, 2 somite stage, Pop2 is expressed in the cardiac crescent. **G** At E 8, 6 somite stage, expression is seen throughout the tubular heart but absent from the outflow region. **H, I** At E 8.5 and E 9.5, respectively, expression is confined to the myocardium of both atrium and ventricle. **J, K** At E 10.5 and E 14.5, respectively, expression is seen throughout the heart, but is absent from the outflow tract. **F′–K′** Transversal sections through the heart of the embryos shown in **F–K**. **F′** Expression in the cardiac crescent is confined to the cardiogenic splanchnic mesoderm (*me*). **G′** In the tubular heart expression is in the myocardium (*mc*) and absent from the endocardium. **H′** Both atria and ventricle (*ve*) display prominent staining for POP3. **I′** At E 9.5 and E 10.5, Pop2 expression is present in the atria (*at*) and both in the trabeculated area and in the compact layer; however, no staining is observed in the outflow tract (*ot*). At E 14.5, Pop2 expression is confined to the subepicardial compact layer of the ventricle. *Ec* Ectoderm; *en* endoderm. *Bars* in **A–E** and **F–K** demarcate approximate plane of section. (Andrée et al. 2000; with kind permission of Academic Press, Orlando

tion. Homozygous Pop1-LacZ mice are viable and until 6 months of age do not reveal any sign of lethality (B. Andrée et al. unpubl. observ.). Further work is required to characterize the function of this gene family in cardiac development.

Acknowledgements. Studies in the authors laboratory are funded by the Deutsche Forschungsgemeinschaft, Sonderforschungsbereich 271: Molecular Genetics of Morphoregulatory Processes, TP B10: "Molecular control of early cardiac organogenesis" and the German-Israeli Foundation (GIF). We would like to acknowledge Martha J. Marvin, Andrew B. Lassar and Mark Mercola for communicating results prior to publishing. The authors gratefully acknowledge the critical reading of the manuscript and helpful comments by Dr. H.H. Arnold.

References

Alexander J, Rothenberg M, Henry GL, Stainier DY (1999) *Casanova* plays an early and essential role in endoderm formation in zebrafish. Dev Biol 215:343–357

Andrée B, Duprez D, Vorbusch B, Arnold H-H, Brand T (1998) BMP-2 induces ectopic expression of cardiac lineage markers and interferes with somite formation in chicken embryos. Mech Dev 70:119–131

Andrée B, Hillemann T, Kessler-Icekson G, Schmitt-John T, Jockusch H, Arnold HH, Brand T (2000) Isolation and characterization of the novel popeye gene family expressed in skeletal muscle and heart. Dev Biol 223:371–382

Antin PB, Taylor RG, Yatskievych T (1994) Precardiac mesoderm is specified during gastrulation in quail. Dev Dyn 200:144–154

Antin PB, Yatskievych T, Dominguez JL, Chieffi P (1996) Regulation of avian precardiac mesoderm development by insulin and insulin-like growth factors. J Cell Physiol 168:42–50

Arai A, Yamamoto K, Toyama J (1997) Murine cardiac progenitor cells require visceral embryonic endoderm and primitive streak for terminal differentiation. Dev Dyn 210:344–353

Auda-Boucher G, Bernard B, Fontaine-Perus J, Rouaud T, Mericksay M, Gardahaut MF (2000) Staging of the commitment of murine cardiac cell progenitors. Dev Biol 225:214–225

Bao ZZ, Bruneau BG, Seidman JG, Seidman CE, Cepko CL (1999) Regulation of chamber-specific gene expression in the developing heart by Irx4. Science 283:1161–1164

Barak Y, Nelson MC, Ong ES, Jones YZ, Ruiz-Lozano P, Chien KR, Koder A, Evans RM (1999) PPAR gamma is required for placental, cardiac, and adipose tissue development. Mol Cell 4:585–595

Barron M, Gao M, Lough J (2000) Requirement for BMP and FGF signaling during cardiogenic induction in non-precardiac mesoderm is specific, transient, and cooperative. Dev Dyn 218: 383–393

Belo JA, Bachiller D, Agius E, Kemp C, Borges AC, Marques S, Piccolo S, De Robertis EM (2000) Cerberus-like is a secreted BMP and nodal antagonist not essential for mouse development. Genesis 26:265–270

Benson DW, Silberbach GM, Kavanaugh-McHugh A, Cottrill C, Zhang Y, Riggs S, Smalls O, Johnson MC, Watson MS, Seidman JG, Seidman CE, Plowden J, Kugler JD (1999) Mutations in the cardiac transcription factor NKX2.5 affect diverse cardiac developmental pathways. J Clin Invest 104:1567–1573

Biben C, Harvey RP (1997) Homeodomain factor Nkx2-5 controls left/right asymmetric expression of bHLH gene eHand during murine heart development. Genes Dev 11:1357–1369

Biben C, Hatzistavrou T, Harvey RP (1998) Expression of NK-2 class homeobox gene Nkx2-6 in foregut endoderm and heart. Mech Dev 73:125–127

Bodmer R (1993) The gene *tinman* is required for specification of the heart and visceral muscles in *Drosophila*. Development 118:719–729

Bodmer R, Jan LY, Jan YN (1990) A new homeobox-containing gene, msh-2, is transiently expressed early during mesoderm formation of *Drosophila*. Development 110:661–669

Boettger T, Stein S, Kessel M (1997) The chicken NKX2.8 homeobox gene. A novel member of the NK-2 gene family. Dev Gene Evol 1:65–70

Bouwmeester T, Kim S, Sasai Y, Lu B, De Robertis EM (1996) *Cerberus* is a head-inducing secreted factor expressed in the anterior endoderm of Spemann's organizer. Nature 382:595–601

Brand T, Andrée B, Schneider A, Buchberger A, Arnold HH (1997) Chicken NKx2-8, a novel homeobox gene expressed during early heart and foregut development. Mech Dev 64:53–59

Bruneau BG, Bao ZZ, Tanaka M, Schott JJ, Izumo S, Cepko CL, Seidman JG, Seidman CE (2000) Cardiac expression of the ventricle-specific homeobox gene *Irx4* is modulated by Nkx2-5 and dHand. Dev Biol 217:266–277

Bruneau BG, Bao ZZ, Fatkin D, Xavier-Neto J, Georgakopoulos D, Maguire CT, Berul CI, Kass DA, Kuroski-De Bold ML, de Bold AJ, Conner DA, Rosenthal N, Cepko CL, Seidman CE, Seidman JG (2001) Cardiomyopathy in *Irx4*-deficient mice is preceded by abnormal ventricular gene expression. Mol Cell Biol 21:1730–1736

Buchberger A, Pabst O, Brand T, Seidl K, Arnold HH (1996) Chick NKx-2.3 represents a novel family member of vertebrate homologues to the Drosophila homeobox gene tinman: differential expression of cNKx-2.3 and cNKx-2.5 during heart and gut development. Mech Dev 56:151–163

Chen CY, Schwartz RJ (1996) Recruitment of the tinman homolog Nkx-2.5 by serum response factor activates cardiac alpha-actin gene transcription. Mol Cell Biol 16:6372–6384

Chen J, Kubalak SW, Chien KR (1998) Ventricular muscle-restricted targeting of the RXRalpha gene reveals a non-cell-autonomous requirement in cardiac chamber morphogenesis. Development 125:1943–1949

Chen JN, Fishman MC (2000) Genetics of heart development. Trends Genet 16:383–388

Chen Z, Friedrich GA, Soriano P (1994) Transcriptional enhancer factor 1 disruption by a retroviral gene trap leads to heart defects and embryonic lethality in mice. Genes Dev 8:2293–2301

Chien K (2000) Genomic circuits and the integrative biology of cardiac disease. Nature 407: 227–232

Chin MT, Maemura K, Fukumoto S, Jain MK, Layne MD, Watanabe M, Hsieh CM, Lee ME (2000) Cardiovascular basic helix loop helix factor 1, a novel transcriptional repressor expressed preferentially in the developing and adult cardiovascular system. J Biol Chem 275:6381–6387

Christoffels VM, Keijser AG, Houweling AC, Clout DE, Moorman AF (2000) Patterning the embryonic heart: identification of five mouse Iroquois homeobox genes in the developing heart. Dev Biol 224:263–274

Cohen-Gould L, Mikawa T (1996) The fate diversity of mesodermal cells within the heart field during chicken early embryogenesis. Dev Biol 177:265–273

Colas J, Schoenwolf GC (2000) Subtractive hybridization identifies chick-cripto, a novel EGF-CFC ortholog expressed during gastrulation, neurulation and early cardiogenesis. Gene 255: 205–217

Colas JF, Lawson A, Schoenwolf GC (2000) Evidence that translation of smooth muscle alpha-actin mRNA is delayed in the chick promyocardium until fusion of the bilateral heart-forming regions. Dev Dyn 218:316–330

Crossley PH, Martin GR (1995) The mouse Fgf8 gene encodes a family of polypeptides and is expressed in regions that direct outgrowth and patterning in the developing embryo. Development 121:439–451

Crossley PH, Martinez S, Martin GR (1996) Midbrain development induced by FGF8 in the chick embryo. Nature 380:66–68

Davis DL, Wessels A, Burch JB (2000) An Nkx-dependent enhancer regulates cGATA-6 gene expression during early stages of heart development. Dev Biol 217:310–322

DeHaan R (1965) Morphogenesis of the vertebrate heart. In: DeHaan R, Ursprung H (eds) Organogenesis. Holt, Rinehart and Winston, New York, pp 377–419

Dersch H, Zile M (1993) Induction of normal cardiovascular development in the vitamin A-deprived quail embryo by natural retinoids. Dev Biol 160:424–433

Dettman RW, Denetclaw W Jr, Ordahl CP, Bristow J (1998) Common epicardial origin of coronary vascular smooth muscle, perivascular fibroblasts, and intermyocardial fibroblasts in the avian heart. Dev Biol 193:169–181

Ding J, Yang L, Yan YT, Chen A, Desai N, Wynshaw-Boris A, Shen MM (1998) Cripto is required for correct orientation of the anterior-posterior axis in the mouse embryo. Nature 395:702–707

Dono R, Scalera L, Pacifico F, Acampora D, Persico MG, Simeone A (1993) The murine cripto gene: expression during mesoderm induction and early heart morphogenesis. Development 118: 1157–1168

Dudley A, Robertson E (1997) Overlapping expression domains of bone morphogenetic protein family members potentially account for limited tissue defects in BMP7 deficient embryos. Dev Dyn 208:349–362

Eisenberg CA, Eisenberg LM (1999) WNT11 promotes cardiac tissue formation of early mesoderm. Dev Dyn 216:45–58

Evans SM (1999) Vertebrate tinman homologues and cardiac differentiation. Semin Cell Dev Biol 10:73–83

Fossett N, Zhang Q, Gajewski K, Choi CY, Kim Y, Schulz RA (2000) The multitype zinc-finger protein U-shaped functions in heart cell specification in the *Drosophila* embryo. Proc Natl Acad Sci USA 97:7348–7353

Frasch M (1995) Induction of visceral and cardiac mesoderm by ectodermal Dpp in the early *Drosophila* embryo. Nature 374:464–467

Fu Y, Yan W, Mohun TJ, Evans SM (1998) Vertebrate tinman homologues XNkx2–3 and XNkx2–5 are required for heart formation in a functionally redundant manner. Development 125: 4439–4449

Gaio U, Schweickert A, Fischer A, Garatt AN, Müller T, Özcelik C, Lankes W, Strehle M, Britsch S, Blum M, Birchmeier C (1999) A role of the cryptic gene in the correct establishment of the left–right axis. Curr Biol 9:1339–1342

Gajewski K, Fossett N, Molkentin JD, Kim Y, Choi CY, Schulz RA (1999) The zink finger proteins Pannier and GATA4 function as cardiogenic factors in *Drosophila*. Development 126:5679–5688

Gannon M, Bader D (1995) Initiation of cardiac differentiation occurs in the absence of anterior endoderm. Development 121:2439–2450

Garcia-Martinez V, Schoenwolf G (1993) Primitive-streak origin of the cardiovascular system in avian embryos. Dev Biol 159:706–719

Gassmann M, Casagranda F, Orioli D, Simon H, Lai C, Klein R, Lemke G (1995) Aberrant neural and cardiac development in mice lacking the ErbB4 neuregulin receptor. Nature 378:390–394

Glinka A, Wu W, Onichtchouk D, Blumenstock C, Niehrs C (1997) Head induction by simultaneous repression of Bmp and Wnt signalling in *Xenopus*. Nature 389:517–519

Grepin C, Nemer G, Nemer M (1997) Enhanced cardiogenesis in embryonic stem cells overexpressing the GATA-4 transcription factor. Development 124:2387–2395

Gritsman K, Zhang J, Cheng S, Heckscher E, Talbot WS, Schier AF (1999) The EGF-CFC protein one-eyed pinhead is essential for nodal signaling. Cell 97:121–132

Grow MW, Krieg PA (1998) Tinman function is essential for vertebrate heart development: elimination of cardiac differentiation by dominant inhibitory mutants of the tinman-related genes, XNkx2–3 and XNkx2–5. Dev Biol 204:187–196

Grunz H (1999) Amphibian embryos as a model system for organ engineering: in vitro induction and rescue of the heart anlage. Int J Dev Biol 43:361–364

Ho E, Shimada Y (1978) Formation of the epicardium studied with the scanning electron microscope. Dev Biol 66:579–585

Hoffman JI (1995a) Incidence of congenital heart disease: I. Postnatal incidence. Pediatr Cardiol 16:103–13

Hoffman JI (1995b) Incidence of congenital heart disease: II. Prenatal incidence. Pediatr Cardiol 16:155–65

Inagaki T, Garcia-Martinez V, Schoenwolf GC (1993) Regulative ability of the prospective cardiogenic and vasculogenic areas of the primitive streak during avian gastrulation. Dev Dyn 197: 57–68

Jaber M, Koch WJ, Rockman H, Smith B, Bond RA, Sulik KK, Ross J Jr, Lefkowitz RJ, Caron MG, Giros B (1996) Essential role of beta-adrenergic receptor kinase 1 in cardiac development and function. Proc Natl Acad Sci USA 93:12974–12979

Jiang Y, Tarzami S, Burch J, Evans T (1998) Common role for each of the cGATA4/5/6 genes in the regulation of cardiac morphogenesis. Dev Gen 22:263–277

Kasahara H, Usheva A, Ueyama T, Aoki H, Horikoshi N, Izumo S (2000) Characterization of homo- and heterodimerization of cardiac Csx/Nkx2.5 homeoprotein. J Biol Chem 276:4570–4580

Kastner P, Grondona JM, Mark M, Gansmuller A, LeMeur M, Decimo D, Vonesch JL, Dolle P, Chambon P (1994) Genetic analysis of RXR alpha developmental function: convergence of RXR and RAR signaling pathways in heart and eye morphogenesis. Cell 78:987–1003

Kastner P, Messaddeq N, Mark M, Wendling O, Grondona JM, Ward S, Ghyselinck N, Chambon P (1997) Vitamin A deficiency and mutations of RXRalpha, RXRbeta and RARalpha lead to early differentiation of embryonic ventricular cardiomyocytes. Development 124:4749–4758

Kirby ML, Waldo KL (1995) Neural crest and cardiovascular patterning. Circ Res 77:211–215

Kishimoto Y, Lee KH, Zon L, Hammerschmidt M, Schulte-Merker S (1997) The molecular nature of zebrafish swirl: BMP2 function is essential during early dorsoventral patterning. Development 124:4457–4466

Kodera Y, Takeyama K, Murayama A, Suzawa M, Masuhiro Y, Kato S (2000) Ligand type-specific interactions of peroxisome proliferator-activated receptor gamma with transcriptional co-activators. J Biol Chem 275:33201–33204

Kokan-Moore N-P, Bolender DL, Lough J (1991) Secretion of Inhibin β_A by endoderm cultured from early embryonic chicken. Dev Biol 146:242–245

Komada M, Soriano P (1999) Hrs, a FYVE finger protein localized to early endosomes, is implicated in vesicular traffic and required for ventral folding morphogenesis. Genes Dev 13: 1475–1485

Kostetskii I, Jiang Y, Kostetskaia E, Yuan S, Evans T, Zile M (1999) Retinoid signaling required for normal heart development regulates GATA-4 in a pathway distinct from cardiomyocyte differentiation. Dev Biol 206:206–218

Kreidberg JA, Sariola H, Loring JM, Maeda M, Pelletier J, Housman D, Jaenisch R (1993) WT-1 is required for early kidney development. Cell 74:679–691

Kuhl M, Sheldahl LC, Park M, Miller JR, Moon RT (2000) The Wnt/Ca2+ pathway: a new vertebrate Wnt signaling pathway takes shape. Trends Genet 16:279–283

Kuo CT, Morrisey EE, Anandappa R, Sigrist K, Lu MM, Parmacek MS, Soudais C, Leiden JM (1997) GATA4 transcription factor is required for ventral morphogenesis and heart tube formation. Genes Dev 11:1048–1060

Kupperman E, An S, Osborne N, Waldron S, Stainier DY (2000) A sphingosine-1-phosphate receptor regulates cell migration during vertebrate heart development. Nature 406:192–195

Kwee L, Baldwin HS, Shen HM, Stewart CL, Buck C, Buck CA, Labow MA (1995) Defective development of the embryonic and extraembryonic circulatory systems in vascular cell adhesion molecule (VCAM-1) deficient mice. Development 121:489–503

Ladd AN, Yatskievych TA, Antin PB (1998) Regulation of avian cardiac myogenesis by activin/TGFbeta and bone morphogenetic proteins. Dev Biol 204:407–419

LeCouter JE, Kablar B, Whyte PF, Ying C, Rudnicki MA (1998) Strain-dependent embryonic lethality in mice lacking the retinoblastoma-related p130 gene. Development 125:4669–4679

Lee K-F, Simon H, Chen H, Bates B, Hung M-C, Hauser C (1995) Requirement for neuregulin receptor erbB2 in neural and cardiac development. Nature 378:394–398

Lee RK, Stainier DY, Weinstein BM, Fishman MC (1994) Cardiovascular development in the zebrafish. II. Endocardial progenitors are sequestered within the heart field. Development 120:3361–3366

Leimeister C, Externbrink A, Klamt B, Gessler M (1999) Hey genes: a novel subfamily of hairy- and Enhancer of split related genes specifically expressed during mouse embryogenesis. Mech Dev 85:173–177

Liberatore CM, Searcy-Schrick RD, Yutzey KE (2000) Ventricular expression of tbx5 inhibits normal heart chamber development. Dev Biol 223:169–180

Lien CL, Wu C, Mercer B, Webb R, Richardson JA, Olson EN (1999) Control of early cardiac-specific transcription of Nkx2-5 by a GATA-dependent enhancer. Development 126:75–84

Lin Q, Schwarz J, Bucana C, Olson EN (1997) Control of mouse cardiac morphogenesis and myogenesis by transcription factor MEF2C. Science 276:1404–1407

Lints TJ, Parsons LM, Hartley L, Lyons I, Harvey RP (1993) Nkx-2.5: a novel murine homeobox gene expressed in early heart progenitor cells and their myogenic descendants. Development 119:419–431

Lough J, Sugi Y (2000) Endoderm and heart development. Dev Dyn 217:327–342

Lough J, Barron M, Brogley M, Sugi Y, Bolender DL, Zhu X (1996) Combined BMP-2 and FGF-4, but neither factor alone, induces cardiogenesis in non-precardiac embryonic mesoderm. Dev Biol 178:198–202

Lyons I, Parsons LM, Hartley L, Li R, Andrews JE, Robb L, Harvey RP (1995) Myogenic and morphogenetic defects in the heart tubes of murine embryos lacking the homeo box gene Nkx2-5. Genes Dev 9:1654–1666

Männer J (1993) Experimental study on the formation of the epicardium in chick embryos. Anat Embryol 187:281–289

Markwald R, Trusk T, Moreno-Rodriguez R (1999) Formation and septation of the tubular heart. Integrating dynamics of morphology with emerging molecular concepts. In: de la Cruz M, Markwald R (eds) Living morphogenesis of the heart. Birkhäuser, Boston, pp 43–84

Marvin MJ, Di Rocco G, Gardiner A, Bush SM, Lassar AB (2001) Inhibition of Wnt activity induces heart formation from posterior mesoderm. Genes Dev 15:316–327

Meyer D, Birchmeier C (1995) Multiple essential functions of neuregulin in development. Nature 378:386–390

Meyers EN, Martin GR (1999) Differences in left-right axis pathways in mouse and chick: functions of FGF8 and SHH. Science 285:403–406

Meyers EN, Lewandoski M, Martin GR (1998) An Fgf8 mutant allelic series generated by Cre- and Flp-mediated recombination. Nat Genet 18:136–141

Mikawa T, Fischman DA (1996) The polyclonal origin of myocyte lineages. Annu Rev Physiol 58:509–521

Mjaatvedt CH, Yamamura H, Capehart AA, Turner D, Markwald RR (1998) The Cspg2 gene, disrupted in the hdf mutant, is required for right cardiac chamber and endocardial cushion formation. Dev Biol 202:56–66

Moens CB, Stanton BR, Parada LF, Rossant J (1993) Defects in heart and lung development in compound heterozygotes for two different targeted mutations at the N-myc locus. Development 119:485–499

Molkentin JD (2000) The Zinc Finger-containing Transcription Factors GATA-4, -5, and -6. Ubiquitously expressed regulators of tissue-specific gene expression. J Biol Chem 275:38949–38952

Molkentin JD, Lin Q, Duncan SA, Olson EN (1997) Requirement of the transcription factor GATA4 for heart tube formation and ventral morphogenesis. Genes Dev 11:1061–1072

Molkentin JD, Lu JR, Antos CL, Markham B, Richardson J, Robbins J, Grant SR, Olson EN (1998) A calcineurin-dependent transcriptional pathway for cardiac hypertrophy. Cell 93:215–228

Molkentin JD, Antos C, Mercer B, Taigen T, Miano JM, Olson EN (2000a) Direct activation of a GATA6 cardiac enhancer by Nkx2.5: evidence for a reinforcing regulatory network of Nkx2.5 and GATA transcription factors in the developing heart. Dev Biol 217:301–309

Molkentin JD, Tymitz KM, Richardson JA, Olson EN (2000b) Abnormalities of the genitourinary tract in female mice lacking GATA5. Mol Cell Biol 20:5256–5260

Montgomery M, Litvin J, Gonzalez-Sanchez A, Bader D (1994) Staging of commitment and differentiation of avian cardiac myocytes. Dev Biol 164:63–71

Monzen K, Shiojima I, Hiroi Y, Kudoh S, Oka T, Takimoto E, Hayashi D, Hosoda T, Habara-Ohkubo A, Nakaoka T, Fujita T, Yazaki Y, Komuro I (1999) Bone morphogenetic proteins induce cardiomyocyte differentiation through the mitogen-activated protein kinase TAK1 and cardiac transcription factors Csx/Nkx-2.5 and GATA-4. Mol Cell Biol 19:7096–7105

Moore AW, McInnes L, Kreidberg J, Hastie ND, Schedl A (1999) YAC complementation shows a requirement for Wt1 in the development of epicardium, adrenal gland and throughout nephrogenesis. Development 126:1845–1857

Morin S, Charron F, Robitaille L, Nemer M (2000) GATA-dependent recruitment of MEF2 proteins to target promoters. EMBO J 19:2046–2055

Morrisey EE, Tang Z, Sigrist K, Lu MM, Jiang F, Ip HS, Parmacek MS (1998) GATA6 regulates HNF4 and is required for differentiation of visceral endoderm in the mouse embryo. Genes Dev 12:3579–3590

Moss JB, Xavier-Neto J, Shapiro MD, Nayeem SM, McCaffery P, Drager UC, Rosenthal N (1998) Dynamic patterns of retinoic acid synthesis and response in the developing mammalian heart. Dev Biol 199:55–71

Nakagawa O, Nakagawa M, Richardson JA, Olson EN, Srivastava D (1999) HRT1, HRT2, and HRT3: a new subclass of bHLH transcription factors marking specific cardiac, somitic, and pharyngeal arch segments. Dev Biol 216:72–84

Nakagawa O, McFadden DG, Nakagawa M, Yanagisawa H, Hu T, Srivastava D, Olson EN (2000) Members of the HRT family of basic helix-loop-helix proteins act as transcriptional repressors downstream of Notch signaling. Proc Natl Acad Sci USA 97:13655–13660

Narita N, Bielinska M, Wilson DB (1997) Wild-type endoderm abrogates the ventral developmental defects associated with GATA-4 deficiency in the mouse. Dev Biol 189:270–274

Nikolova M, Chen X, Lufkin T (1997) Nkx2.6 expression is transiently and specifically restricted to the branchial region of pharyngeal-stage mouse embryos. Mech Dev 69:215–218

Pabst O, Schneider A, Brand T, Arnold HH (1997) The mouse Nkx2-3 homeodomain gene is expressed in gut mesenchyme during pre- and postnatal mouse development. Dev Dyn 209: 29–35

Park M, Wu X, Golden K, Axelrod JD, Bodmer R (1996) The wingless signaling pathway is directly involved in *Drosophila* heart development. Dev Biol 177:104–116

Parlow MH, Bolender DL, Kokan-Moore NP, Lough J (1991) Localization of bFGF-like proteins as punctate inclusions in the pre-septation myocardium of the chicken embryo. Dev Biol 146: 139–147

Pereira FA, Qiu Y, Zhou G, Tsai MJ, Tsai SY (1999) The orphan nuclear receptor COUP-TFII is required for angiogenesis and heart development. Genes Dev 13:1037–1049

Pfeffer PL, De Robertis EM, Izpisua-Belmonte JC (1997) Crescent, a novel chick gene encoding a Frizzled-like cysteine-rich domain, is expressed in anterior regions during early embryogenesis. Int J Dev Biol 41:449–458

Reecy JM, Yamada M, Cummings K, Sosic D, Chen C-Y, Eichele G, Olson EN, Schwart RJ (1997) Chicken Nkx-2.8: a novel homeobox gene expressed in early heart progenitor cells and pharyngeal pouch-2 and -3 endoderm. Dev Biol 188:296–311

Reecy JM, Li X, Yamada M, DeMayo FJ, Newman CS, Harvey RP, Schwartz RJ (1999) Identification of upstream regulatory regions in the heart-expressed homeobox gene Nkx2–5. Development 126:839–849

Reifers F, Walsh EC, Leger S, Stainier DY, Brand M (2000) Induction and differentiation of the zebrafish heart requires fibroblast growth factor 8 (fgf8/acerebellar). Development 127: 225–235

Reiter JF, Alexander J, Rodaway A, Yelon D, Patient R, Holder N, Stainier DY (1999) Gata5 is required for the development of the heart and endoderm in zebrafish. Genes Dev 13:2983–2995

Reiter JF, Kikuchi Y, Stainier DY (2001) Multiple roles for Gata5 in zebrafish endoderm formation. Development 128:125–135

Riley PR, Gertsenstein M, Dawson K, Cross JC (2000) Early exclusion of Hand1-deficient cells from distinct regions of the left ventricular myocardium in chimeric mouse embryos. Dev Biol 227:156–168

Rodriguez Esteban C, Capdevila J, Economides AN, Pascual J, Ortiz A, Izpisua Belmonte JC (1999) The novel Cer-like protein Caronte mediates the establishment of embryonic left-right asymmetry. Nature 401:243–251

Roebroek AJ, Umans L, Pauli IG, Robertson EJ, van Leuven F, Van de Ven WJ, Constam DB (1998) Failure of ventral closure and axial rotation in embryos lacking the proprotein convertase Furin. Development 125:4863–4876

Rosenquist GC (1970) Cardia bifida in chick embryos: anterior and posterior defects produced by transplanting tritiated thymidine-labeled grafts medial to the heart-forming regions. Teratology 3:135–142
Rosenthal N, Xavier-Neto J (2000) From the bottom of the heart: anteroposterior decisions in cardiac muscle differentiation. Curr Opin Cell Biol 12:742–746
Saga Y, Miyagawa-Tomita S, Takagi A, Kitajima S, Miyazaki J, Inoue T (1999) MesP1 is expressed in the heart precursor cells and required for the formation of a single heart tube. Development 126:3437–3447
Salomon DS, Bianco C, De Santis M (1999) Cripto: a novel epidermal growth factor (EGF)-related peptide in mammary gland development and neoplasia. Bioessays 21:61–70
Schlange T, Andree B, Arnold H, Brand T (2000a) Expression analysis of the chicken homologue of CITED2 during early stages of embryonic development. Mech Dev 98:157–160
Schlange T, Andree B, Arnold HH, Brand T (2000b) BMP2 is required for early heart development during a distinct time period. Mech Dev 91:259–270
Schlange T, Schnipkoweit I, Andrée B, Ebert A, Zile M, Arnold H, Brand T (2001) Chick CFC controls Lefty1 expression in the embryonic midline and nodal expression in the lateral plate. Dev Biol 234:376–389
Schneider VA, Mercola M (1999) Spatially distinct head and heart inducers within the *Xenopus* organizer region. Curr Biol 9:800–809
Schneider VA, Mercola M (2001) Wnt antagonism initiates cardiogenesis in *Xenopus laevis*. Genes Dev 15:304–315
Schott JJ, Benson DW, Basson CT, Pease W, Silberbach GM, Moak JP, Maron BJ, Seidman CE, Seidman JG (1998) Congenital heart disease caused by mutations in the transcription factor NKX2-5. Science 281:108–111
Schultheiss TM, Xydas S, Lassar AB (1995) Induction of avian cardiac myogenesis by anterior endoderm. Development 121:4203–4214
Schultheiss T, Burch J, Lassar A (1997) A role for bone morphogenetic proteins in the induction of cardiac myogenesis. Genes Dev 11:451–462
Searcy RD, Vincent EB, Liberatore CM, Yutzey KE (1998) A GATA-dependent nkx-2.5 regulatory element activates early cardiac gene expression in transgenic mice. Development 125:4461–4470
Sepulveda JL, Belaguli N, Nigam V, Chen CY, Nemer M, Schwartz RJ (1998) GATA-4 and Nkx-2.5 coactivate Nkx-2 DNA binding targets: role for regulating early cardiac gene expression. Mol Cell Biol 18:3405–3415
Shawlot W, Min Deng J, Wakamiya M, Behringer RR (2000) The cerberus-related gene, Cerr1, is not essential for mouse head formation. Genesis 26:253–258
Shen M, Schier A (2000) The *EGF-CFC* gene family in vertebrate development. Trends Genet 16: 303–309
Shi Y, Katsev S, Cai C, Evans S (2000) BMP signaling is required for heart formation in vertebrates. Dev Biol 224:226–237
Slack JMW (1991) From egg to embryo. Regional specification in early development, 2nd edn. Cambridge University Press, Cambridge
Smith SJ, Kotecha S, Towers N, Mohun TJ (2000) *Xenopus* hand2 expression marks anterior vascular progenitors but not the developing heart. Dev Dyn 219:575–581
Solloway MJ, Robertson EJ (1999) Early embryonic lethality in Bmp5;Bmp7 double mutant mice suggests functional redundancy within the 60A subgroup. Development 126:1753–1768
Srivastava D, Olson E (2000) A genetic blueprint for cardiac development. Nature 407:221–226
Stainier DY, Lee RK, Fishman MC (1993) Cardiovascular development in the zebrafish. I. Myocardial fate map and heart tube formation. Development 119:31–40
Stanley EG, Biben C, Allison J, Hartley L, Wicks IP, Campbell IK, McKinley M, Barnett L, Koentgen F, Robb L, Harvey RP (2000) Targeted insertion of a lacZ reporter gene into the mouse Cer1 locus reveals complex and dynamic expression during embryogenesis. Genesis 26:259–264

Sucov HM, Dyson E, Gumeringer CL, Price J, Chien KR, Evans RM (1994) RXR alpha mutant mice establish a genetic basis for vitamin A signaling in heart morphogenesis. Genes Dev 8:1007–1018

Sugi Y, Lough J (1994) Anterior endoderm is a specific effector of terminal cardiac myocyte differentiation of cells from the embryonic heart forming region. Dev Dyn 200:155–162

Sugi Y, Lough J (1995) Activin-A and FGF-2 mimic the inductive effects of anterior endoderm on terminal cardiac myogenesis. Dev Biol 168:567–574

Tam PP, Parameswaran M, Kinder SJ, Weinberger RP (1997) The allocation of epiblast cells to the embryonic heart and other mesodermal lineages: the role of ingression and tissue movement during gastrulation. Development 124:1631–1642

Tanaka M, Yamasaki N, Izumo S (2000) Phenotypic characterization of the murine Nkx2.6 homeobox gene by gene targeting. Mol Cell Biol 20:2874–2879

Tevosian S, Deconinck A, Tanaka M, Schinke M, Litovsky S, Izumo S, Fujiwara Y, Orkin S (2000) FOG-2, a cofactor for GATA transcription factors, is essential for heart morphogenesis and development of coronary vessels from epicardium. Cell 101:729–739

Tokuyasu K (1990) Co-development of embryonic myocardium and myocardial circulation. In: Clark E, Takao A (eds) Developmental cardiology: morphogenesis and function. Futura, Mt Kisco, pp 205–218

Torres MA, Yang-Snyder JA, Purcell SM, DeMarais AA, McGrew LL, Moon RT (1996) Activities of the Wnt-1 class of secreted signaling factors are antagonized by the Wnt-5A class and by a dominant negative cadherin in early *Xenopus* development. J Cell Biol 133:1123–1137

Tran CM, Sucov HM (1998) The RXRalpha gene functions in a non-cell-autonomous manner during mouse cardiac morphogenesis. Development 125:1951–1956

Tzahor E, Lassar AB (2001) Wnt signals from the neural tube block ectopic cardiogenesis. Genes Dev 15:255–260

Weber H, Symes CE, Walmsley ME, Rodaway AR, Patient RK (2000) A role for GATA5 in *Xenopus* endoderm specification. Development 127:4345–4360

Wei Y, Mikawa T (2000) Fate diversity of primitive streak cells during heart field formation in ovo. Dev Dyn 219:505–513

Wu H, Lee S, Gao J, X L, Iruela-Arispe M (1999) Inactivation of erythropoietin leads to defects in cardiac morphogenesis. Development 126:3597–3605

Wu X, Golden K, Bodmer R (1995) Heart development in *Drosophila* requires the segment polarity gene *wingless*. Dev Biol 169:619–628

Xavier-Neto J, Neville CM, Shapiro MD, Houghton L, Wang GF, Nikovits W, Stockdale FE, Rosenthal N (1999) A retinoic acid-inducible transgenic marker of sino-atrial development in the mouse heart. Development 126:2677–2687

Xavier-Neto J, Shapiro MD, Houghton L, Rosenthal N (2000) Sequential programs of retinoic acid synthesis in the myocardial and epicardial layers of the developing avian heart. Dev Biol 219:129–141

Xu C, Liguori G, Adamson ED, Persico MG (1998) Specific arrest of cardiogenesis in cultured embryonic stem cells lacking Cripto-1. Dev Biol 196:237–247

Xu C, Liguori G, Persico MG, Adamson ED (1999) Abrogation of the Cripto gene in mouse leads to failure of postgastrulation morphogenesis and lack of differentiation of cardiomyocytes. Development 126:483–494

Yamada M, Szendro PI, Prokscha A, Schwartz RJ, Eichele G (1999) Evidence for a role of Smad6 in chick cardiac development. Dev Biol 215:48–61

Yan Y-T, Gritsman K, Ding J, Burdine RD, Corrales JD, Price SM, Talbot WS, Schier AF, Shen MM (1999) Conserved requirement for EGF-CFC genes in vertebrate left-right axis formation. Genes Dev 13:2527–2537

Yatskievych TA, Ladd AN, Antin PB (1997) Induction of cardiac myogenesis in avian pregastrula epiblast: the role of the hypoblast and activin. Development 124:2561–2570

Yelon D, Ticho B, Halpern ME, Ruvinsky I, Ho RK, Silver LM, Stainier DY (2000) The bHLH transcription factor hand2 plays parallel roles in zebrafish heart and pectoral fin development. Development 127:2573–2582

Yokouchi Y, Vogan KJ, Pearse RV 2nd, Tabin CJ (1999) Antagonistic signaling by Caronte, a novel Cerberus-related gene, establishes left-right asymmetric gene expression. Cell 98:573–583

Yoshida K, Taga T, Saito M, Suematsu S, Kumanogoh A, Tanaka T, Fujiwara H, Hirata M, Yamagami T, Nakahata T, Hirabayashi T, Yoneda Y, Tanaka K, Wang WZ, Mori C, Shiota K, Yoshida N, Kishimoto T (1996) Targeted disruption of gp130, a common signal transducer for the interleukin 6 family of cytokines, leads to myocardial and hematological disorders. Proc Natl Acad Sci USA 93:407–411

Zhang H, Bradley A (1996) Mice deficient for BMP2 are nonviable and have defects in amnion/chorion and cardiac development. Development 122:2977–2986

Zhu L, Marvin MJ, Gardiner A, Lassar AB, Mercola M, Stern CD, Levin M (1999a) Cerberus regulates left-right asymmetry of the embryonic head and heart. Curr Biol 9:931–938

Zhu X, Sasse J, Lough J (1999b) Evidence that FGF receptor signaling is necessary for endoderm-regulated development of precardiac mesoderm. Mech Ageing Dev 108:77–85

Subject Index

Printing: Mercedes-Druck, Berlin
Binding: Stein + Lehmann, Berlin

GPSR Compliance

The European Union's (EU) General Product Safety Regulation (GPSR) is a set of rules that requires consumer products to be safe and our obligations to ensure this.

If you have any concerns about our products, you can contact us on ProductSafety@springernature.com

In case Publisher is established outside the EU, the EU authorized representative is:

Springer Nature Customer Service Center GmbH
Europaplatz 3
69115 Heidelberg, Germany

Batch number: 10755466

Printed by Printforce, the Netherlands